ILLUSTRATED GUIDE TO PRACTICAL SOLID STATE CIRCUITS...
WITH EXPERIMENTS AND PROJECTS

ILLUSTRATED GUIDE TO PRACTICAL SOLID STATE CIRCUITS ... WITH EXPERIMENTS AND PROJECTS

Robert C. Genn, Jr.

Illustrated by E. L. Genn

PRENTICE-HALL, INC.
(Business and Professional Division)
ENGLEWOOD CLIFFS, N.J.

Prentice-Hall International, Inc., *London*
Prentice-Hall of Australia, Pty. Ltd., *Sydney*
Prentice-Hall Canada Inc., *Toronto*
Prentice-Hall of India Private Ltd., *New Delhi*
Prentice-Hall of Japan, Inc., *Tokyo*
Prentice-Hall of Southeast Asia Pte. Ltd., *Singapore*
Whitehall Books, Ltd., *Wellington, New Zealand*
Editora Prentice-Hall do Brasil, Ltda., *Rio de Janeiro*

© 1983 by
Prentice-Hall, Inc.
Englewood Cliffs, N.J.

All rights reserved. No part of this book may be reproduced in any form or by any means, without permission in writing from the publisher.

Editor: George E. Parker

Library of Congress Cataloging in Publication Data

Genn, Robert C.
   Illustrated guide to practical solid state circuits—with experiments and projects.

   Includes index.
   1. Electronic circuits. 2. Solid state electronics.
I. Genn, E. L. II. Title.
TK7867.G38 1983   621.3815'3   83-4414

ISBN 0-13-450643-X

Printed in the United States of America

How This Book Will Help You Work with State-of-the Art Solid State Devices and Systems

It is safe to say that at least 80 percent of today's electronics technicians are employed in jobs requiring a knowledge of solid state circuits. And there is every reason to believe that solid state circuits, particularly integrated circuits (IC's), will become even more widely used in the future. With this book you will acquire the technical knowledge and experience needed to master the IC's used in today's—as well as in tomorrow's—electronics industry. The discussions in the following chapters begin with basics, then turn to specific devices, and finally, to working systems.

To be prepared to grow with the world's largest growth industry, you'll need hands-on experience. The last chapter contains 30 projects that have been selected because they use up-to-date components and are within the capability of the technician just starting in the electronics field. An important part of the presentation of the projects is the availability of circuit components. Therefore, the solid state components are all Motorola Company products that are up to date, easily obtained, and can be purchased at reasonable prices.

Not only is this handbook clearly indexed and easy to understand, in addition it will help you keep on top of the many remarkable changes taking place—or about to take place—in the solid state industry. Without this information, the electronics technician or experimenter is severely handicapped. A sound grasp of the newer solid state technologies is vital to a successful career in electronics, both now and in the foreseeable future.

To start, the essential information pertaining to electronics basics is presented in Chapter 1. Then you will acquire a complete working knowledge of modern linear and digital circuits in Chapter 2. You will learn microcomputer basics in Chapter 3, and go on to actual state-of-the-art microcomputer units in Chapter 4. For example, Motorola's MC6805R2 8-bit microcomputer unit with analog-to-digital converter is explained in detail.

An illustrated guide such as this one may be used in varying degrees by different readers. For instance, if you are new to a particular technique, you can start with basics and study at your own pace. You'll be under no pressure to keep up, and there are no slow learners to hold you back. This book will help cut the time normally involved in mastering today's solid state circuits.

On the other hand, some may desire a particular solid state device to fulfill a certain system requirement and will make minimal use of the section on how solid state circuits work and how to use them. For these applications, this illustrated guide includes over 100 Motorola solid state devices, with pin configurations for the most widely used ones.

There are a lot of good jobs in electronics, but the best jobs go to the people who have experience and ability. This book can help you cut the time you must spend to get both. You gain practical experience by doing—by actually building modern electronic circuits, running tests, and verifying specs.

You will find that the emphasis has been placed on work-related problems, shop hints that will save time when building and troubleshooting, plus simplified, practical ways to learn the modern servicing techniques that are essential to your increasing success as an electronics technician.

This comprehensive handbook contains all the essential information the person interested in modern solid state needs to know. It is written for the technician who is searching for concise, up-to-date guidelines and more economical ways to troubleshoot and maintain today's equipment, designed using current solid state technology. You will find that every chapter stresses simplified methods that can be used for years to come. The integrated circuits used in this book are currently the mainstay in solid state electronics, and no workbench would be complete without the information contained in the following pages.

Robert C. Genn, Jr.

Other books by the author:

Digital Electronics: A Workbench Guide to Circuits, Experiments and Applications
Practical Handbook of Solid State Troubleshooting
Manual of Electronic Servicing Tests and Measurements
Practical Handbook of Low-Cost Electronic Test Equipment
Workbench Guide to Electronic Troubleshooting

Table of Contents

How This Book Will Help You Work with State-of-the-Art Solid State Devices and Systems vii

1. A BRIEF REVIEW OF BASIC ELECTRONICS FUNDAMENTALS WITH PRACTICAL APPLICATIONS 1

    Working with Voltage ... Solving Resistance Problems ... Current and Its Relationships ... How to Make dc Measurements ... Guide to AC Measurements ... What You Should Know About Circuit Loading During Measuring ... Using Capacitors ... Guide to Inductance ... Working with Solid State Rectifiers ... Practical Applications Using Transistors ... Today's and Tomorrow's Integrated Circuits.

2. UNDERSTANDING TODAY'S SOLID STATE CIRCUITS 33

    Introduction to Linear and Digital Integrated Circuits ... TTL CMOS and ECL— Why? ... Principles of CMOS Devices and Circuitry ... MOS Power Transistors (MOSFET's and VMOSFET's ... CMOS Handling Precautions ... Basic Logic Elements ... Fundamentals of Multivibrators and Flip-Flops ... Principles of Register and Counter IC's ... Understanding Memory IC's ... Working with Decoders/Encoders ... Essentials of Display Decoders.

3. SIMPLIFIED DIGITAL BASICS 79

> Understanding Numbering Systems Used in Modern Computer Units . . . The ASCII Alphanumeric Code System . . . Introduction to Today's Computer-on-a-Chip . . . Basic Microcomputer Address, Data, and Control Lines . . . MPU Signal Descriptions . . . Programming Fundamentals . . . How and Why Addressing Modes Are Used in a Microcomputer System . . . Summary.

4. PRACTICAL GUIDE TO STATE-OF-THE-ART ON-CHIP MICROCOMPUTERS 103

> About Today's Microcomputer Units . . . The M6805 Family of MCU's . . . How the RAM'S Operate . . . Understanding ROM's . . . Input/Output Signals . . . Input/Output Ports . . . Registers . . . Timer Circuitry and How It Works . . . Counter ($09:OA) . . . Output Compare Register ($OB:OC) . . . Input Capture Register ($OD:OE) . . . Timer Control and Status Register ($08) . . . Addressing Modes . . . Software . . . Introduction to Programming Basics.

5. COMPLETE GUIDELINES FOR THE CIRCUIT BUILDER, USING MODERN SOLID STATE DEVICES 133

> Practical Techniques for Working with Resistors, Capacitors, and Inductors . . . How to Select and Use Linear Multitesters . . . Breadboarding Solid State Linear Circuits . . . Test Equipment for Digital Systems . . . Troubleshooting Digital Circuits . . . DC Power Supplies for Linear Circuits . . . DC Power Supply Requirements for Digital Circuits.

6. HOW TO WORK WITH PRACTICAL AUDIO CIRCUITS USING SOLID STATE DEVICES 155

> Introduction to Audio Systems Using IC's . . . Discrete Power Transistors . . . Audio Amplifier Design Fundamentals . . . Power Requirements . . . Distortion in Audio

TABLE OF CONTENTS xi

Systems ... Practical Decibel Measurements ... Power Measurements ... Impedance Measurements ... Audio Preamplifiers ... Audio Power Amplifiers ... Troubleshooting Breadboarded Audio Amplifiers.

7. MODERN REGULATED SOLID STATE POWER
SUPPLIES 181

Regulator Power Supplies ... Basic Concept of Regulation ... Line Voltage Regulation ... Load Current Control ... Using Protection Techniques ... Switchmode Regulators ... Basic Switchmode Configurations ... Practical Guide to Troubleshooting Solid State Power Supply Circuits ... Voltage Regulator for Photovoltaic Power Systems.

8. HOW TO INTERFACE TO CONTEMPORARY
IC APPLICATIONS 205

Analog-to-Digital/Digital-to-Analog Conversion Techniques ... Understanding Phase Shift and Compensation ... Output Interfacing ... LED Indicator/CMOS Driver Circuits ... Driving High-Voltage/Current Peripherals ... Testing a Peripheral Driver ... Bus Interface ... MPU Memory Interface ... Peripheral Interface Using Drivers and Receivers in Computer Applications.

9. PRACTICAL GUIDE TO MODERN POWER
TRANSISTORS, THYRISTORS, AND
OPTOELECTRONIC DEVICES 235

Darlington Transistors: the IC's of the Power Field ... Power MOS Fet's ... TMOS and Bipolar Power: Advantages and Disadvantages ... Introduction to Thyristors ... Unijunction Transistors (UJT) ... Bilateral Triggers (DIAC) ... Programmable UJT's ... Silicon Bidirectional Switches (SBS) ... Optically Coupled Triac Drivers ... Optoisolators ... Using Opto Coupler/Isolators for TTL to MOS Interface ... How Semiconductor Phototransistors Operate ... Photo Darlington Amplifiers ...

Practical Mounting Guide for Power Semiconductors ... Guidelines for Electrical Insulation of Power Semiconductors.

10. BASIC FUNDAMENTALS FOR RADIO FREQUENCY SOLID STATE DEVICES AND CIRCUITS 255

Frequency Spectrum ... Frequency Versus Wavelength ... RF Semiconductors ... Impedance (at rf Frequencies) ... Matching Networks ... Testing Solid State Radio Frequency Systems ... Controlled Quality Factor (Q) Transistors ... Introduction to Modern Resonant Circuits ... Practical Solid State RF Circuits and Devices ... Radio Frequency Measurements ... Special RF Measurement Precautions ... A Collection of Formulas Frequently Needed in RF Work.

11. HOW TO SELECT AND USE HIGH FREQUENCY DIODES, TRANSISTORS, INTEGRATED CIRCUITS, AND MODULES 273

Introduction to Electronic Tuning and Control Applications ... Tuning Diode ... Mixer Diode ... Hot-Carrier Diodes ... Pin Switching Diodes ... Selecting the Right Transistor For Your Application ... Maximum Ratings ... Collector Current ... Total Power Dissipation ... Operating and Storage Junction Temperature Range ... Thermal Characteristics ... OFF Characteristics ... ON Characteristics ... Dynamic Characteristics ... CB/Amateur High Frequency Transistors ... Driving rf Output Transistors ... FM Mobile Communications Transistors ... Marine Radio Transistors/Modules.

12. 30 ELECTRONIC PROJECTS USING STATE-OF-THE-ART SOLID STATE DEVICES 293

Project 12.1: Low-Cost Rectifier Circuit for Solid State Projects ... Project 12.2: Practical 5-Volt, 3-Ampere Regulator ... Project 12.3: Overvoltage Protection Circuit ...

TABLE OF CONTENTS

Project 12.4: Using a Power Transistor as a Rectifier ... Project 12.5: Lab-Type Power Supply for Shop Use ... Project 12.6: Simple Fixed Output Voltage Regulator (MC78 Series) ... Project 12.7: How to Double the Output Voltage ... Project 12.8: Easy-to-Make High-Impedance Microphone ... Project 12.9: Using a MC1741 OP AMP to Interface a High Impedance to a Low Impedance ... Project 12.10: Building a Voltage Level Detector Using a MC1741C OP AMP ... Project 12.11: How to Connect and Test a High Performance Dual Operational Amplifier ... Project 12.12: How to Construct a Dual Flasher ... Project 12.13: Audio Amplifier ... Project 12.14: Lamp Dimmer Using a Power MOSFET ... Project 12.15: Variable Timing Circuit (Clock) ... Project 12.16: Contact Debouncer ... Project 12.17: Square-Wave Generator ... Project 12.18: Touch-Controlled Flip-Flop ... Project 12.19: CMOS Quad 2-Input NAND Gate (MC14011B) ... Project 12.20: Constructing an Inverter Using a MC14011B ... Project 12.21: Building an AND Gate Using a MC14011B ... Project 12.22: Using the MC14011B to Build an OR Gate ... Project 12.23: Combination Gates ... Project 12.24: Using the MC14011B to Build a NOR Gate ... Project 12.25: Building an Exclusive-OR Gate Using a MC14011B ... Project 12.26: Building a 4-Input NAND Gate Using Two MC14011B's ... Project 12.27: Divide-by-4 Counter Using Dual J-K Flip-Flops (MC14027B) ... Project 12.28: Building a 4-Bit Serial Shift Register Using Two Dual J-K Flip-Flops (MC14027B's) ... Project 12.29: Phototransistor Light-Operated Relay ... Project 12.30: Diode Transistor Coupler (Optoisolator).

Index 239

ILLUSTRATED GUIDE TO PRACTICAL
SOLID STATE CIRCUITS...
WITH EXPERIMENTS AND PROJECTS

CHAPTER ONE

A Brief Review of Basic Electronics Fundamentals with Practical Applications

Have you forgotten many of your electronics basics? Still have areas you are not sure about? This chapter will help you sort out the pieces, one at a time. In a new, practical approach, using the material in the following pages overcomes the tedious task of reviewing basic electronics. It combines essential information with ready-reference data—material you can use on the job.

Each section contains information that is of prime importance for understanding and servicing circuits containing solid state devices; you will find that no important factor in modern-day electronics fundamentals has been left out or passed over. Basic electrical theory is explained in depth and examples are presented to illustrate all principles.

Working with Voltage

The desire of electricity to flow is called "electromotive force" and is initialed "emf." The unit for emf is the volt. Typical examples: a zinc-carbon dry cell (such as those used to power flashlights, toys, etc.) usually has an emf of 1½ volts; a wet cell has 2.1 volts; most household electricity has an emf of about 115 volts.

Today, literally hundreds of different kinds of batteries are available in a wide range of voltages. Nevertheless, their output voltages do not reverse polarity during use and they all are referred to as direct current (abbreviated dc) voltage sources. On the other hand, the type of voltage that reverses its polarity regularly, such as your house electricity, is referred to as an alternating voltage and is generated by an alternating current (abbreviated ac) source, such as an alternator.

The waveform of a single cycle of a pure ac voltage in which the amplitude at each instant is proportional to the sine of the angle as it changes from 0 to 360° (one cycle) is referred to as a *sine wave* voltage (see Figure 1-1).

We've all been taught that the rms (abbreviation for root mean square) value of a sine wave of voltage is 0.707 times

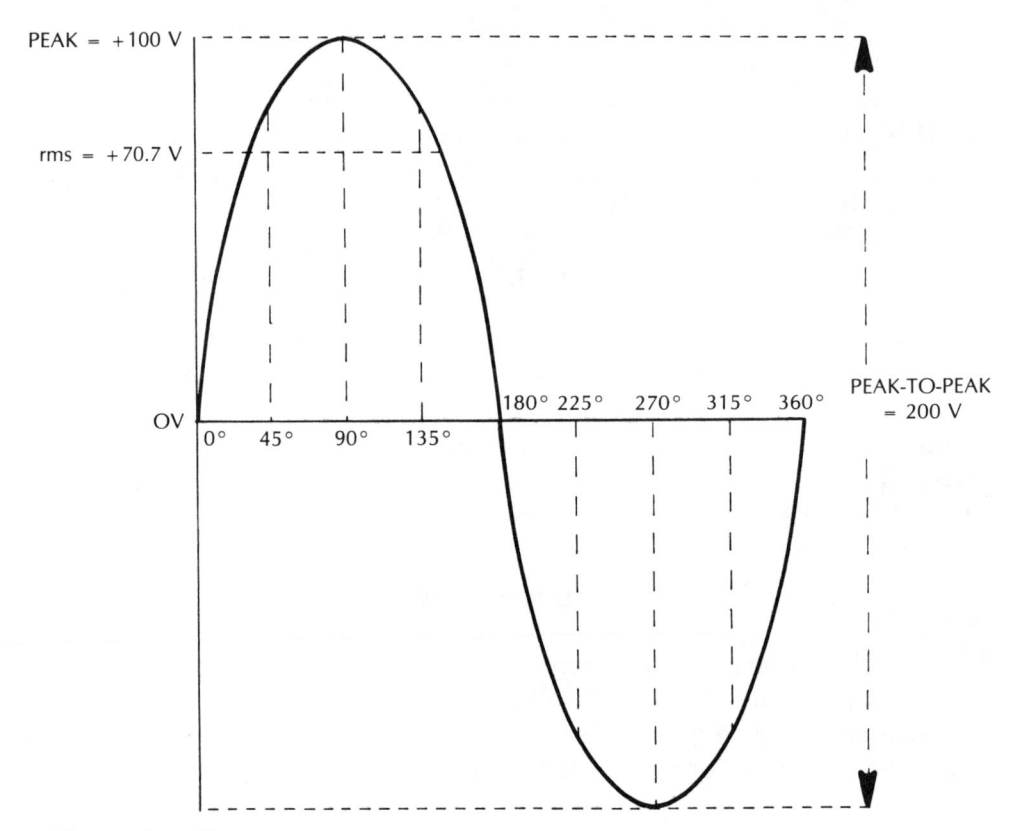

Figure 1-1: Sine wave.

REVIEW OF BASIC ELECTRONICS FUNDAMENTALS 3

TO CONVERT

FROM	TO
PEAK-TO-PEAK	AVERAGE
PEAK-TO-PEAK	EFFECTIVE
PEAK-TO-PEAK	PEAK
PEAK	AVERAGE
PEAK	EFFECTIVE
EFFECTIVE	AVERAGE
AVERAGE	PEAK
PEAK	PEAK-TO-PEAK
EFFECTIVE	PEAK-TO-PEAK
AVERAGE	PEAK-TO-PEAK

YOUR KNOWN VOLTAGE

YOUR CONVERTED VOLTAGE

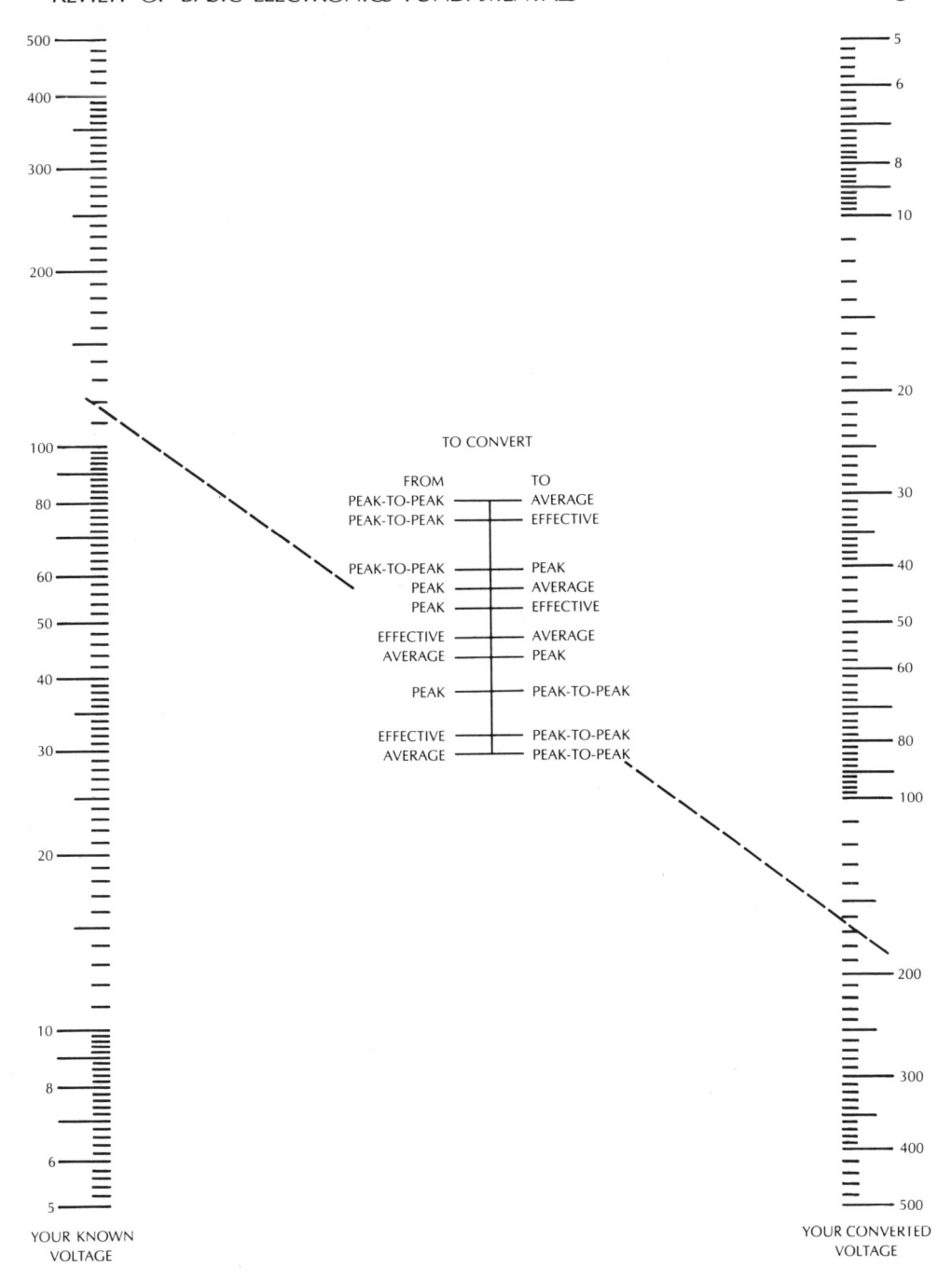

Figure 1-2: Chart for converting sine wave voltages (or currents). From *Workbench Guide to Electronic Troubleshooting*, **page 199.**

its peak value. Or, the peak value is 1.414 times the rms value. You'll probably remember fighting these problems during your first few weeks of studying basic electronics. By using a straightedge and the chart shown in Figure 1-2, you will find that converting sine wave voltages (or currents) to whatever value you need is an almost instantaneous procedure. As an example of what you can do with the chart, let's say that you want to know the peak value of the 115-volt (rms) line voltage you desire to connect to a certain solid state component (such as a diode, etc.).

Your first step is to start from 115 on the left-hand scale (see Figure 1-2), then draw a straight line through rms (listed as "effective" on the chart) on the center scale and extend the line to cross the right-hand scale, as shown by the dashed line on the chart. Your converted voltage is about 163 volts. If you use the formula for peak voltage calculations, you will come up with 162.2 volts, so, as you can see, we are off only 0.8 volt, which is close enough for use with solid state components such as diodes, general purpose transistors, and the like. In general, you will find all your answers to be about eight to 14 percent off calculated value. When solving this type of problem, a programmable calculator or computer is much more accurate and should be used with many of today's circuit designs.

Solving Resistance Problems

Any dissimilar metals immersed in a suitable electrolyte (salt water or lemon juice, for example) may produce an electron flow, and this flow is what we call *current*. The amount of current flow depends on several factors: the emf generated by the do-it-yourself electrochemical generator, and the *resistance* in the circuit that completes the current path from one dissimilar metal to the other. Or, to put it another way, anything that reduces the flow of direct current from one point to another is called *resistance*. Also, resistance in any circuit will always transform part of the electrical energy in that circuit into heat energy. Resistance (R) is measured in units called *ohms* and is symbolized by the Greek letter omega (Ω). This symbol is frequently used in schematic diagrams.

Resistance comes in many degrees and forms. Multiple resistors and resistor arrays in single or dual in-line packages

REVIEW OF BASIC ELECTRONICS FUNDAMENTALS

are available (see Figure 1-3 A). Such arrays are very useful in a variety of systems and can mate with IC sockets. Composition resistors (see Figure 1-3 B) are made with carbon particles mixed with a binder. They generally are very inexpensive, and are easily identified because of a color-coding technique that permits the value of the resistor to be determined simply by observation.

Commercially available composition resistors usually are provided in 5, 10, or 20% tolerance values. Figure 1-3 (B)

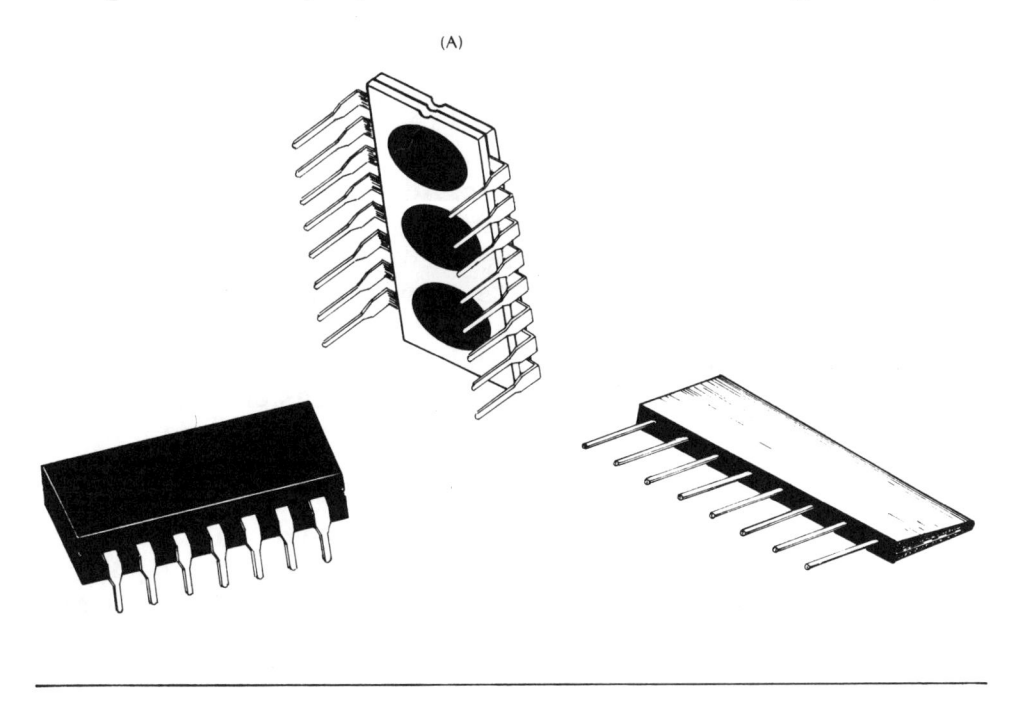

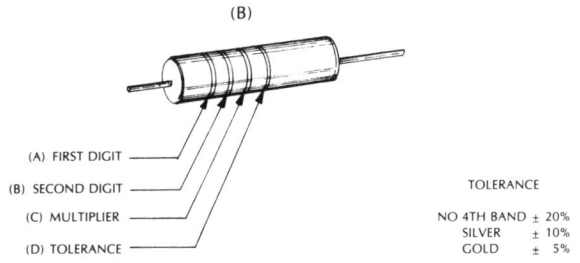

(A)

(B)

(A) FIRST DIGIT
(B) SECOND DIGIT
(C) MULTIPLIER
(D) TOLERANCE

TOLERANCE

NO 4TH BAND ± 20%
SILVER ± 10%
GOLD ± 5%

R = (1st digit) (2nd digit) (multiplier) ± tolerance

Figure 1-3: (A) Multiple resistor and arrays in dual in-line and single in-line packages. (B) Composition resistor color code and bands.

shows the bands around a resistor that can be used to determine the value of the resistor and its tolerance. Notice, band D is the tolerance-indicating band: gold 5%, silver 10%, and no fourth band ±20%. The first three bands (A, B, and C) determine the value of the resistor. The third band (C) tells you how many zeros to add to the first two numbers. For example, a 220,000 ±10% resistor would have band A red, band B red, and band C yellow. Band D would be silver. Also, the resistor may actually be 2,200 ohms higher or lower than the color-coded value of 220,000; i.e., it may range from 222,200 to 217,800 ohms.

Very often it is desirable to connect resistors in series to increase the total resistance, or in parallel to create some value less than the smaller of the parallel resistors. As you probably know, it is a very simple matter to connect resistors in series to construct a larger value resistance. Add the values and that is all there is to it—as far as the mathematical calculation is concerned. For example, if you connect a 10,000- and a 15,000-ohm resistor in series, the total resistance would be 25,000 ohms, or 25k Ω (k = 1,000). But you'll remember that the combined value of resistors in parallel is equal to the reciprocal of the sum of the reciprocals of the individual values. A much easier way to solve this type of problem is to use the chart shown in Figure 1-4. All you need is a straightedge to lay across the scales.

When you use the chart shown in Figure 1-4 to determine the resultant value of resistors in parallel, you can read the scale numbers in ohms, or megohms, depending on which you're working with, and provided you read the same units on all scales. As an example of how to use the chart, let's assume that you have a 6- and a 12-ohm resistor and want to know the value when they are paralleled. First, place a straightedge at 6 on the outer left-hand scale and run it diagonally up to 12 on the right-hand section of the chart (see dashed line on Figure 1-4). Notice that you read 4 on the center scale—your answer.

To work with three or more resistors, you only have to solve the problem for two, as just explained, then use the resulting answer as a single value with the next resistor, and so on. Furthermore, to extend the range of the scales, simply multiply or divide all scales by the same number. For instance, if you divide by 10, the two scales labeled Component Value become 0.5 to 10 rather than 5 to 100, as shown. Of course, the center scale (of the left-hand three scales) would now be 0.26 to 5.

REVIEW OF BASIC ELECTRONICS FUNDAMENTALS 7

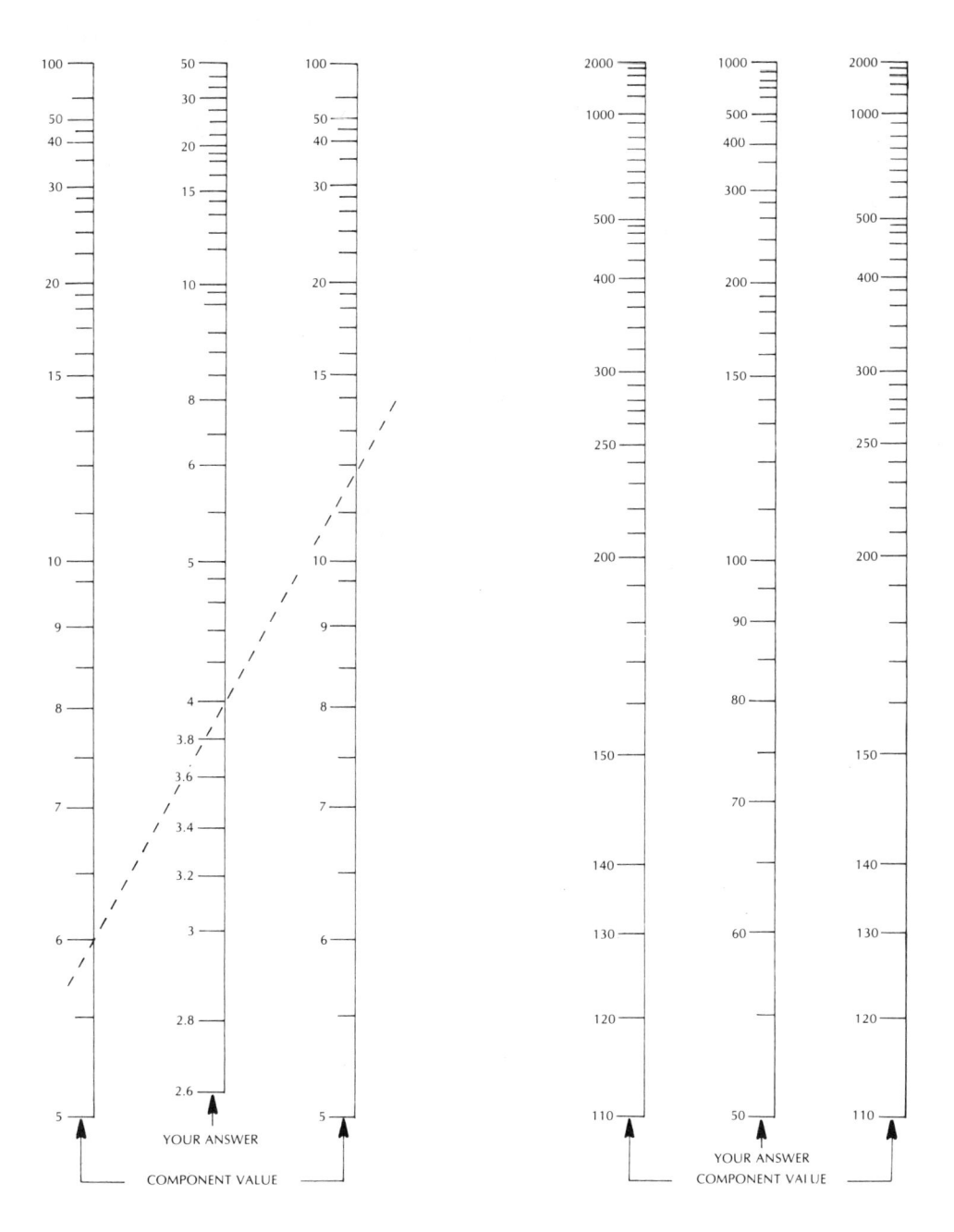

Figure 1-4: Chart for resistors in parallel problems. From *Workbench Guide to Electronic Troubleshooting*, page 186.

When selecting composition resistors for various projects, it is important that you choose resistors with the proper power dissipation rating. In general, you will find that you can purchase composition resistors in five different ratings: ⅛, ¼, ½, 1, and 2 watts. If you are in doubt about the power dissipation rating required, simply remember (and use) the standard formula for power; i.e., power dissipation is equal to the maximum current you expect to pass through the resistor, times the voltage you expect to measure across the resistor—or you can use any of these well-known formulas:

$$P = IE \qquad P = I^2 R \qquad P = E^2/R$$

Temperature, however, is also a factor when working with resistors. If you are placing a certain resistor in a particularly hot spot on a chassis, don't forget that as temperature increases, so does resistance. A good rule of thumb is this: when in doubt, use the next highest power dissipation rating.

Current and Its Relationships

Most of us think of current as being analogous to water flowing in a pipe but, in reality, a flow of current is any movement of electric charge carriers (electrons, protons, ions, or holes). If we speak of electric current in a wire (a flow of electrons), it is usually from negative to positive. As explained before, direct current (such as from a battery) flows in one direction only, whereas alternating current periodically reverses direction (for example, common house current). There is a direct relationship between current, voltage, and resistance. In equation form, it is called Ohm's Law and is expressed as

$$\text{current (I)} = \frac{\text{voltage (E)}}{\text{resistance (R)}}.$$

Current is measured in units called "amperes," abbreviated as A.

You can convert sine wave currents to average rms, peak, and peak-to-peak values just as easily as you can make voltage conversions of the same type, by using the chart (shown in Figure 1-2) and the same procedure as previously explained for voltage conversions. As an example of how to use the chart for this purpose, suppose you wanted to know the peak value of

REVIEW OF BASIC ELECTRONICS FUNDAMENTALS

an ac current flowing through a diode in a *series circuit*. A diode is a solid state device that allows current to flow in one direction only. It is symbolized as shown in Figure 1-5.

The voltage you would measure across the diode is very small when current is flowing through it. In this example, we will assume it is zero. To find the peak value of the current produced by the 5 V source through a 1,000-ohm resistor, we must first determine the rms value of the current. This can be done by using Ohm's Law:

$I_{rms} = E_{rms}/R_{ohms} = 5/1000 = 0.005$ ampere (5 milliamperes).

Using the chart:

> Effective (rms) current converted to peak current
> = about 7.07 milliamperes.

Still referring to series circuits, Kirchoff's voltage law states that the *algebraic* sum of all the voltages in any *closed* loop in a circuit is equal to zero. That is to say that if a circuit has a single battery and three resistors all connected in series, the sum of the voltages you would measure across all the resistors will be equal to the battery voltage. See Figure 1-6.

Now let's examine current flow in a *parallel* circuit. Kirchoff's circuit rules state that the sum of the currents into a particular junction equals the sum of the currents out of the same junction. To put this in simpler terms, if a circuit has a single battery and three parallel resistors, the sum of the

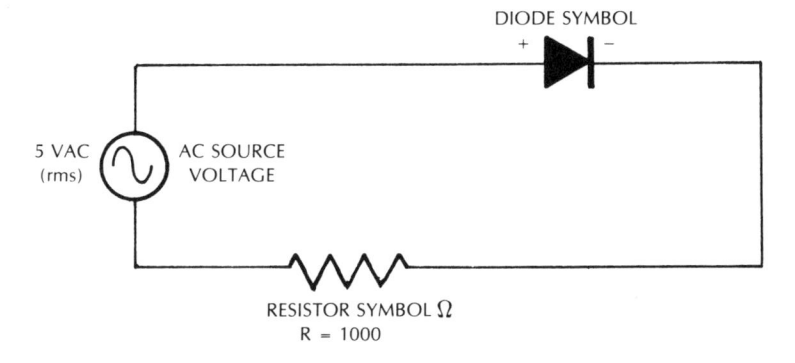

Figure 1-5: Example series circuit. Electron current will flow only when the anode (+) of the diode is positive and the cathode (−) is negative. When the polarity is reversed, no current will flow.

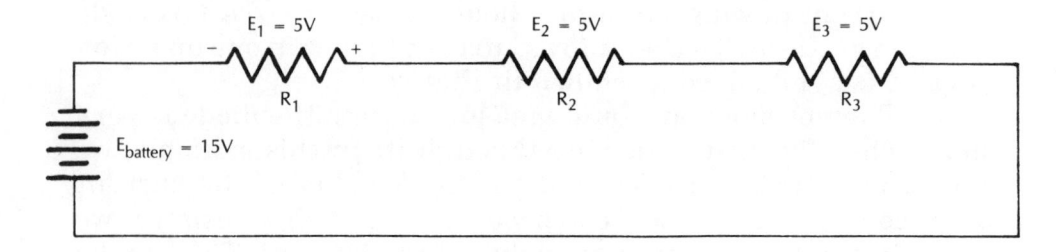

Figure 1-6: Example showing Kirchoff's voltage law. All three resistors are the same value.

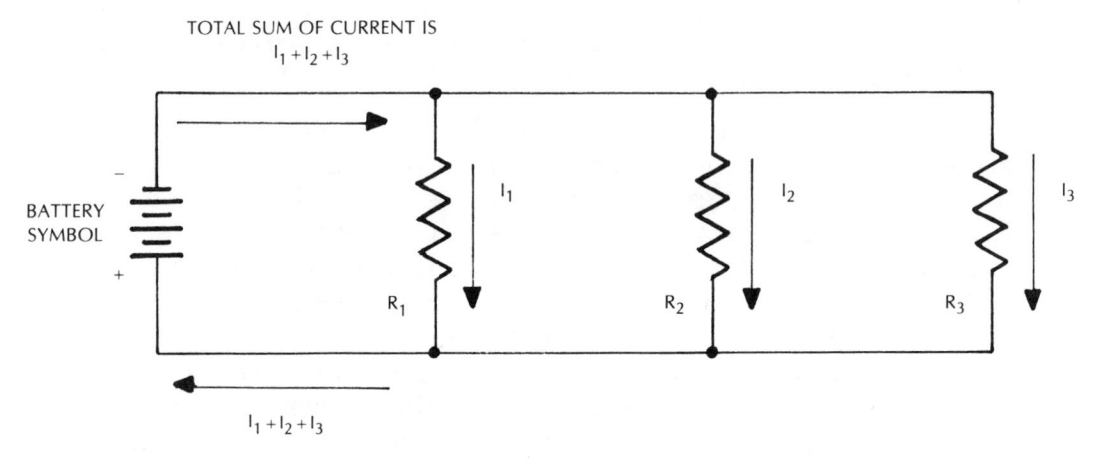

Figure 1-7: Example parallel circuit. The total current flowing out of the negative terminal of the battery equals the sum of the currents flowing through the three parallel resistors.

current flowing through the resistors will equal the current flowing into and out of the battery. See Figure 1-7.

How to Make dc Measurements

An instrument used to measure current is an ammeter. Nevertheless, when you are working with solid state circuits, you will usually need to measure much less than one ampere; this is often done by using a digital multimeter (DMM). Typically, your measurements will be in milliamperes (thousandths of an ampere) and microamperes (millionths of an ampere). As a general rule, a DMM is limited to measuring no more than

1,000 or 2,000 milliamperes (1 or 2 amperes). Moreover, when a DMM is set to its current function, it really measures the voltage developed by a current flowing in a known shunt (or shunts) resistance. Even though the DMM is operating as a voltmeter, the current function display is calibrated in milliamperes or amperes.

There are both internal and external shunts. Internal shunts are located inside the DMM case; external shunts, outside the case. An external shunt is a very low resistance conductor connected across the input terminals of the DMM (or other type of current-measuring device), to carry most of the current being measured. When this is done, it is usually referred to as "extending the range" of the ammeter, DMM, etc.

To extend the current range of a DMM, all you have to do is shunt (parallel) the meter input leads with the proper value of resistance. If the resistance of the shunt is low, a large current through it will develop only a small voltage drop. For example, as shown in Figure 1-8, a shunt resistor of 0.001 ohm will produce a voltage across the resistance of 0.001 volt for each ampere of current that flows through it.

If your DMM will display a reading of 0.001 volt when connected across a resistor of 0.001 ohm, as shown in Figure 1-8, all you have to remember is that for each ampere flowing through the resistor there will be a voltage reading of 0.001 volt. Therefore, with a reading of 0.001, you know that 1 ampere of current is being supplied to the load. As another example, if you read 0.010 volt, this represents 10 amperes flowing through the 0.001-ohm shunt, and obviously, in this case, that is the current supplied to the load.

There are two important things to note here: (1) for all practical purposes, you are required to shunt your meter (DMM or whatever) with a short circuit, and (2) the current-measuring meter and its external shunt resistor *must* be placed in series with the load.

Now comes the matter of wattage. Since an external meter shunt necessarily offers some resistance to the current flowing through it, it will develop heat that the shunt must dissipate. You can calculate the heat loss by using Ohm's laws for dc circuits. These are

$$W = I^2 R = EI = E^2/R$$

where W is watts, I is current, E is voltage, and R is the resist-

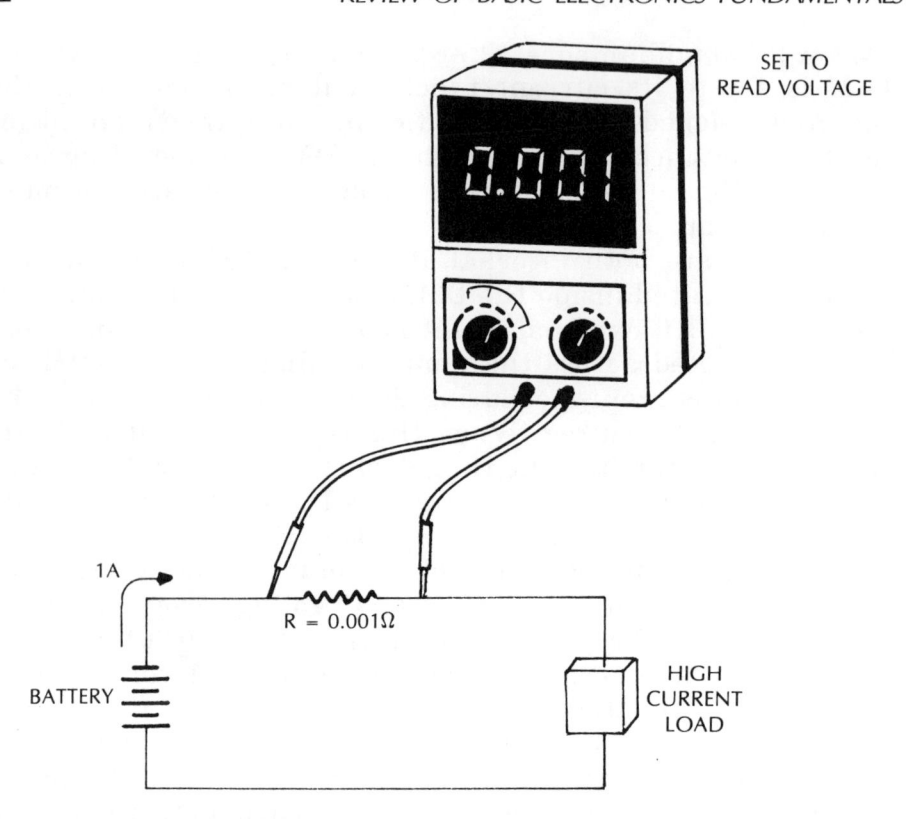

Figure 1-8: Using a DMM to measure current through a very small shunt resistor.

ance of the shunt. Thus, a current of 10 amperes would develop 25 watts of heat in 0.25 ohms of resistance. But the same 10 amperes flowing through a 0.001-ohm shunt would develop only 0.1 watt of heat.

Now, let's go a bit farther and ask the question, "What would be the voltage measured across the 0.001-ohm shunt when it is dissipating 0.1 watt of heat?" Transposing the formula $W = E^2/R$ to find voltage:

$$E = \sqrt{WR} = \sqrt{(0.1)(0.001)} = \sqrt{0.0001}$$

$$= 0.01 \text{ V or 10 millivolts.}$$

One last question. What if your DMM does not indicate readings such as 0.001 volt? To get around this problem, simply

REVIEW OF BASIC ELECTRONICS FUNDAMENTALS

construct a shunt of higher resistance, using the procedure we have just discussed. For example, if you build a 0.01-ohm shunt, a reading of 0.017 volt represents a current flow of 1.7 amperes through the load.

Guide to AC Measurements

Since all digital multimeters measure dc voltages, an internal ac converter is used to convert ac voltages to dc. The dc is then fed to an analog-to-digital (A/D) converter that, in turn, feeds the display section. In most cases, the scale of such an instrument is calibrated to read rms values. By the way, the rms value of a sine wave of current or voltage is 1.11 times the average of the instantaneous values. For example, the average value of a 100-volt peak sinusoidal voltage (or current of 100 amps) is approximately 69 volts, if we use the chart shown in Figure 1-2. This comes out to be about 8% off the calculated value (63.7 V).

WAVEFORM	VOLTAGE VALUE				
	PEAK VOLTAGE	AVERAGE VOLTAGE	ACTUAL METER READING	AR × CF = TRUE READING	CORRECTION FACTOR
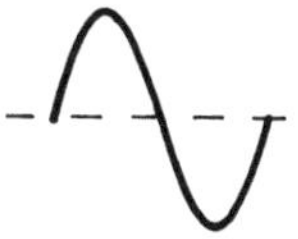 sine wave	10	6.38	7.07	7.07	1
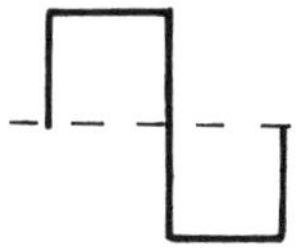 square wave	10	10	11.1	10	0.9
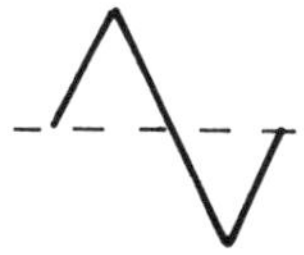 triangle wave	10	5	5.55	5.77	1.04

Figure 1-9: Correction factors for average reading meter scaled to read rms. From *Workbench Guide to Electronic Troubleshooting*, page 22.

If you measure a voltage or current of some other waveform—say a square or triangular wave rather than a sinusoidal one—and your meter is scaled to read rms, in all probability your readings will be incorrect. For instance, let's assume that you measure the output voltage of a function generator set in the sine, square, and triangular waveform modes. With a constant peak value of 10 volts from the function generator during each measurement, you will measure 7.07 for the sine wave, 11.1 volts for the square, and 5.55 volts for the triangular. The only correct rms reading occurred when you were measuring the sine wave voltage. You should use correction factors to find the true values for the others. Figure 1-9 shows the correction factors for an average reading meter scaled to read rms.

What You Should Know About Circuit Loading During Measurement

The multimeter is the workhorse for almost the entire electronics industry. Moreover, it is the principal—and sometimes only—test instrument used by most experimenters. Nevertheless, many technicians are at a loss to explain why most low-cost meters give different readings on different ranges when they are placed across the same circuit component. When you connect a voltmeter across a circuit to make a voltage measurement (ac or dc), it shunts the circuit. If the voltmeter has a low internal resistance, it will bypass an appreciable amount of current. In effect, what you have done is decrease the actual resistance of the circuit being tested by paralleling it with another resistance (the meter). As a result, the voltage reading will also decrease. Under these conditions, it is said that the voltmeter is "loading the circuit." To see why circuit loading will cause you to have different readings on different ranges—and erroneous readings—let's consider the effects of circuit loading by a meter on the different ranges.

To begin, the amount of circuit loading depends on the ohms-per-volt rating of the meter you are using. This rating is equal to the reciprocal of the meter's full-scale current in amperes, or the quotient of the meter's resistance divided by the full-scale voltage reading. To find the resistance of a voltmeter, multiply the ohm-per-volt (Ω/V) rating by the voltage. Table 1-1 shows the meter resistance for three scales of a

Ω/V		R_{METER} = FULL SCALE READING		
10,000	×	10V	=	100,000V
10,000	×	50V	=	500,000V
10,000	×	250V	=	2,500,000V

Table 1-1: Meter resistance of different scales of the same ac voltmeter (see text).

10,000-ohms-per-volt ac voltmeter, calculated by using the formula ohms-per-volt times full-scale voltage.

For negligible loading when making voltage readings, your voltmeter resistance should be at least 20 times greater than the resistance of the circuit being measured. This means that you cannot measure the voltage across a resistance of more than 5,000 ohms when using the full-scale voltage of 10 volts shown in Table 1-1 without realizing that there will be an erroneous reading. Circuit loading is less a problem in low-resistance circuits than in high-resistance circuits because, as you can see by this example, shunting effect is less in a low-resistance circuit. Whenever possible, it is best to use a voltmeter with an ohms/volt rating as large as practical. By referring to Table 1-1, you can see that the higher the voltage range, the larger the meter resistance. It follows that the higher range you use during a measurement, the less loading you will have. It also should be apparent by now why there will be different readings on different scales of a low ohms/volt voltmeter.

Using Capacitors

As you may remember, a capacitor is a component used in electronic circuits, which basically consists of two conducting surfaces separated by an insulating material (dielectric) such as paper, ceramic, plastic, or air. Still, we should not think of capacitance solely in terms of its manifestation in capacitors.

We can define capacitance as the property of a capacitor that determines how much charge can be stored in it for a given potential difference across its terminals. But it seems as if there is more to it than that. Theoretically, all conductive surfaces not electrically connected to each other exhibit capacitance between each other. To put this statement into practical terms, almost everything you work on in the shop has capacitance: coax, twin lead, ribbon cable, printed circuit boards, semiconductor junctions, and, as we have said, almost any electronic circuit you can name.

The traditional textbook formulas given for calculating capacitance are not practical for use by most of today's technicians. It is far too easy, using one of the newer digital capacitance meters, simply to measure the exact value (using laboratory standards, ±0.2%) of a capacitor. Digital capacitance meters provide a wonderful way to determine the value of a capacitor. But, in a practical sense, most capacitors are used in a circuit operating at a given frequency. Capacitive reactance is the key to these problems.

The traditional formula for calculating capacitive reactance is $X_C = 1/(2\pi fC)$, where X_C is in ohms, π is equal to 3.14159292, f is frequency in hertz, and C is capacitance in farads. Going a bit further, resistance-capacitance (RC) combinations exhibit what is known as a "time constant." The formula for determining the time constant is t = RC, where t is the time in seconds, R is the resistance in ohms, and C is the capacitance in farads. Incidentally, microfarads and ohms yield time constants in microseconds, or, if you use picofarads and megohms, you will also obtain answers in microseconds.

Signal losses may be determined by using the formula for capacitive reactance, or the chart in Figure 1-11 (assuming no other losses in the attenuator design); and the RC time constant formula can be used to calculate such things as the components used for the 75-microsecond de-emphasis used in FM radio.

Example:

The reactance of a 0.01-microfarad capacitor at 54 megahertz is

$X_C = 1/(2\pi fC) = 1/6.2831858 \times 54 \times 0.01 = 0.2947313$ ohms.

As you can see, almost no signal loss would occur when using this capacitor at this frequency.

REVIEW OF BASIC ELECTRONICS FUNDAMENTALS

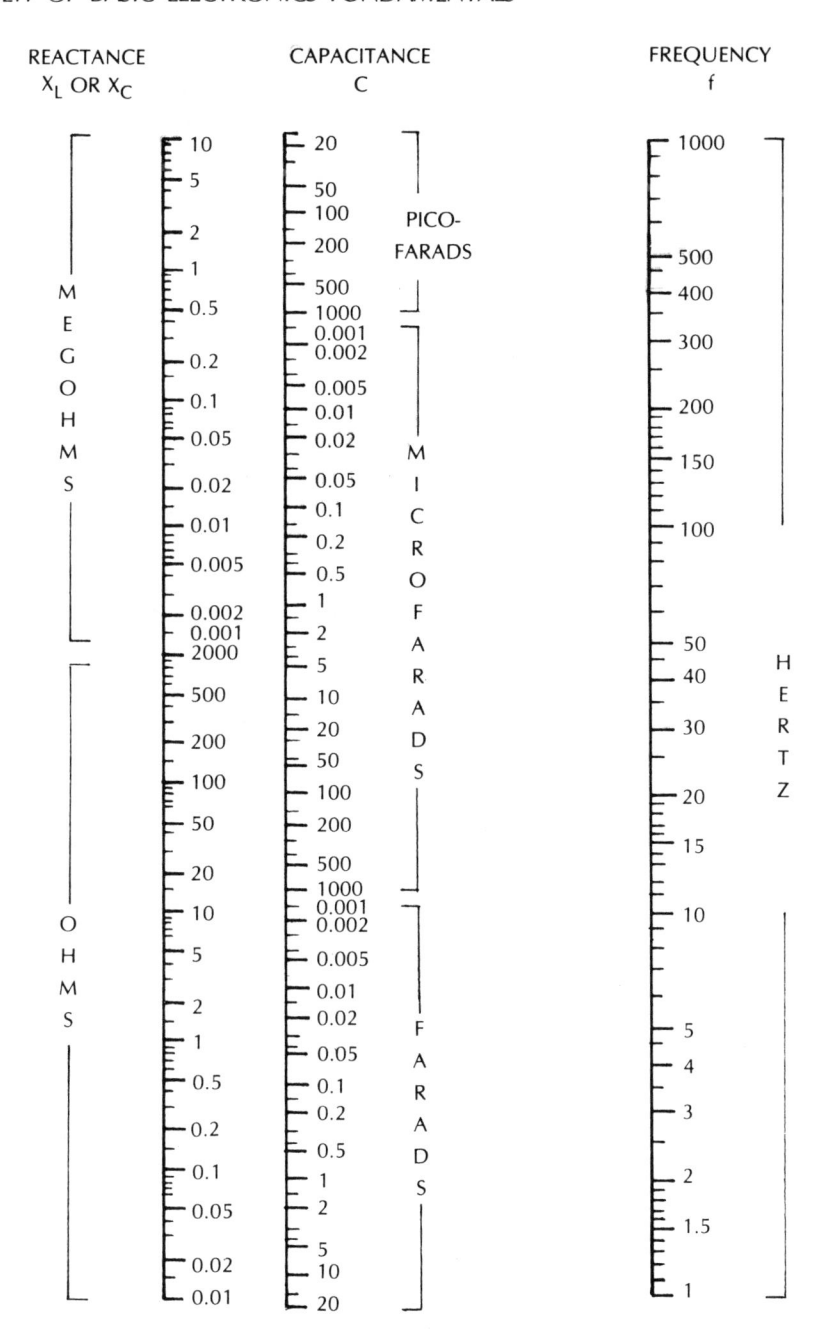

Figure 1-11: Chart for determining the capacitive reactance of a certain capacitor at a given frequency.

Your answer is 0.002 microfarads. Next, what is the equivalent reactance of a 0.01-microfarad capacitor at 1600 hertz? First, place your straightedge across the chart, starting at 1600 hertz (the first mark above the number 150), then across the 0.01 microfarads. Read 0.1 megohm on the reactance scale.

Guide to Inductance

Every conductor has inductance, even though the conductor may not have been designed as an inductor, if (and only if) the current flowing through the conductor is changing to some higher or lower value. Inductance in any circuit is the property that opposes any change in the existing current in that circuit. The unit of inductance may be stated in henrys, millihenrys, or microhenrys.

When an alternating voltage is applied to an inductance, a *back* electromotive force (emf) is generated in the inductance. This emf is proportional to the rate at which the current changes (the more rapid the change, the greater the back emf developed), and this, in turn, is proportional to the frequency of the alternating voltage. This opposition to the flow of alternating current is called *inductive reactance*, symbolized by X_L. Reactance is measured in ohms, and the formula for inductive reactance is

$$X_L = 2\pi fL$$

where X_L is inductive reactance measured in ohms, f is frequency in hertz, L is inductance in henrys, and π is equal to 3.1459296.

Example:

The reactance of a certain 15-microhenry inductor at 54 megahertz is

$X_L = 2\pi fL = 6.2831858 \times 54 \times 15 = 5,089.3804$ ohms.

Working with Solid State Rectifiers

By definition, a solid state rectifier is a device which, by virtue of its lack of symmetry, converts an alternating current into a unidirectional or direct current. The device for accom-

REVIEW OF BASIC ELECTRONICS FUNDAMENTALS

V_{RRM} VOLTS	1.0 PLASTIC	3 METAL	3 METAL	6 PLASTIC	12 METAL	12 PLASTIC	15 METAL	25 PLASTIC	25 METAL	50 METAL
50	1N4001	1N4719	1N4997	MR750	MR1120 1N119,A,B	1N1199C	1N3208	MR2500	1N3491	MR5005
100	1N4002	1N4720	1N4998	MR751	MR1121 1N1200,A,B	1N1200C	1N3209	MR2501	1N3492	MR5010
200	1N4003	1N4721	1N4999	MR752	MR1122 1N1202,A,B	1N1202C	1N3210	MR2502	1N3493	MR5020
400	1N4004	1N4722	1N5000	MR754	MR1124 1N1204,A,B	1N1204C	1N3212	MR2504	1N3495	MR5040
600	1N4005	1N4723	1N5001	MR756	MR1126 1N1206,A,B	1N1206C	1N3214	MR2506	MR328	—

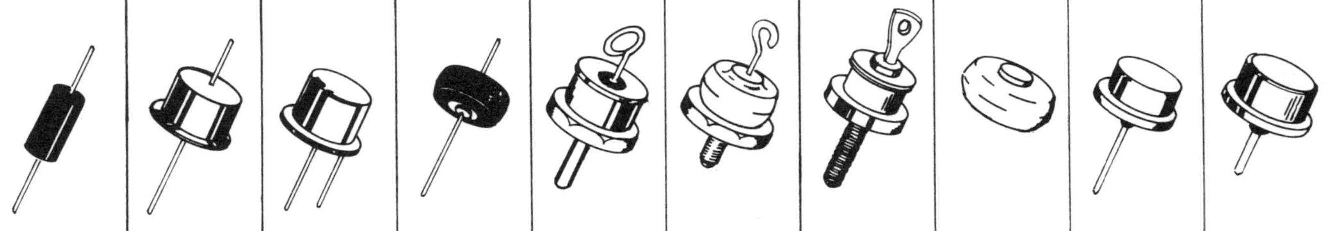

I_O, AVERAGE RECTIFIED FORWARD CURRENT (Amperes)

Figure 1-12: Modern general purpose rectifiers.

plishing this is a semiconductor diode rectifier. The diode rectifier does its job of converting ac to dc by offering current a low resistance path in one direction (the "forward" direction), and a high resistance path in the other direction (the "reverse" direction). Some typical modern packages and forward current values for a few general purpose rectifiers offered by the Motorola Company are shown in Figure 1-12.

When one of these devices is connected with a positive voltage on its anode and a negative voltage on its cathode, it conducts in the forward direction. If the diode connections are reversed (offering a high resistance path to current flow), the diode does not conduct (instead, it blocks) the current. If an ac voltage is applied to the diode, it conducts only during the time that its anode is positive. Or, to put it another way, it simply blocks one-half (the negative half-cycle) the ac voltage cycle. Figure 1-13 is an example showing the voltage-current char-

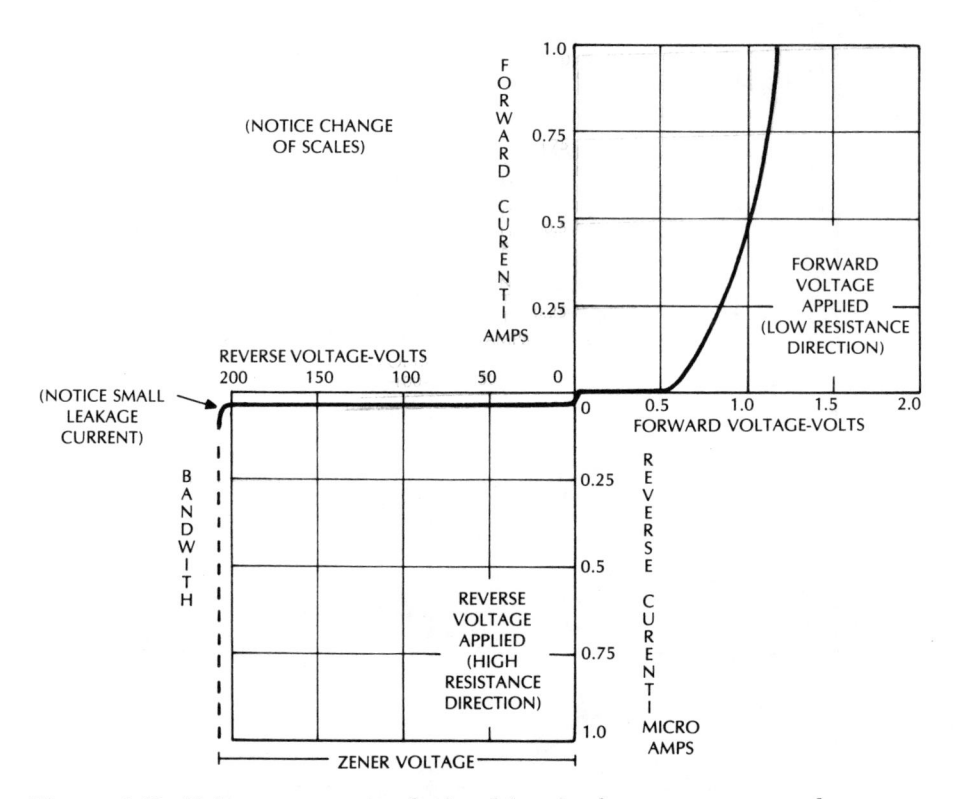

Figure 1-13: Voltage-current relationship of a low-power general purpose solid state diode.

REVIEW OF BASIC ELECTRONICS FUNDAMENTALS

acteristics of a typical general purpose rectifier. The most useful parameters are labeled for your convenience.

Granting that, in practice, most diodes are used in simple rectifying applications, you will find several uses where special purpose rectifiers and diodes are of paramount importance. Table 1-2 lists several special types and applications.

Conventional silicon diodes are sine wave diodes and they work fine for a 60-hertz sine wave or pulse. However, when these standard types are used in applications that utilize high frequency pulses (such as some TV flyback-derived power supplies), they will not turn off fast enough. Because of this,

TYPE	USE
Fast recovery rectifiers	Used in designs requiring a power rectifier having maximum switching time (on-off time) ranging from 200 nanoseconds to 750 nanoseconds.
Schottky rectifiers	Used in low-voltage high-frequency power supplies and as very fast clamping diodes. These devices feature switching times less than 10 nanoseconds.
Rectifier bridges	These devices are used when the performance of four individual diodes is required, and cost about the same as only two. Also, reliability is good — comparable to that of a single diode. Some of the modern versions have fast recovery time — less than 200 nanoseconds.
Tunnel diodes	Used in applications such as UHF and microwave oscillators and amplifiers.
Zener diodes	Basically automatic variable resistance devices. Manufactured in both plastic (low-cost) and glass (highest reliability). Applications: voltage reference, regulation, and limiting. Available tolerances from 10% to as close as 1% (where strict control is required).
Varactor diode	Used in resonant and oscillator circuits.
Pin diodes	UHF switching and microwave circuits.
Switching diodes	Used in applications that require rapid switching or modulation at nanosecond speed. Usually for the control of low-power microwave frequencies.

Table 1-2: Special rectifier and diode applications.

they will draw a very high reverse current. An example of what can happen if you use the wrong diode is that the diode will burn out in less than a minute. A good rule is this: when in doubt, use a fast-recovery diode. Even where there are very sharp pulses, a fast-recovery diode will turn off completely and will not draw reverse current.

A word about zener diodes: if you are working with a modern power supply regulator circuit and find the voltage across a zener diode is not correct to within ±5%, check it or replace the diode.

Practical Applications Using Transistors

Transistors are a very popular electronic component—an element you will use a lot, if you haven't already. They come in all shapes and sizes, from almost invisible ones constructed on semiconductor chips up to large ones that handle high voltages (V_{CC}, 28 volts) and high powers (input power 20 watts, and output power 100 watts). Figure 1-14 shows a variety of transistors and module packages offered by Motorola.

For the moment, let's restrict ourselves to discussing bipolar transistors (the general name for npn and pnp transistors). Transistors of this type may have a positive voltage applied to the collector (npn), or a negative voltage on the collector (pnp). In npn transistors, the polarity of the applied voltage that causes current to flow from the emitter to the collector (assuming a common emitter configuration) is called *forward bias*. Note: The word "current," in this case, means electron flow. The forward bias is applied between the *emitter* and the *base* lead of the transistor which, in turn, controls the current in the output circuit—the collector circuit. Figure 1-15 shows a common emitter transistor amplifier circuit.

Notice that the forward bias polarity can be determined by arrow direction. If the arrow does not point in towards the base, it is an npn transistor. (This transistor should have a positive voltage on its base—see Figure 1-15.) On the other hand, if the schematic diagram shows the arrow pointing inward toward the base, the base should be negative and the transistor is a pnp. Note: All transistors must have at least the minimum rated voltage (bias voltage) applied to their base or they will cut off (there will be no current flow).

REVIEW OF BASIC ELECTRONICS FUNDAMENTALS 25

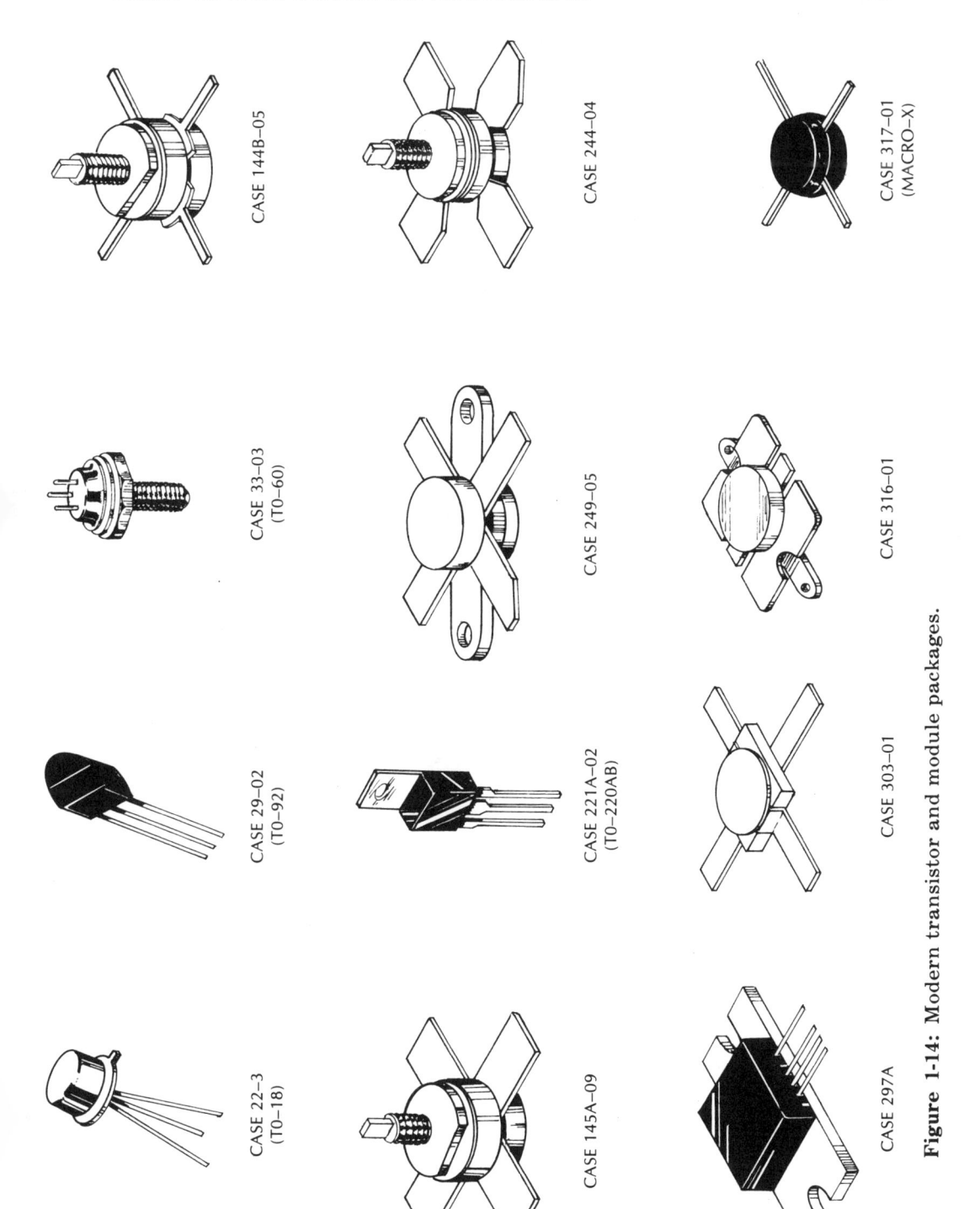

Figure 1-14: Modern transistor and module packages.

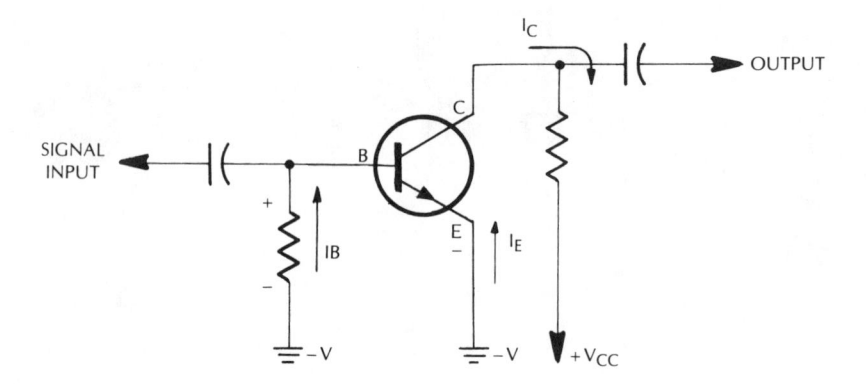

Figure 1-15: Common emitter transistor circuit. Current will flow into the base (B) and emitter (E) and out the collector (C), when the correct polarity is applied to the transistor leads, as shown.

These transistors always have a reverse bias on the collector. This means that whatever the collector material is (n or p material), the collector voltage polarity will be the opposite. For example, an npn transistor should have a positive voltage on its collector and a pnp should have a negative voltage on its collector.

We have said that in order to have a current flow in the collector circuit, all transistors must have a bias voltage applied to their base. There are several different circuits used to develop this required bias. However, due to the fact that we will cover the full range of transistor circuits in the following chapters, we will not go into the details of such circuits now. But it is a good idea to review transistor circuit computations at this time. As an example of how to calculate the various currents flowing in a transistor amplifier, we will use the common-emitter circuits shown in Figure 1-16.

First, calculating the base current (I_b):

$I_b = 5/100,000 = 0.00005 = 0.05$ milliampere.

The voltage drop between the emitter and base is assumed to be zero. Next, the collector current (I_c) is found by multiplying the base current by the beta (current gain) of the transistor. You

REVIEW OF BASIC ELECTRONICS FUNDAMENTALS

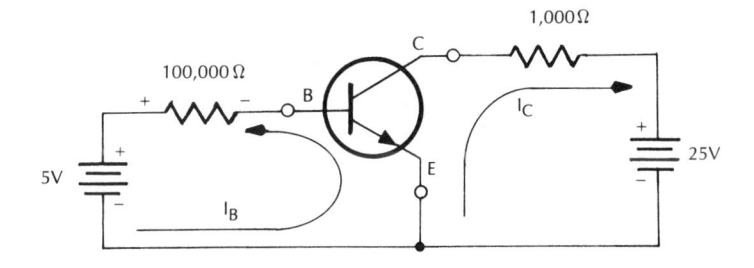

Figure 1-16: An npn transistor using fixed bias and wired to operate in the common-emitter mode. Arrows show direction of electron flow.

can find the current gain (β) of a transistor in manufacturers' spec sheets, data books, etc. Using our assumed value:

$$I_c = (\beta)(I_b) = 100 \times 0.05 \text{ milliampere} = 5 \text{ milliampere}.$$

Now that we know the collector current, we can calculate the voltage across the 1000-ohm resistor in the collector circuit:

$$E = IR = 5 \text{ milliampere} \times 1000 = 5V.$$

Therefore, the difference between the battery voltage (25V) and the voltage you measure across the 1000-ohm resistor is 20 volts. This 20 volts is what you should measure between the emitter and collector of the transistor.

Example:

Find the I_b, I_c, and the voltage between the emitter and collector V_{CE}, when the transistor's $\beta = 80$ (see Figure 1-17).

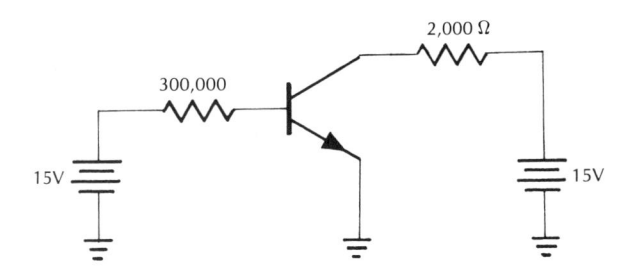

Figure 1-17: Schematic for example problem (see text).

Your work should look like this:

$I_b = 15/300{,}000 = 0.05$ milliampere.
$I_c = (\beta)(I_b) = 80 \times 0.05$ milliampere $= 4$ milliampere.
$V_{CE} = 15 - (2000 \times 0.004) = 7V$.

There are several variations from the circuit shown in Figure 1-17: the common collector (also called the *emitter follower*) and the common base. To calculate the various voltages and currents in these circuits, you can use the same general procedure as we have used to calculate these values in the common-emitter circuit. Just remember, electron flow is always from negative to positive.

Today's and Tomorrow's Integrated Circuits

Remarkable advances have been made in the integrated circuit industry in the past few years. For example: Bell Laboratories' 100,000 transistors on a single chip (an integrated circuit ready for use is a packaged chip); Hewlett-Packard's 450,000 transistor central processing unit (**CPU**) on a single chip; and there are VLSI (very large scale integration) devices that contain up to 600,000 transistors per chip. If what the industry tells me becomes a reality, we can expect even more miniaturization in the next few years—*millions of transistors* on a chip. These new devices will be introduced to us as "Ultra-Large-Scale Integration" (ULSI), if things work out as now planned.

Integrated circuits (**IC's**) are sometimes described in terms of package type (metal, ceramic, or plastic), and, at other times, according to operating mode (*linear* or *digital*). In case you have forgotten, linear operation is the mode in which the output of the **IC** is proportional to the input and, frequently, the **IC** reproduces a faithful, amplified version of the input. This type of operation is most often also sinusoidal. Some examples of linear **IC's** are operational amplifiers, balanced modulator-demodulators, voltage regulators, intermediate frequency (IF) amplifiers, TV sound circuits, TV video IF amplifiers, Darlington amplifiers, and differential amplifiers.

REVIEW OF BASIC ELECTRONICS FUNDAMENTALS   29

Digital **IC**'s work in an "on-off" mode; i.e., bistable operation. The circuits inside the digital **IC** operate as a switch (either on or off) and, most important, can make logical decisions. Digital **IC**'s include the following types: gates, shift registers, adders, flip-flops, multiplexers, counters, BCD-to-decimal decoders, and microcomputer chips such as Motorola's MC6805R2, an 8-bit microcomputer unit with A/D converter (see Chapters three and four). Figure 1-18 shows three microcomputer **IC** packages.

It would be very tiresome to draw the complete **IC** circuits (such as for the packages shown in Figure 1-18) in a schematic diagram, especially if several **IC**'s are being used in a complete system. The symbol shown in Figure 1-19 represents a complete **IC** and will be used throughout this book.

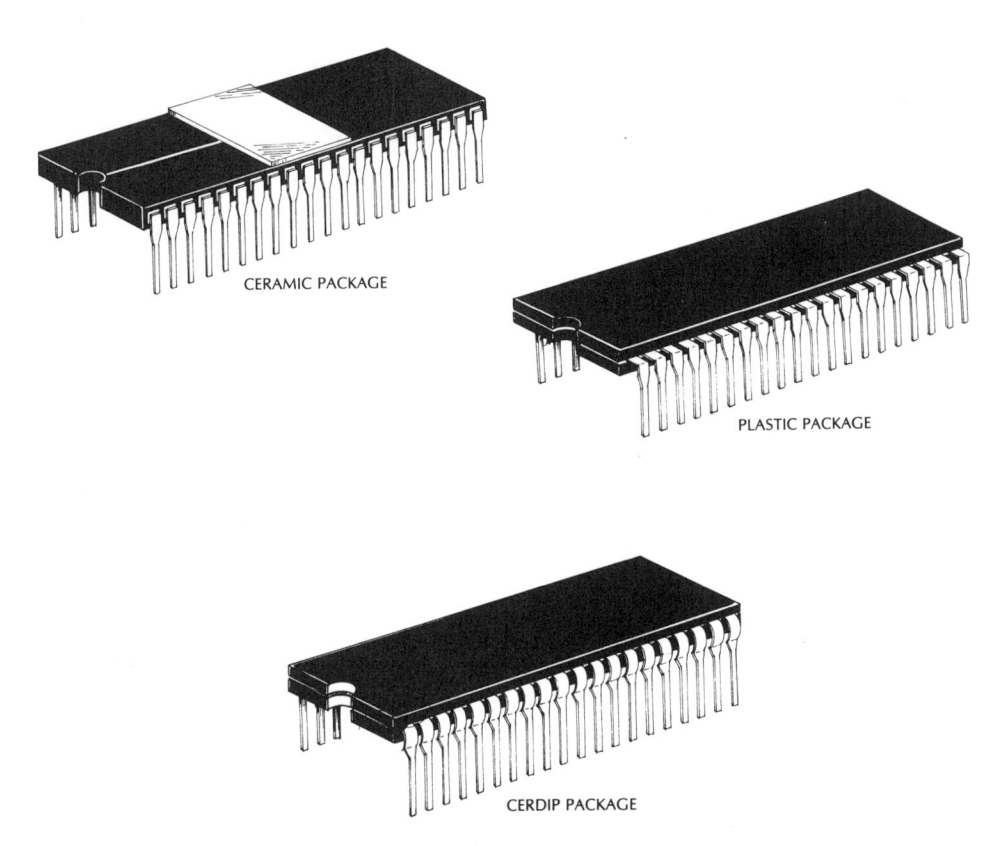

CERAMIC PACKAGE

PLASTIC PACKAGE

CERDIP PACKAGE

Figure 1-18: Microcomputer IC packages.

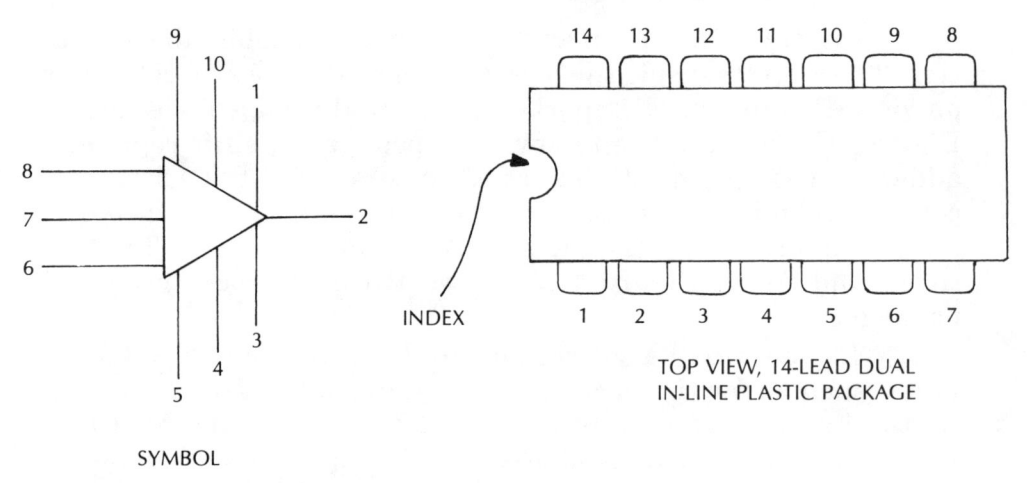

Figure 1-19: Example symbol and pin layout for an IC.

You will also find that some digital IC's will contain digital symbols within the pin layout package. Figure 1-20 shows some common digital IC symbols. In general, two, four, or six of these symbols (gates, flip-flops, operational amplifiers, etc.) are drawn inside the illustration of the package of a certain IC.

The use of four operational amplifier symbols inside a single IC package is illustrated in Figure 1-21. This *linear* IC is known as a *quad differential input operational amplifier*, identification numbers LM124, LM224, LM324, and LM2901, or, simply, the LM124 series.

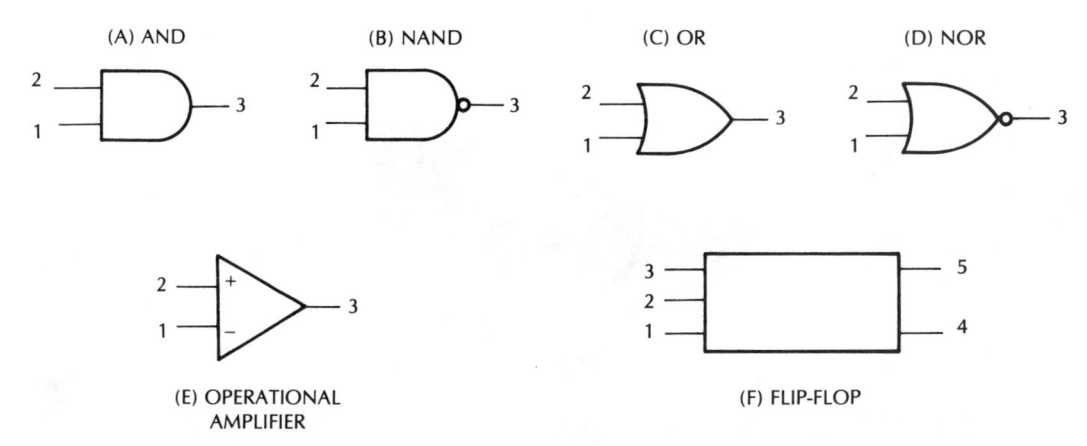

Figure 1-20: Typical interpackage showing digital symbols.

REVIEW OF BASIC ELECTRONICS FUNDAMENTALS 31

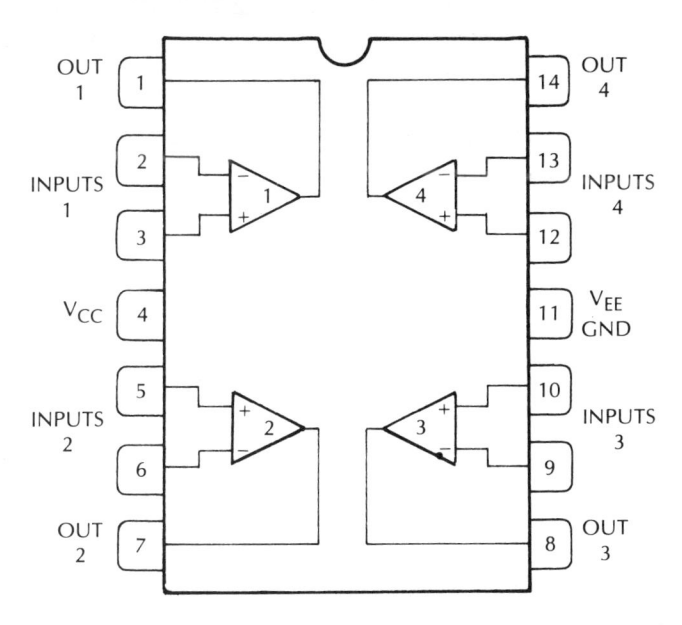

Figure 1-21: Single IC package showing pin connections and internal operational amplifier symbols for Motorola's LM124 OP AMP series.

CHAPTER TWO

Understanding Today's Solid State Circuits

This chapter provides the practical guidance you'll need to work with today's digital, as well as linear, equipment. For example: solid state speech-processing systems (talking calculators, games, toys, etc.); microprocessors (found in everything from kitchen equipment to sophisticated laboratory data-acquisition systems); and such things as video cassette recorders, video disc players, and analog-to-digital converters.

You'll find that this chapter contains information that will enable you to use modern procedures when working with digital circuits. When you are working with state-of-the-art hardware (a term used to describe the physical or hard components that make up a computer system), the main requirement is to understand computer-oriented technology. You must work using two points of view—a logic structure in one, and an electronic circuit in the other. This will not be difficult if you follow the procedures set forth in the following pages.

Introduction to Linear and Digital Integrated Circuits

In the last chapter, it was stated that a *linear* amplifier is an amplifier that faithfully reproduces the input waveform on its output. There are two terms you will encounter that are sometimes confusing: *linear* and *analog*. There is a difference. An analog circuit is one in which the output varies as a continuous function (often an electrical analogy of a mathematical problem) of the input. But a linear circuit has an output that varies in *direct proportion* to the input; i.e., we want the output to be an exact reproduction of the input (except for possible amplification, noise reduction, etc.). As an example of a linear IC circuit, we will use Motorola's general purpose transistor array CA3054. The pin connections and transistor diagram are shown in Figure 2-1. The CA3054 has six npn transistors that comprise the amplifiers and are useful from dc to 120 megahertz (MHz).

The CA3054 consists of a single slice of silicon substrata on which the integrated circuit is built; hence, it is called a *monolithic integrated circuit*. The monolithic construction of this IC provides close electrical and thermal matching of the amplifiers, which makes it particularly useful in dual-channel applications where matched performance of two channels is required. Referring to Figure 2-1, you will see that the CA3054 is a *dual independent differential amplifier*. Pins 2 and 13 are the two inputs for the upper IC differential amplifier and pins 6 and 9 are the inputs to the differential amplifier on the bottom half of the IC.

In case you are not familiar with a differential amplifier, it is a circuit that amplifies the *difference* between two input signals, but effectively suppresses all *like* voltages or currents on these inputs. When properly connected, a differential amplifier (such as either of these) will produce an output only when there is a difference in signals at the inputs.

Differential amplifiers are particularly useful (especially in oscilloscopes, electronic meters, recording instrument amplifiers, and for data transmission in electrically noisy environments) because signals common to both inputs (for example, pins 2 and 13 in Figure 2-1) are eliminated or greatly reduced. Inputs such as these are known as a *common mode signal*.

Frequently, a differential amplifier is used as an input circuit of an operational amplifier (for instance, Motorola's LM124 series). The ability of the differential amplifier to prevent common mode signals from passing on to the next amplifier (the OP AMP), is one of its most useful characteristics.

UNDERSTANDING TODAY'S SOLID STATE CIRCUITS 35

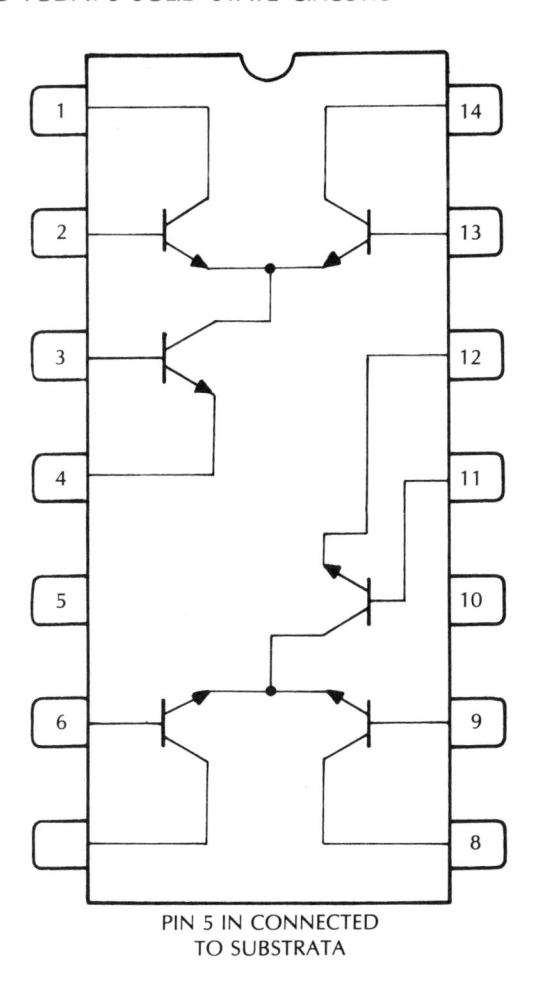

PIN 5 IN CONNECTED
TO SUBSTRATA

Figure 2-1: Pin connections for Motorola's linear integrated general purpose transistor array CA3054.

Of course, as we have said, it is a linear amplifier; therefore all desirable signals are amplified replicas of the difference signal (the desired signal), and these are fed to the next stage.

Let's now turn our attention to digital IC's. These IC's are built containing electronic circuits that use numerical and logical data for their operation. The input and output signals are in the form of electrical pulses, ordinarily and ideally of the perfect square wave type. Digital circuits are nothing more than switching circuits, as we know. This switching (on and off) produces electrical pulses that are called *bits*. An *off* condition may be referred to as a 0 bit (for example, zero voltage or ground), while the *on* condition (generally about 2 to 15 volts) may be represented as a 1 bit. For proper operation, digital circuits should always be in one or the other of these two states or conditions, either *high* (1), or *low* (0). These states (levels) are also called *logic levels*.

When checking logic levels, absolute amplitudes are unimportant. As an example, in some transistor-transistor logic (TTL) data sheets, when the amplitude of the digital pulse is 0.8 ±0.15 volts, it's at logic 0 level. When the amplitude of the pulse is 2.1 ±0.25 volts, it's at logic 1 level. The series of logic levels (pulses), as shown in Figure 2-2, is referred to as a *pulse train* and frequently will represent *binary* data in a digital circuit. Note: The binary number system uses only the 0 and 1 digits.

In summary, digital circuits operate with pulses. The voltage levels assigned to a logic level (1 or 0) determine whether the circuit is in an on or off condition—also called *high* (1), *low* (0), and *true* (1), *false* (0). The binary number system with a base 2 is frequently used to convey information from point to point in a digital system where a pulse represents 1 and the absence of a pulse represents 0.

Three elements used in digital circuits (called integrated circuits, **IC**, or logic circuits) are **AND** gates, **OR** gates, and inverters. They are shown symbolically in Figure 2-3. Exactly how these gates work will be discussed in the following pages.

On the other hand, a linear circuit is a system in which the output voltage is, for all practical purposes, directly propor-

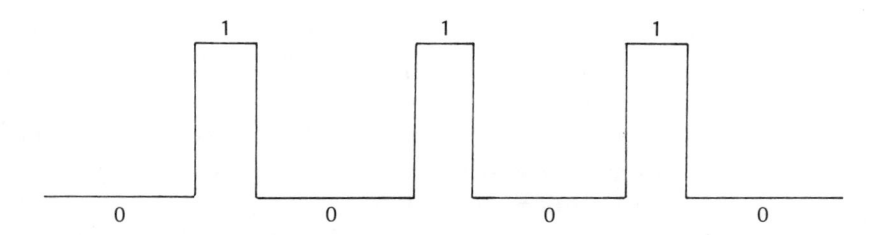

Figure 2-2: Voltage representation of bits in a pulse train.

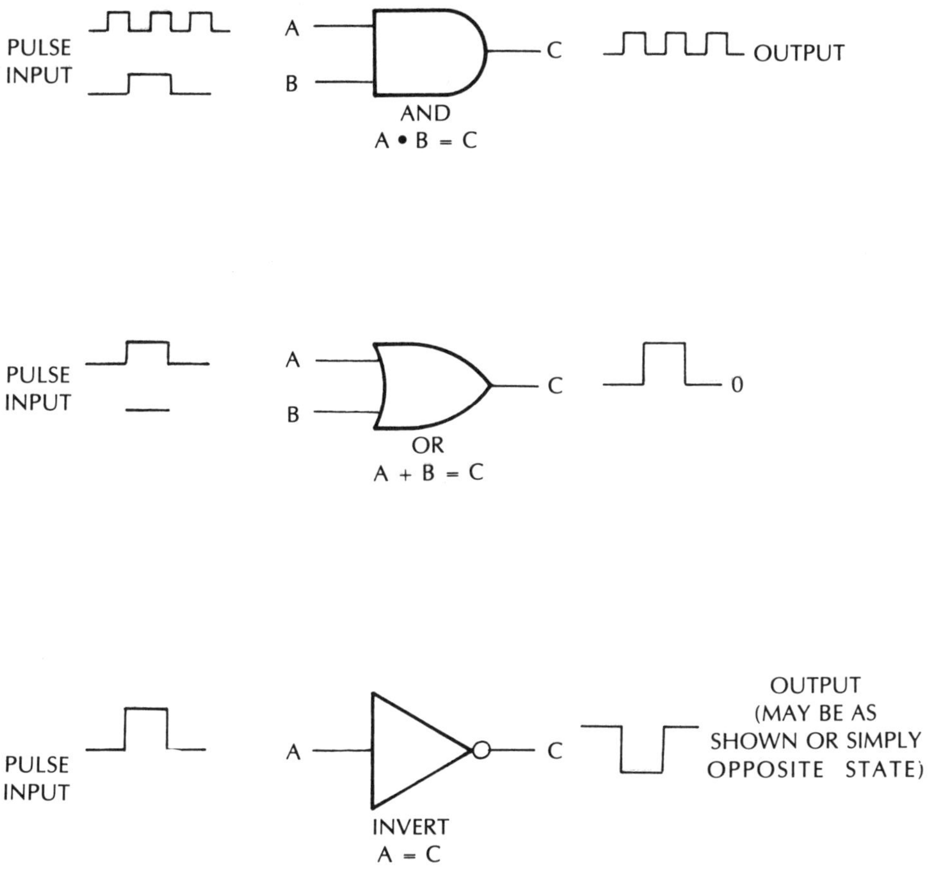

Figure 2-3: Three symbols of the basic elements of IC logic. To determine input-output signal results, read the dot (·) as AND, the plus (+) as OR, and the bar over a legend as signal inversion.

tional to the input voltage. This is only true provided you do not overdrive the circuit and do not exceed the frequency-reproducing capabilities of the device.

Another example of a linear amplifier IC is shown in Figure 2-4. This device is an 8-watt audio power amplifier (TDA2002). It is a class B (a linear amplifier) power amplifier designed for automotive and general purpose audio applications.

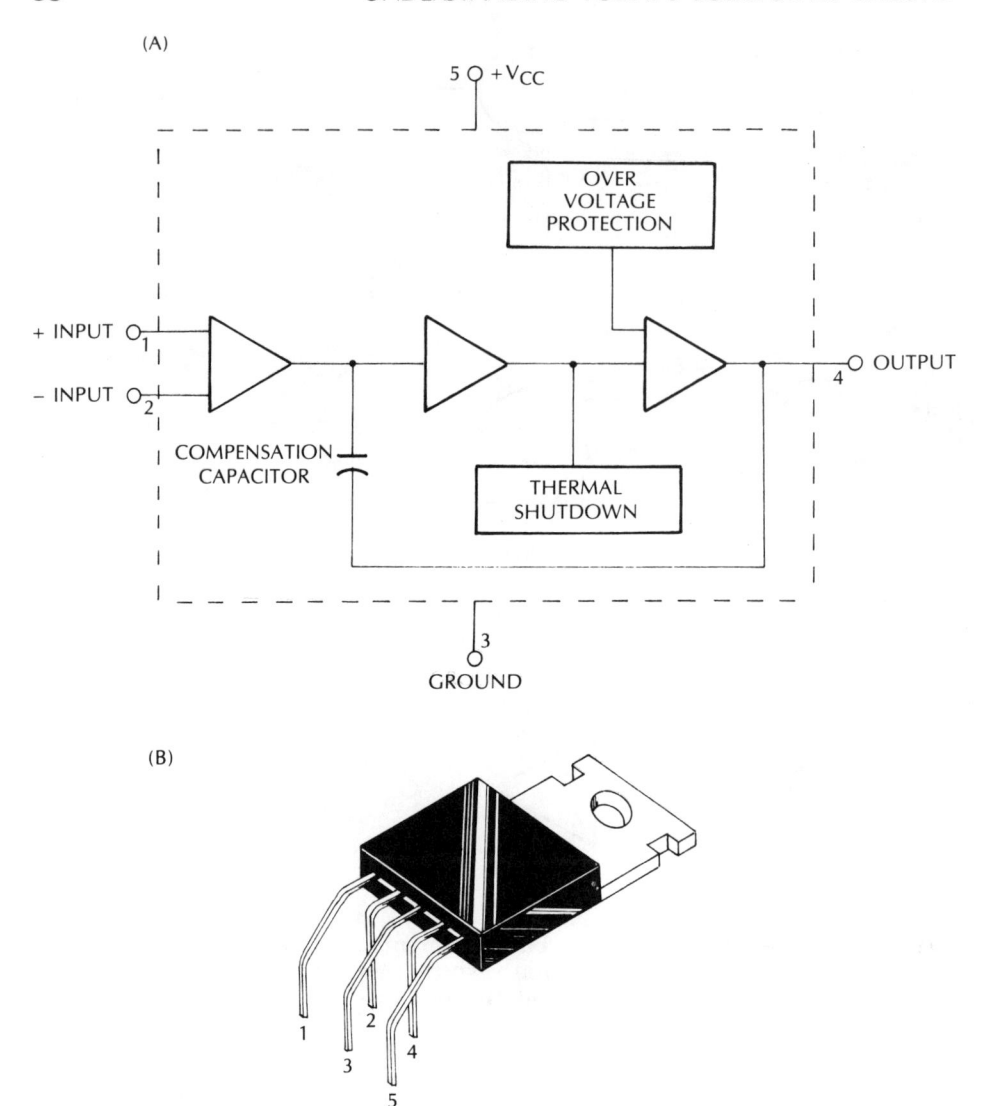

Figure 2-4: An example of a linear IC, an 8-watt audio power amplifier (TDA2002). (A) is a block diagram of the device and (B) shows the actual package of the IC.

TTL CMOS and ECL—Why?

Transistor-transistor logic refers to the type of electronic components used to construct certain integrated circuits. This type of logic is manufactured using multiple-emitter transistors

UNDERSTANDING TODAY'S SOLID STATE CIRCUITS 39

in the input circuit, and the output transistors are switching transistors, which greatly reduces matching problems. See Figure 2-5 for a schematic drawing of a **TTL IC NAND** gate.

The 7400 series is currently inexpensive and quite popular with many experimenters. This is mostly due to the low cost (a few cents) and to the fact that it has a comparatively high operating speed, fairly low power dissipation, large multiple load capabilities (fan-out), and excellent electrical noise margin. Incidentally, the example **TTL** gate shown in Figure 2-5 has only two emitters on input transistor Q1. This is not always the case. A four-input **TTL** gate will have four emitters of a single transistor as the input element. On the other hand, a 7400 series hex inverter (the 7405) has only one emitter on the input transistor. The main point is that **TTL** circuits use bipolar transistors in their design.

Compared to bipolar types, metal-oxide semiconductor (**MOS**) devices offer the advantage of low power requirements, high input impedance, and high gain, but these advantages are

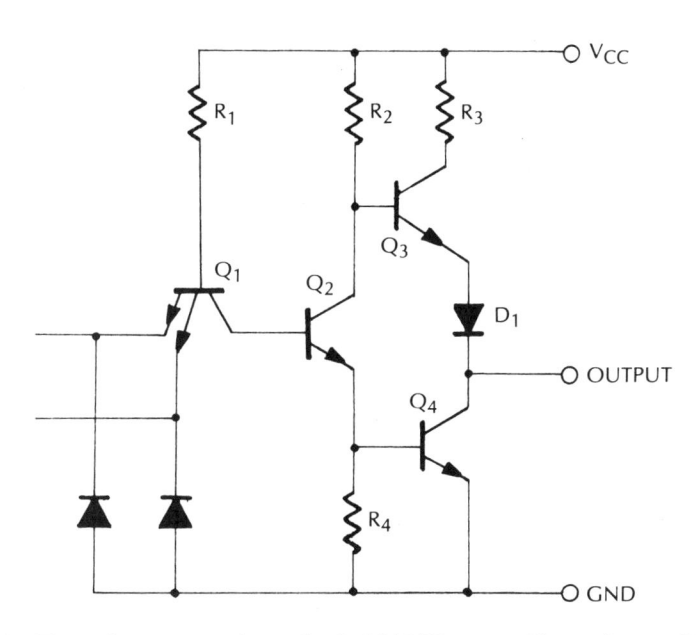

Figure 2-5: Transistor-transistor logic NAND gate. Note the multiple-emitter transistor on the input and the switching transistors on the output. This type of construction is called TTL logic.

offset for many applications by a limited power capability, and they are easily damaged by sudden changes in voltage (voltage transients) and short-term overloads.

An area of **MOS** technology that uses P-channel and N-channel metal oxide field effect transistors (**MOSFET's**), fabricated on the same chip (integrated circuit before packaging) in a complementary switching arrangement, is called **CMOS**. An illustration of a typical **CMOS** circuit configuration is shown in Figure 2-6.

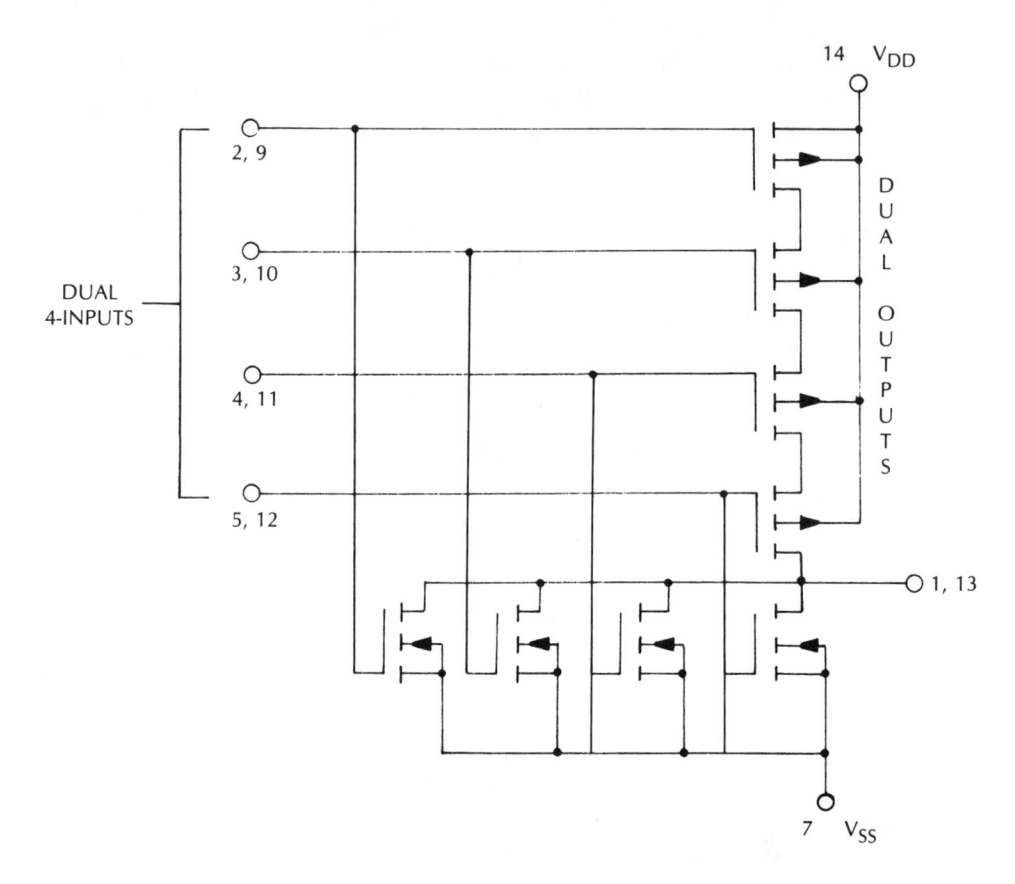

Figure 2-6: A dual 4-input NOR gate (MC14002UB) constructed with P and N channel enhancement mode devices in a single monolithic structure (complementary MOS). This circuit schematic shows dual inputs (pins 2, 3, 4, 5, and 9, 10, 11, 12) and dual outputs (pins 1 and 13): i.e., the schematic is one-half the IC but shows all pin numbers.

UNDERSTANDING TODAY'S SOLID STATE CIRCUITS 41

The **CMOS** logic family uses voltage-sensitive **MOSFET**'s as opposed to current-sensitive bipolar transistors used in the **TTL** family of IC's, resulting in an ultra-low power requirement. To see why **CMOS** requires so little power, let's examine one single **CMOS** gate extracted from the dual 4-input **NOR** gates shown in Figure 2-6. The gate (an inverter) we are going to examine is shown in Figure 2-7, with pin numbers of the MC14002UB for easy identification.

First, notice that a complementary (the "C" in **CMOS**) pair of enhancement-mode **MOSFET**'s comprise this inverter. Although the rest of the **CMOS** logic elements appear more complex, they all are complementary **MOSFET**'s. The MC14002 IC's are constructed with P- and N-channel enhancement mode devices, as we have said. A gate such as this is normally off, displaying a high resistance between drain (labeled D in the drawing) and source (S). To turn it on, you must have a sufficiently large voltage between the gate and source (for example, pins 5 and 7). In this circuit, there are two **MOSFET**'s: the upper one is a P-channel **MOSFET**, and the lower is an N-channel device. The two form a series circuit between V_{DD} (pin 14) and V_{SS} (pin 7, usually ground).

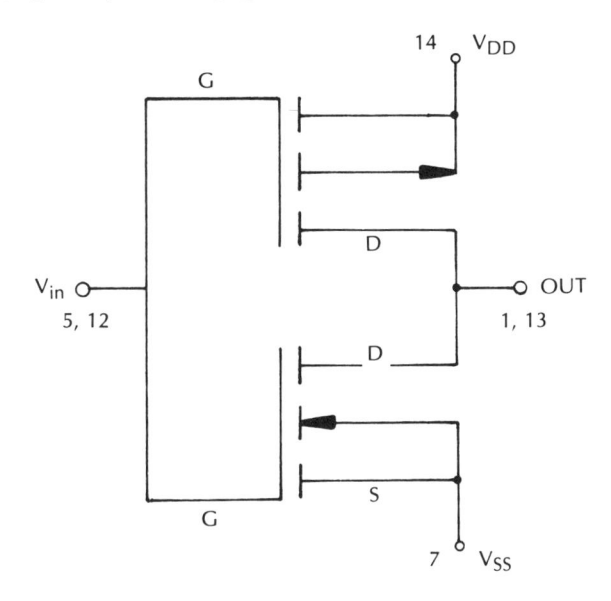

Figure 2-7: Basic CMOS gate (an inverter) extracted from the schematic drawing of the dual 4-input NOR gate shown in Figure 2-6.

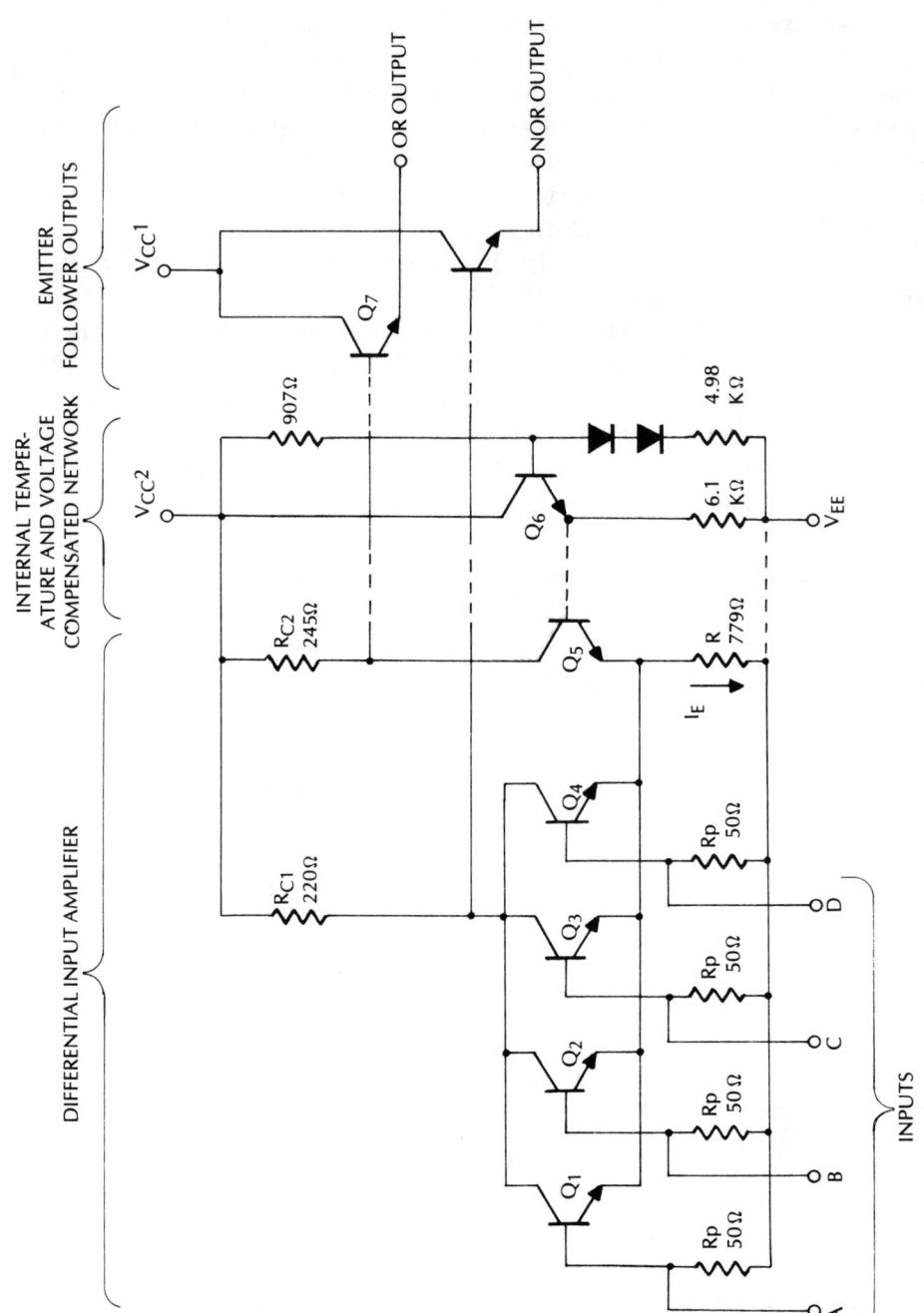

Figure 2-8: A MECL basic gate (Motorola's 10,000 family).

If the voltage on pin 5 (input) is low, the upper **MOSFET** is on and the lower **MOSFET** is off. On the other hand, if the input is high, the upper **MOSFET** is off and the lower one is on. In other words, because this is a series circuit—between V_{DD} and V_{SS} in this case—an opposition to current flow is offered in either state. But during the input transition (input going towards V_{DD} or V_{SS}), some current will flow through the series circuit composed of the two **MOSFET**'s. That's why **CMOS** requires so little supply current; no change in input level, no supply current drawn.

In regard to available solid state digital devices, you can also use Motorola's *emitter-coupled logic* (**MECL**). As in the standard **TTL** family, emitter-coupled logic (**ECL**) uses transistors in its design, but **ECL** is a form of digital logic that eliminates transistor storage time as a speed-limiting characteristic, permitting very high speed operation. As you would probably assume, "emitter-coupled" refers to the manner in which transistor amplifiers (differential amplifiers) are electronically coupled within the **IC**'s.

The differential input amplifiers of this logic family provide high-impedance inputs and voltage gain within the circuit. By referring to Figure 2-8, you'll notice that the output of this basic **MECL** gate is an emitter follower. This circuit restores the logic levels and provides low output impedance for good line driving and multiple load (fan-out) capability.

Motorola has offered **MECL** circuits in four logic families: MECL I, MECL II, MECL III, and MECL 10,000. The MECL III logic family is listed as the fastest standard logic available, designed to provide the high speed required by computers, communications, and instrument systems.

Principles of CMOS Devices and Circuitry

In the last section, it was pointed out that the best solution to the **TTL** power requirement problem is to use **CMOS** **IC**'s instead. However, technicians must be ever-conscious of whether they are working with **TTL**, **CMOS**, or other logic families because these are classifications of fabrication techniques, as has been explained. These procedures for manufacturing an **IC** determine critical electrical characteristics such as operating voltage levels, input and output impedance (important when interfacing **IC**'s), drive capabilities (fan-in and fan-out) and, of course, operating speed.

CMOS devices offer a wider range of operating voltage levels and higher input impedances. These **IC**'s are an attractive option if you are building relatively low speed circuits that are to operate from standard battery power sources of 6V, 9V, and 12V. At this point you may be wondering, "Why not combine **CMOS** and **TTL IC**'s in the same project?" You'll find that, because of interfacing problems, using both these devices in a single circuit isn't easy. For example, a **TTL** quad-2-input **NAND** gate (N7403) has a normalized fan-out from a single gate of a maximum of ten **TTL** loads, whereas a **CMOS** quad-2-input **NAND** gate (MC14011B) has only a maximum capability of driving two low-power **TTL** loads. Also, the N7403's recommended operating voltage (V_{CC}) is a minimum of 4.5V to a maximum of 5.5V. The MC14011B has a recommended voltage (V_{DD}) of −0.5 to +18VDC (reference to V_{SS}).

Sometimes you will have trouble with logic level mismatching. In fact, in many cases, every time you interface between the two different families it may require some special circuitry, even if you are only driving a low-power **TTL** with a **CMOS**. What this all means is that you should take care when using both **CMOS** and **TTL** in the same project. Nevertheless, the need for combining **TTL** and **CMOS** circuits frequently arises; therefore interface problems are discussed in great detail in Chapter Eight.

Most technicians are familiar with **TTL** family **AND, OR,** and **NAND** gates, but many are not at ease with the newer **CMOS** gates. Figure 2-9 shows a basic **CMOS AND** gate circuit.

To see how this circuit works, suppose you apply a logic low (a near V_{SS} potential) to input B (gate electrode of Q3). At this point, the series circuit composed of Q1 and Q3 will be turned off. On the other hand, while it is turning off Q1 and Q3 (N-channel **MOSFET**'s), it is turning on Q4, a P-channel **MOSFET**. Next, notice Q4 and Q2 are in parallel. This means that these two transistors are both acting as a closed switch (conducting). That in turn causes the gates of the outputs (Q5 and Q6) to be at a potential very near V_{DD}. It turns the P-channel **FET** (Q6) off, and the N-channel (Q5) on. The output

UNDERSTANDING TODAY'S SOLID STATE CIRCUITS

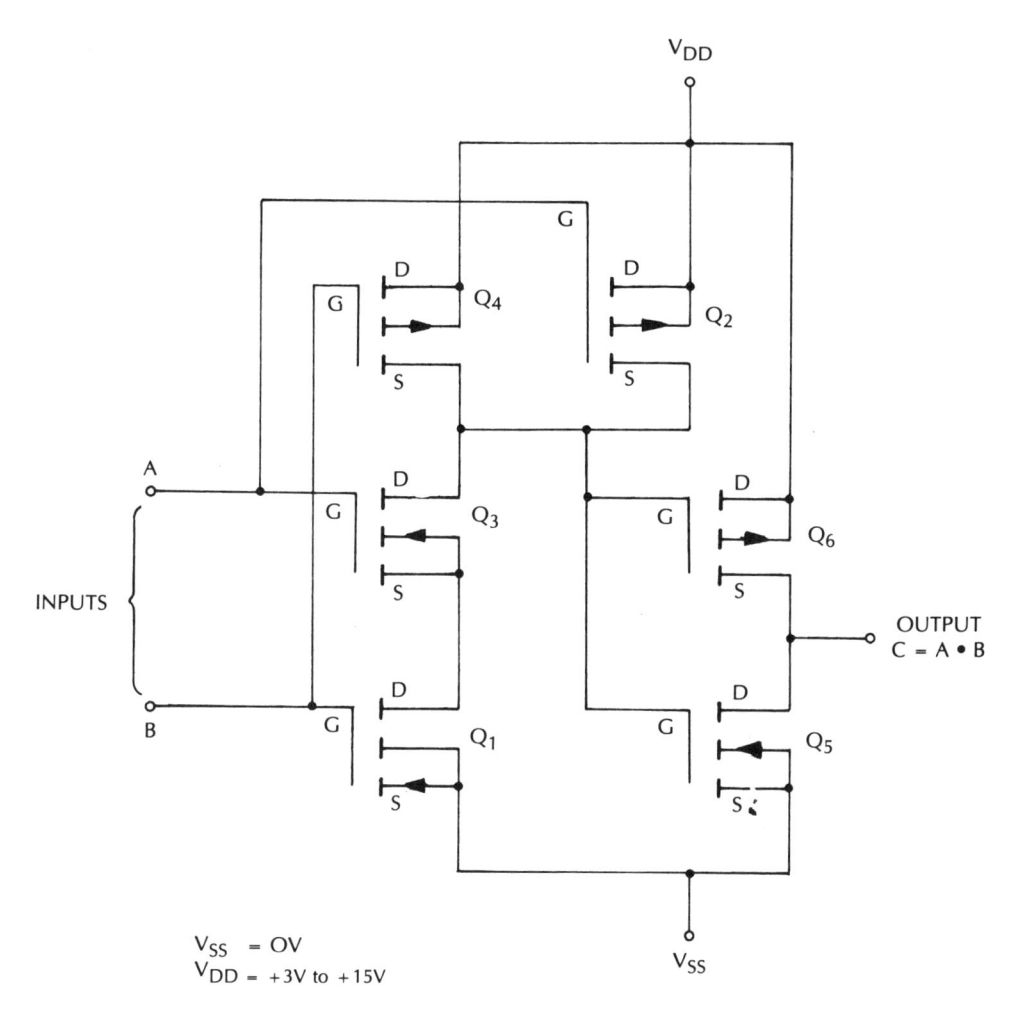

Figure 2-9: Basic CMOS 2-input AND gate. Typically, such a gate will be found in an IC such as a MC14081, a quad-2-input AND gate (a pin-for-pin replacement for the corresponding CD4000 series).

is now near V_{DD} (a logic high). The result: input A and input B arrive at the output C as input A and B, or $C = A \cdot B$. Figure 2-10 is the schematic of a single **CMOS OR** gate circuit. Typically, four of these gates would be found in an **IC** such as a MC1407B, a quad-2-input **OR** gate.

46   UNDERSTANDING TODAY'S SOLID STATE CIRCUITS

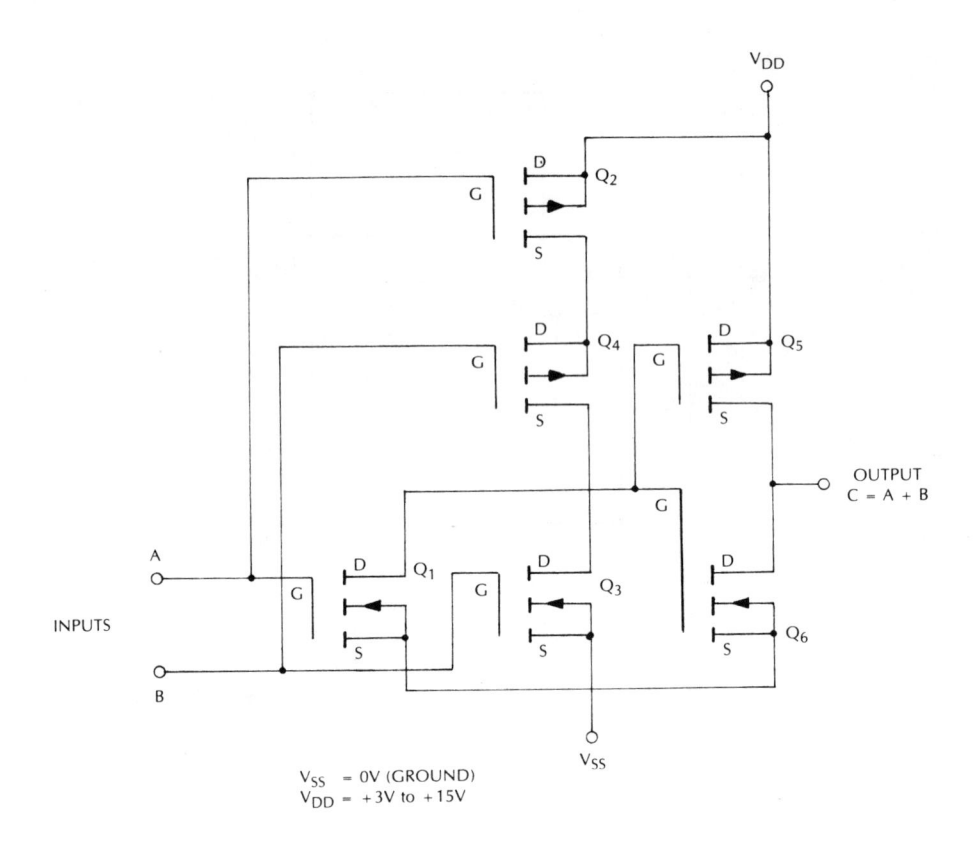

$V_{SS} = 0V$ (GROUND)
$V_{DD} = +3V$ to $+15V$

Figure 2-10: Basic CMOS 2-input OR gate. A gate such as this would be the internal structure of an IC such as the MC1407B, a quad-2-input OR gate.

      To get some idea of the function of the **CMOS OR** gate circuit, let's assume that you place a logic high (V_{DD}) on input A. Notice that this applies V_{DD} to both Q1 and Q2 gates. Q1 is an N-channel **FET**, so it switches on. Q2 is a P-channel **FET**, therefore it switches off. Looking at the schematic very closely, you'll see that Q1 and Q3 are in a parallel circuit and Q2 is part of a series circuit that includes Q4. What this means is that when Q1 is on, the gates of output **FET**'s Q5 and Q6 will drop down to the V_{SS} potential, and turning off Q2 cuts the gates of those output **FET**'s from V_{DD}. The output **FET**'s now have a potential on these gates of V_{SS}, which means Q5 is on and Q6 is off. The output (C) is now very near V_{DD}.

If we go through the circuit again with a logic high (V_{DD}) applied to the input B, it will turn out to be exactly the same; i.e., the output C will be at a logic high. Furthermore, if you apply a logic low (V_{SS}) to the input (A or B), you will produce a logic low on the output. In other words, if you connect both inputs to V_{SS}, it should make your **IC** output go to a logic low or some voltage near V_{SS}. Connect both inputs to V_{DD} and the output should be at a logic high (V_{DD} or some potential near V_{DD}). This means that the circuit performs as an **OR** gate.

Although other **CMOS** logic elements are more complex, they all use complementary **MOSFET**'s, and the preceding circuit description captures the most essential characteristics of the **CMOS** logic family. However, **CMOS** does have an important rival, the **NMOS**, which is a faster operating logic family. In fact, **CMOS** is only a fourth as fast as **TTL**. But this problem is beginning to disappear with a new generation of **CMOS** chips. Both Motorola and National Semiconductor have introduced a family of **CMOS** chips that are functionally similar to standard **TTL LC** parts. This family is designated 74HCxx.

MOS Power Transistors (MOSFET's and VMOSFET's)

MOS power **FET**'s are being designed into more and more electronic equipment, offering better performance, often requiring less support circuitry and, in the long run, costing less. These devices have other advantages when compared to bipolar transistors. Their input impedance is higher, generally usable bandwidth is greater, and linearity is excellent. One particularly interesting thing about **VMOSFET**'s is that they have a negative temperature coefficient; i.e., there is a continuous reduction in the amount of current flow through the device as it gets hotter and hotter (assuming a fixed level input to its gate).

Both **MOSFET**'s and **VMOSFET**'s (the "V" has to do with the **FET**'s physical structure—a V-shaped channel is etched in the internal material during manufacture) are widely used in radio frequency (rf) applications. In the preceding pages it was brought out that some **FET**'s are *single gates*, others are *dual gates*, and so on. If you refer to a manufacturer's data manual (such as Motorola's RF Data Manual), you will also find that **FET**'s are classified as *junction* **FET**'s (example: the symmetrical 2N3823 designed for VHF amplifier and mixer applications) and **MOSFET**'s (example: a 3N201-N-channel dual gate **MOSFET**, same applications as the 2N3823), plus a *double-diffused* **MOSFET** (example: MFE521, used in television tuners).

Notice that in our example of an **MOSFET** (the 3N201), it was said that this **IC** is an N-channel device. There are other classifications: insulated gate and, according to their mode of operation, *depletion, enhancement depletion/enhancement, symmetrical* (the 2N3823), and *asymmetrical*. Most **FET**'s are asymmetrical.

The 2N5484 N-channel junction field-effect transistor (**JFET**) is described as a depletion mode (type A) **JFET** and is used for applications in VHF/UHF amplifiers. First of all, this is an asymmetrical **FET**. What does this mean? An **FET** is termed asymmetrical when the designated source and drain terminals can not be interchanged without causing a loss of performance. Of course, as one would expect, symmetrical means that you can change them. In other words, interchanging the source and gate terminals won't affect the operation.

Next, depletion mode. Two distinct types of construction are used in **FET**'s: depletion and enhancement. They not only are built differently, there is a difference in operation as well. Because this book is about how to use **FET**'s—not how to build them—we will restrict our discussion to their operation. A depletion mode **FET** is one that exhibits substantial device current (ID_{ss}) with zero gate-to-source bias (V_{Gs} = OV). If the negative V_{Gs} is made high enough, drain current (I_D) will be almost cut off.

There is no such thing as a minimum or maximum positive gate bias voltage for a depletion mode **JFET** or **MOSFET**, since all gate voltage is in the negative region. All **JFET**'s usually are operated only in the depletion mode. In other words, they must always use negative bias voltages.

Now that you understand the operation of a depletion **FET** (it must always be with negative gate voltages), the enhancement type is easy. Basically, it's just the opposite; the polarity of the gate bias voltage is reversed. When V_{GS} is zero, there is only leakage current flowing between drain and the source—the device is cut off.

In summary, what's important to the project builder is that (1) zero bias on a depletion type **FET** will cause considerable current flow through the transistor, and (2) zero bias on an enhancement type **FET** will (for all practical purposes) cut it off. It should be pointed out that, except for the bias voltage, all **FET**'s (**JFET**'s and **MOSFET**'s) are connected in a circuit with the source as common (ground), drain as the anode, and, of course, the gate is the control electrode. Figure 2-11 shows typical drain characteristics for a depletion **MOSFET**, an enhancement **MOSFET**, and a depletion/enhancement **MOSFET**.

From what has been said, it should be obvious that if the **FET** is using a negative bias it should be operated in the depletion mode, and that a positive bias means it should be operated in the enhancement mode. If you are working in the depletion/enhancement mode, the gate bias should be zero. You will find **MOSFET**'s manufactured for all three modes but, as you will remember, not **JFET**'s. All **JFET**'s are usually operated in the depletion mode only. Operating them in any other mode will lower their input impedance, which in turn may seriously load any preceding circuit.

Although the **VMOS** power **FET** is not as well known as the conventional **MOSFET**, it does offer the user some advantages. So far, you will find that some manufacturers have used them in place of bipolar power transistors in switching power supplies and audio amplifiers, and they are excellent for use in rf switching applications.

If you wish to experiment with a **VMOS** power **FET**, you might like to use it in place of a conventional power transistor when building a power supply, especially where you want to operate the device at high switching rates. Many bipolar transistors—particularly the sturdy ones—have a bandwidth of 2 or 3 MHz. You'll find that a **VMOS** power **FET** can operate up to 30 MHz and above.

Or you might like to try using one of these **FET**'s in an audio amplifier project. In this case, simply use the **VMOS** in place of a bipolar in the output stage. See Chapter Twelve for a schematic diagram of projects where you can use **VMOS** devices.

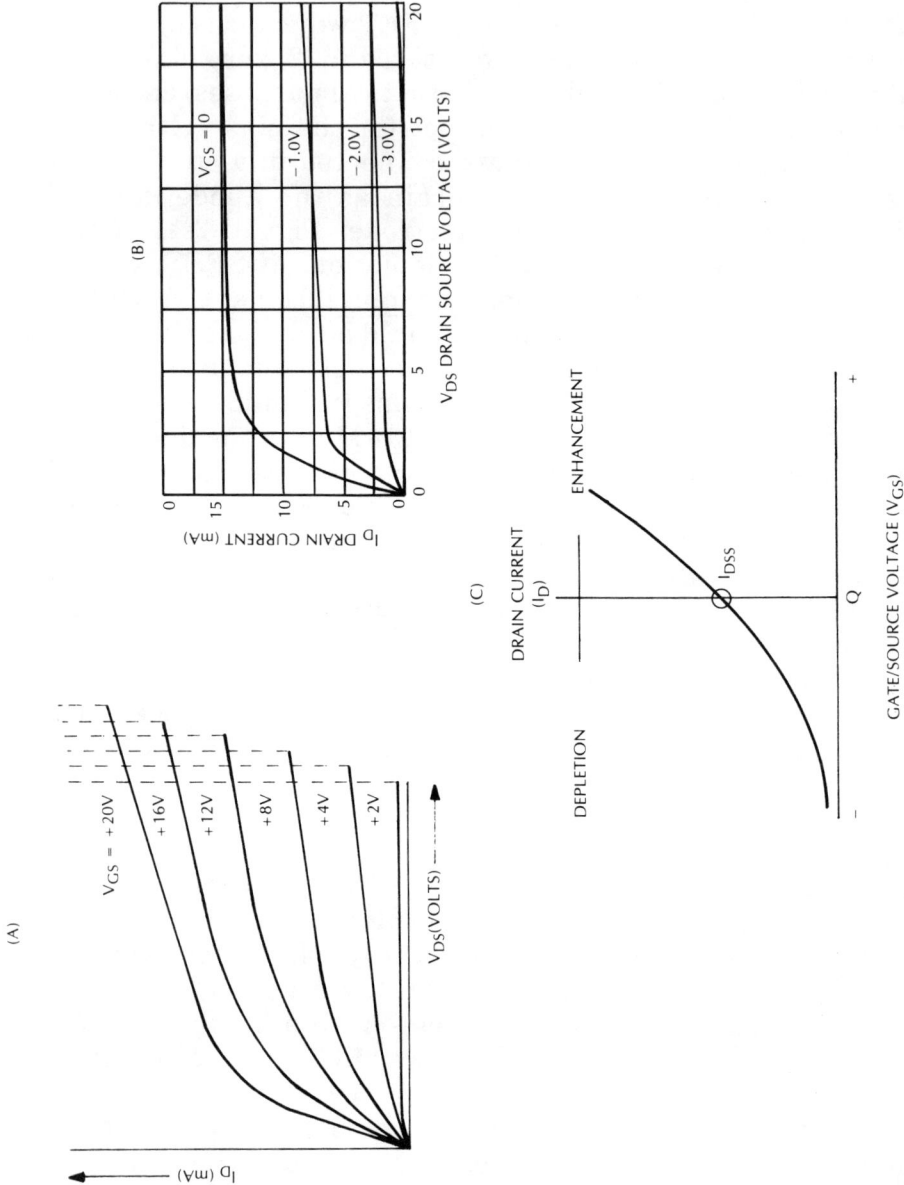

Figure 2-11: (A) enhancement MOSFET, (B) depletion MOSFET, (C) depletion/enhancement MOSFET.

VMOS's that you should be able to purchase at most electronics supply stores (for example, Radio Shack) are the VN66AF, VN64GA, and VN10KM.

CMOS Handling Precautions

There was a time when the handling of **CMOS IC**'s was a very touchy procedure. However, today's **CMOS** devices are generally protected against over-voltages (they have circuitry to protect the inputs against damage due to high static voltages or electric fields). Don't let this statement mislead you about the care you must exercise when working with components vulnerable to static damage. Incidentally, devices other than **CMOS IC**'s with which you must use extreme caution are double heterostructure lasers, tunnel diodes, and crystal diodes. In fact, almost any components that have an ultra-small active area require extreme care in handling. Technicians who regularly work with devices such as these are aware of the fact that solid state components are susceptible to damage from static discharge.

Manufacturers sometimes ship microwave crystal diodes in small lead envelopes (to protect against electrostatic damage). This lead covering does not prevent externally charged objects—for instance, your body—from discharging through the diode once it is out of its case. The question is, once you have this lead-packaged diode in your hands, how do you remove the diode for placement in the equipment you are working on? Simply place both hands in contact with chassis ground (generally the aluminum equipment frame, etc.) and keep them in contact during the opening of the package. This is an example of the care you must exercise when working with components vulnerable to static discharge.

Here are a few important points that you should remember during handling and troubleshooting **CMOS IC**'s:

1. Do not remove **CMOS IC**'s from their original container until you are ready to use them.
2. Do not leave **CMOS IC**'s on your workbench without using some type of shorting method. Place all leads in conducting foam or wrap aluminum foil around them. **CMOS** devices must not be placed in conventional

plastic "snow," Styrofoam, ordinary plastic bags, or plastic trays.
3. Never wear garments made of nylon, or any other material (such as wool) that may develop a static charge, when working with **CMOS IC's**.
4. When working with PC boards that have **CMOS IC's**, it's best to terminate all leads with a 1-megohm resistor before removing the board from the chassis; i.e., don't leave the leads floating. Also, do not leave any **CMOS** inputs open even in a circuit. Tie them in parallel with the input from the gate, etc.
5. Use an isolated soldering iron (you can use a grounding strap connected to the soldering iron tip). Always check to see that all test equipment is properly grounded.
6. Watch the input signal voltage limits:
   a. A safe input voltage is always less than (or equal to) V_{DD} and more than (or equal to) V_{ss}. Break this rule and you probably will lose an IC.
   b. *Never* apply an input signal without applying power to the **IC**. Forgetting this bit of advice could cost you an **IC**.

Basic Logic Elements

About seven or eight digital circuits are the very backbone of today's digital electronics. These are gates, multivibrators, registers, counters, memories, decoders/encoders, optoelectric display systems and, of course, modern microprocessor units. All kinds of digital circuits—from a simple handheld space game to the most sophisticated computer system—depend on the operation of one or more of these basic logic circuits. The electronics technician or experimenter who does not have a working knowledge of the Boolean logic equations, schematic symbols, and truth tables describing the characteristics of these circuits is severely handicapped.

The remaining part of this chapter deals with each of these important digital circuits separately, introducing their schematic symbols and truth tables, and describing their essential operating characteristics. However, because of the importance and complexity of modern microprocessor fundamentals, this subject will not be covered at this time. Instead, the next two chapters will be devoted entirely to helping you acquire knowledge and skills you'll need to work with the basic building blocks of microprocessors and microcomputers.

Fundamentals of Multivibrators and Flip-Flops

Digital circuits are entirely useless without a good, clean source of trigger pulses. Multivibrators are oscillators that primarily produce pulse waveforms that are important to digital electronics. There are numerous multivibrator IC's. For example, there is the **TTL** family, 74121 monostable, and the 74122 retriggerable monostable. Two in the **CMOS** family are the more recent MC14528B dual retriggerable/resettable monostable and the MC14538B dual precision retriggerable/resettable monostable. Notice that each of our example IC multivibrators is referred to as *monostable*. Actually, there are three main types of multivibrators:

1. Astable Multivibrator (Free-Running):

This circuit has two momentarily stable states. It switches rapidly from one state to the other (either a logic high or low). The output signal is a square-edge waveform used as a clock pulse in digital applications: computers, digital games, etc. Figure 2-12 illustrates a simple transistor astable multivibrator.

The simple basic circuit shown in Figure 2-12 lacks the precision of the more sophisticated IC's described in the beginning of this section (Motorola's MC14538B), but is certainly much easier to analyze. To being with, because of slight manufacturing differences, transistors Q1 and Q2 will not conduct the same amount of current when power is first applied to the circuit.

Let's assume that Q1 conducts more current than Q2. Under this condition, the collector voltage of Q1 will initially keep Q2 in a cut-off mode of operation because, as you can see, it is directly wired to the base of Q2. Nevertheless, as Q1 conducts more and more, its collector voltage will drop more and

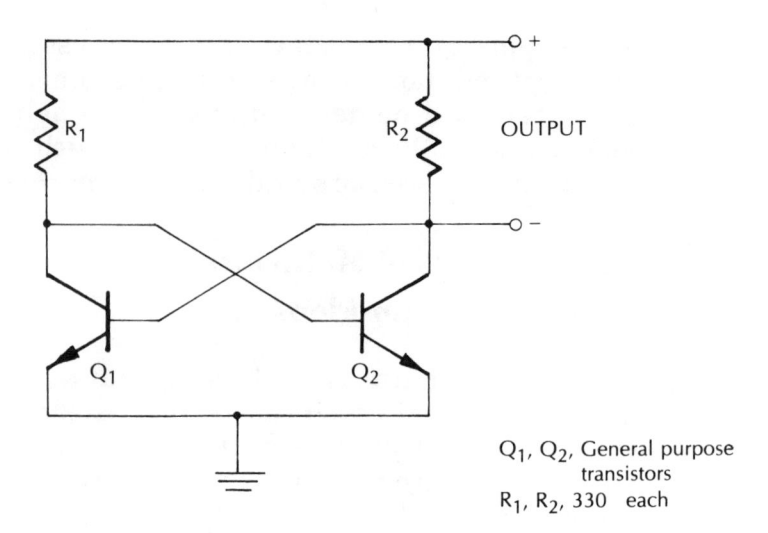

Figure 2-12: A basic bipolar transistor astable multivibrator circuit.

Q₁, Q₂, General purpose transistors
R₁, R₂, 330 each

more, and Q2 will be allowed to start conducting. Now that Q2 is turned on, the initial high collector voltage (notice, this collector voltage is also being applied to the base of Q1), will cut Q1 off. But as the current through Q2 increases, the collector voltage decreases. In fact, it will reach a point where Q1 is again conducting and Q2 is turned off.

In summary, the circuit switches back and forth (between Q1 and Q2) between two states (on and off), providing a square waveform output. Incidentally, you can take the output signal from either the collector of Q2, as shown, or the collector of Q1. Now, with this basic understanding of how multivibrators work, let's examine the other types.

2. Monostable Multivibrator IC:

Also called *mono, one-shot, single-shot,* or *start-stop*. This is a circuit that has only one stable state. However, it can be triggered to change to another state (logic high or low) for a predetermined period of time. At the end of this time, it will return to its original stable state. You will find that monostable multivibrators are used in applications where delaying or reshaping of pulses is desired. Figure 2-13 shows a

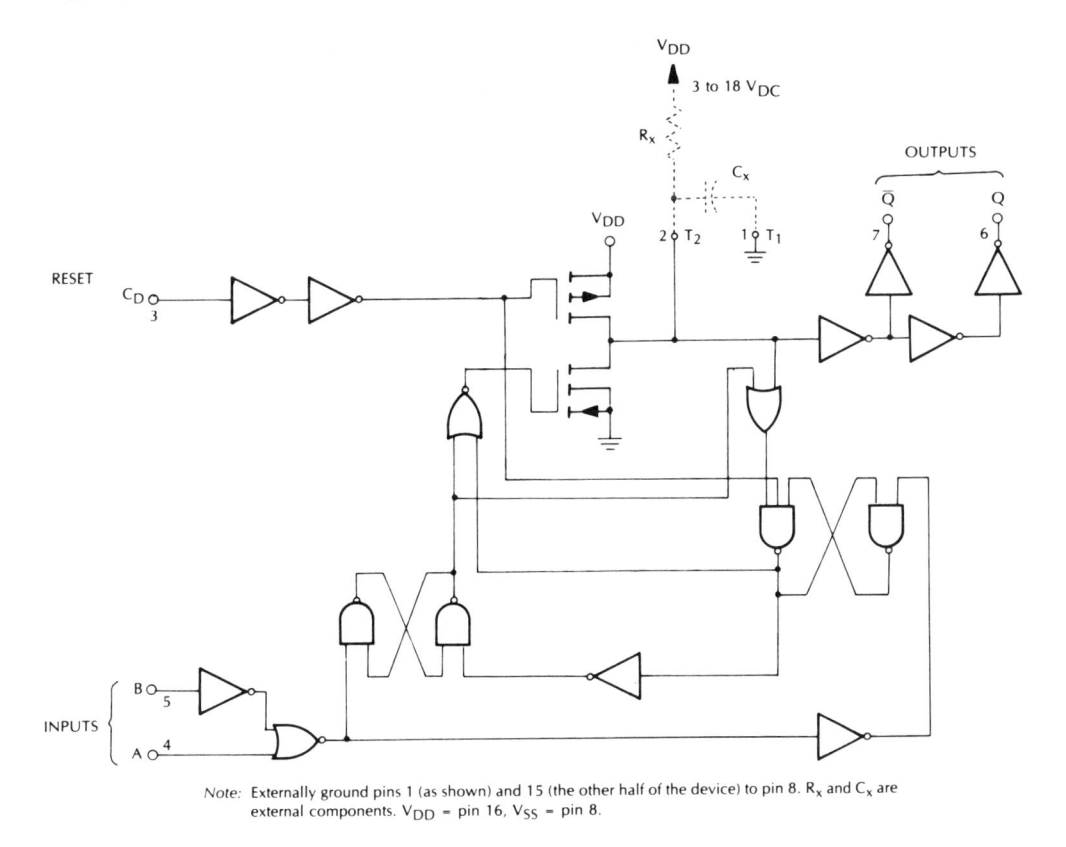

Figure 2-13: Logic diagram for one-half of a CMOS dual monostable multivibrator.

logic diagram for one-half of a **CMOS** dual monostable multivibrator (MC14528B).

As previously mentioned, a mono can be triggered to change state for a predetermined period of time. The MC1428B can be triggered from either edge of an input pulse. The output will be initiated on the positive or negative going edge of the input waveform, depending upon which input is used, A or B (see Figure 2-13).

The time delay of the mono is mainly dependent upon the time constant of the external timing components C_x and R_x, shown as dashed lines in Figure 2-13. Varying these two components will produce an output pulse over a wide range of widths, depending also on which value V_{DD} you used (5, 10, or 15 volts).

You can choose either a negative or a positive going output pulse. A positive Q output is available at pin 6, while a complementing $\bar{Q}$ output is at pin 7. When pin 6 is high, inverted output pin 7 is low and vice versa. To place the **IC** into operation, you could apply a positive going pulse on input A and B (pins 4 and 5) of one-half the device, a negative going pulse on input C_D (reset), and a negative going pulse on the other half of the IC

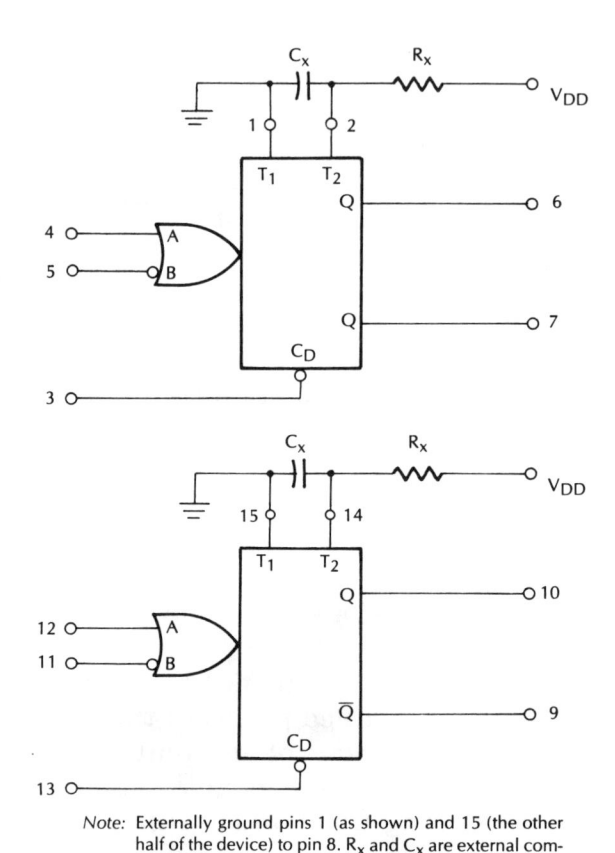

Note: Externally ground pins 1 (as shown) and 15 (the other half of the device) to pin 8. R_x and C_x are external components. V_{DD} = pin 16, V_{SS} = pin 8.

Figure 2-14: A block diagram of the MC14528B dual monostable multivibrator.

(pins 11 and 12). A block diagram of the entire **IC** is shown in Figure 2-14.

3. *Bistable or Flip-Flop IC:*

This integrated circuit is basically a memory device used for storing data in the form of a logic high or low. All flip-flop circuits have two stable states controlled by an input trigger and/or clock pulse. For example, a clocked flip-flop will be triggered (caused to change states) only if the trigger pulse and clock pulse are presented at the same time. Two of the most popular flip-flops in modern digital systems are the delay (or D) and the J-K types. Figure 2-15 A shows a block diagram for a **CMOS** dual D type and 2-15 B shows a **CMOS** dual J-K type produced by Motorola.

The D type shown can be used as an edge-clocked flip-flop. In this case it is in a memory mode while the clock input (C, see Figure 2-15) is at logic low, and in memory mode while the clock input is at logic high. *The only time the outputs (Q, $\overline{Q}$, see Figure 2-15) can change state is during the brief interval of time it takes the clock signal to make a transition from logic low to logic high or, in some cases, from high to low.*

However, there are two other inputs: direct set (S) and direct reset (R). These two inputs influence the outputs without regard to any clocking operations, and their main purpose is to let you set the outputs to either desired logic level before or after a positive-going edge of the clock pulse occurs. This is shown in the truth table for the **IC** as X = don't care (see Table 2-1).

INPUTS				OUTPUTS		
C*	D	R	S	Q	$\overline{Q}$	
⌿	0	0	0	0	1	
⌿	1	0	0	1	0	
⌉	X	0	0	Q	$\overline{Q}$	NO CHANGE
X	X	1	0	0	1	
X	X	0	1	1	0	
X	X	1	1	1	1	

X = Don't care.
* = Level change.

Table 2-1: Truth table for a type D flip-flop (MC14013B).

Note that the truth table indicates an invalid operating mode, one in which reset = set = 1 (a logic high). Under this particular set of circumstances, the Q and $\bar{Q}$ outputs both go to a logic 1, thus destroying the complemented relationship you should find existing between the two. This D flip-flop is only representative of the modern triggered D flip-flop family now available to the experimenter in both **TTL** and **CMOS IC** packages. Nevertheless, whether the device is a **TTL** such as 74274, 74175, 7474, a **CMOS** 4013 or 4042, or the one just described, basically they all work the same. They may have but a single output (Q), or no preset, but once you understand one, you should have no trouble understanding the others.

Today it would be next to impossible for an electronics technician to perform his duties without an understanding of the so-called J-K flip-flops. Table 2-2 shows the truth table for the MC14027B dual J-K flip-flop included in Figure 2-15 (see Figure 2-15 B).

As shown by the truth table, this **IC** is an edge-triggered flip-flop featuring independent J, K, clock (C), set (S), and reset (R) inputs for each flip-flop. Notice that the multiple inputs provide more control conditions for operating the flip-flop. A logic low on the set input and a logic high on the reset will set the Q outputs to logic Q = 0, $\bar{Q}$ = 1. On the other hand, a 1 on the S and a 0 on the R inputs will reverse the output; i.e., Q = 1 and $\bar{Q}$ = 0.

INPUTS						OUTPUTS*		
C†	J	K	S	R	Q_n‡	Q_{n+1}	Q_{n+1}	
╱	1	X	0	0	0	1	0	
╱	X	0	0	0	1	1	0	
╱	0	X	0	0	0	0	1	
╱	X	1	0	0	1	0	1	
╲	X	X	0	0	X	Q_n	Q_n	NO CHANGE
X	X	X	1	0	X	1	0	
X	X	X	0	1	X	0	1	
X	X	X	1	1	X	1	1	

X = Don't care.   † = Level change.
‡ = Present state.   * = Next state.

Table 2-2: Truth table for a dual J-K flip-flop MC14027B. A block diagram of this IC is shown in Figure 2-15 (B).

UNDERSTANDING TODAY'S SOLID STATE CIRCUITS   59

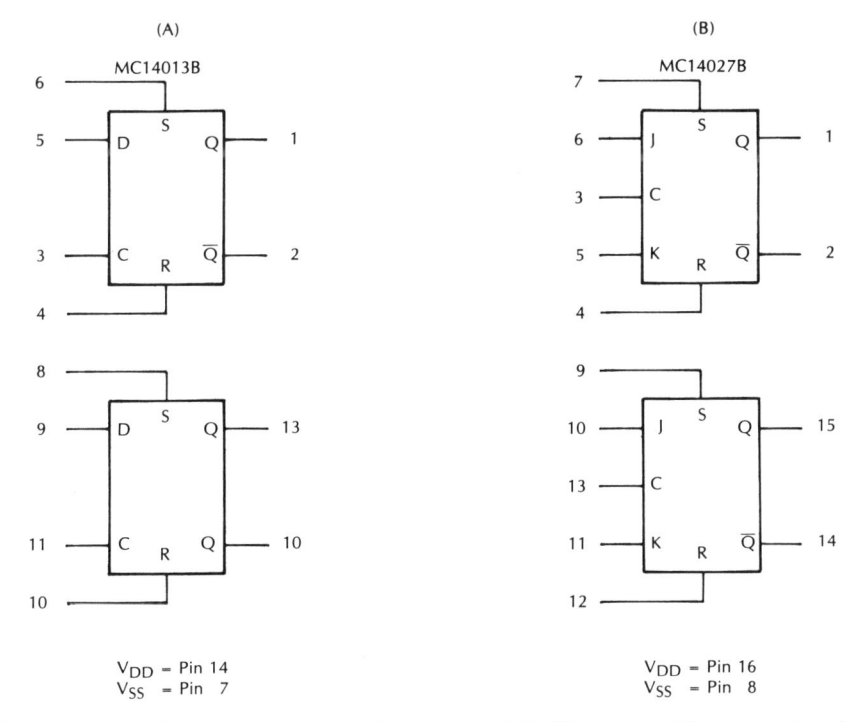

Figure 2-15: Block diagram for dual D and J-K flip-flops. The outputs of these devices are labeled Q and $\overline{Q}$. This notation conforms to modern standards for flip-flops of all kinds. The inputs are always on the left of the drawing—for example, J and K in illustration B.

Basically, you will find that flip-flop IC's are not symbolically designated and can be recognized only by their number and title, as shown by the block diagram in Figure 2-15 A and B. Essentially, the J-K master/slave flip-flop is set with a positive-going edge of a clock pulse, just as is the edge-trigger we just explained. In this case, the first flip-flop (there are at least two in the IC) is called the *master* and is triggered with the positive-going edge of the clock pulse, and the information is transferred to the next flip-flop—called the *slave*—on the negative-going edge of the clock pulse, as shown in Figure 2-16.

Generally, you will find the sequence of operation for a master/slave flip-flop is as follows:

1. Isolate the slave from master.
2. Enter information from an internal (within the IC) AND gate's inputs to master.

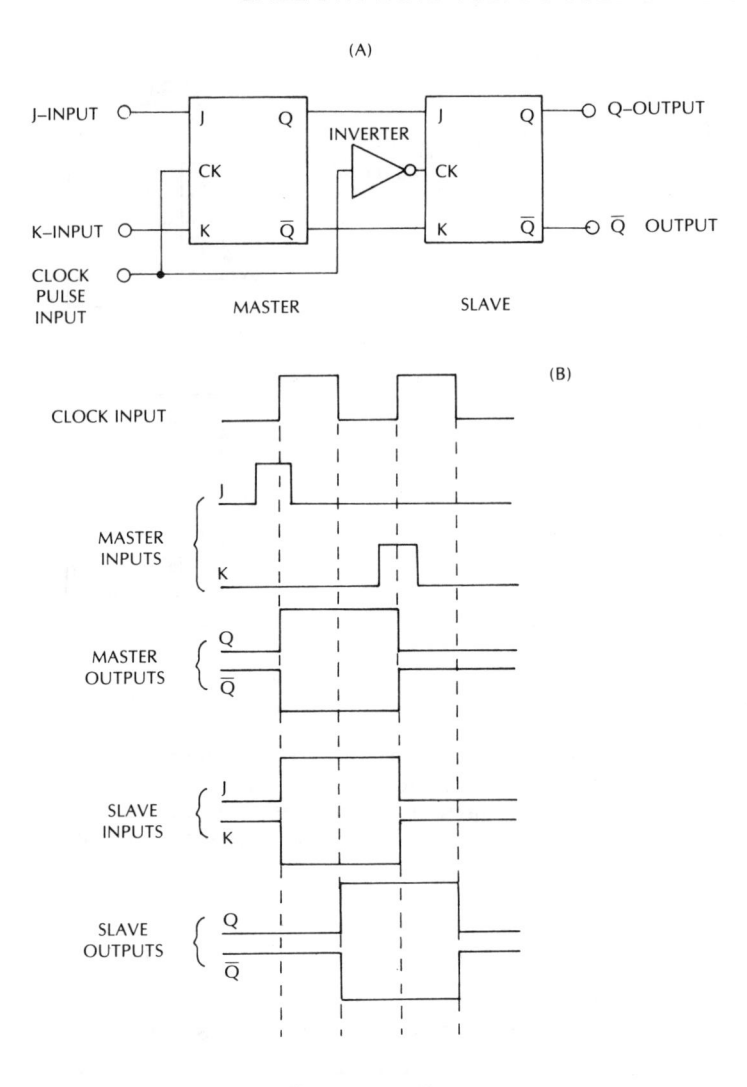

Figure 2-16: J-K master/slave flip-flop. (A) is a logic diagram showing two single flip-flops, master and slave. (B) is a timing chart illustrating inputs, clock pulse, and resulting changes.

3. Disable **AND** gate inputs.

4. Transfer information from master to slave.

Figure 2-17 shows the pin configurations and block diagram for a **TTL** family 7472, J-K master/slave flip-flop that would follow this sequence of operations.

UNDERSTANDING TODAY'S SOLID STATE CIRCUITS　　　　　　　　　　　61

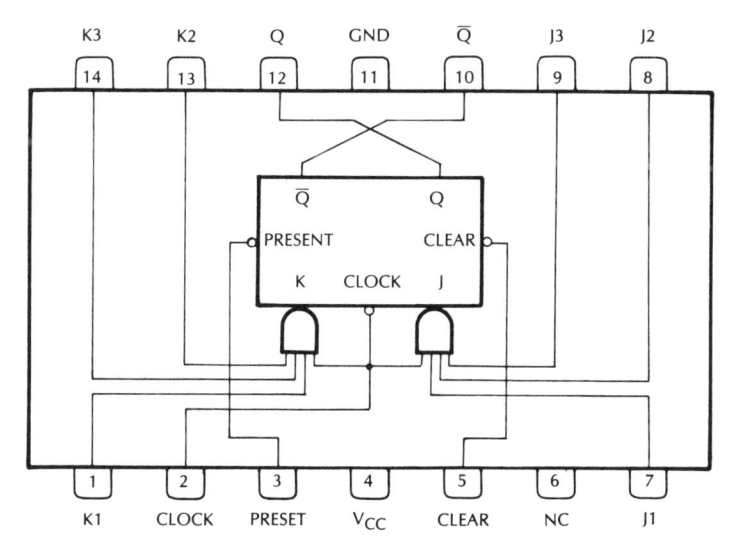

Figure 2-17: TTL family J-K master/slave flip-flop, 7472. Notice, this IC also has a preset and clear. Although these inputs were not shown in Figure 2-16, they are usually found on these types of devices.

Principles of Register and Counter IC's

In this section, you will learn how the logic gates and flip-flops presented in previous sections are arranged in an IC to produce registers and counters. Because registers are constructed using flip-flops, they are used for short-term storage of logic highs and lows. In fact, the number of flip-flops determines the amount of data (usually one computer word) per unit that can be stored. A register may store 4 bits (a bit is a state such as 1 or 0), 8 bits, 16 bits, etc., which means it contains 4, 8, or 16 flip-flops. Furthermore, a register IC must not only be able to store data, but it must also move the data upon command. Figure 2-18 contains a logic diagram for a **CMOS** family 8-bit static shift register, the MC14014B synchronous, or the MC14021B asynchronous, both having parallel input/serial output.

This register (a shift register) is a device that uses logic flip-flops for storage and, as we have said, contains eight flip-flops. *Note:* Flip-flops 4 and 5 are not shown. There are two different types of shift registers you will encounter. These are the dynamic type in which information is stored by means of

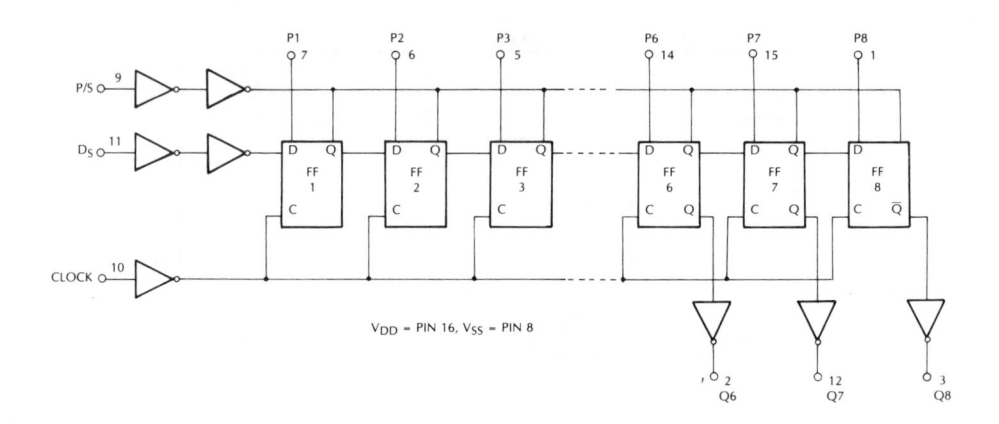

Figure 2-18: Logic diagram for a MC14014B or MC14021B 8-bit static shift register.

temporary charge storage techniques, and the static units such as the **IC** shown in Figure 2-18. The major difference between the two is that as long as you apply power, a static shift register will hold its stored data, whereas a dynamic shift register has a comparatively short holding time. Most true *dynamic* **MOS** shift registers will hold information for about one-tenth of a second at best. What this means is that you must have a recycling circuit to update the contained information within the allowed holding time, or you'll lose all stored information.

You will remember that we said the MC14014 and 21 both have parallel input/serial output. Many **IC**'s use parallel inputs/parallel outputs, serial inputs/serial outputs, and a combination of both. The 4-bit register illustrated in Figure 2-19 shows the basic operations of serial inputs/outputs and parallel inputs/outputs.

First, let's look at the serial input, shifting, and serial output of this 4-bit register. The outputs (Q, Q̄) of one flip-flop (FF1) feed into the inputs J, K of the next flip-flop, traveling (shifting) from left to right (called a *shift-right mode*). If one of the flip-flops has a Q output of a logic high, the following flip-flop has a logic high on its J input and as soon as the next clock pulse (CLK input) appears, that flip-flop will turn on. Of course, an

UNDERSTANDING TODAY'S SOLID STATE CIRCUITS

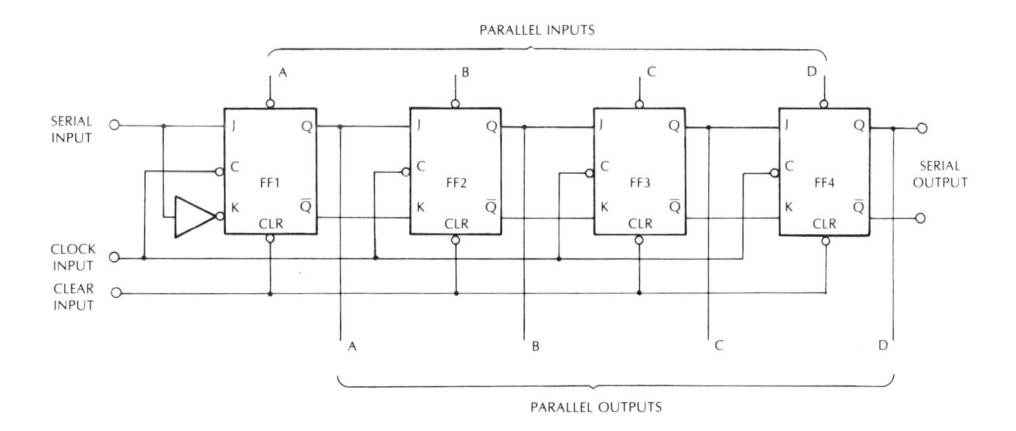

Figure 2-19: Block diagram of a 4-bit register showing serial input/output and parallel inputs/outputs.

inverted Q ($\bar{Q}$) output of the logic high input will be applied to the K input of the following flip-flop and turn it off when the next clock pulse arrives.

This stepping action must occur one step at a time until a certain number is loaded into the register. For example, Table 2-3 shows the steps required to load the binary number 1010 (decimal number 10) into the 4-bit register. Notice, if you keep

CLOCK PULSE	SERIAL INPUT	REGISTER CONTENTS			
		FF1	FF2	FF3	FF4
NO PULSE	0	0	0	0	0
1	0	0	0	0	0
2	1	1	0	0	0
3	0	0	1	0	0
4	1	1	0	1	0
5	0	0	1	0	1
6	0	0	0	1	0
7	0	0	0	0	1
8	0	0	0	0	0

REGISTER LOADED ON 4th CLOCK PULSE → (row 4)

REGISTER CLEAR ON 8th CLOCK PULSE → (row 8)

Table 2-3: Pulse-stepping procedure of the serial input/output 4-bit shift register shown in Figure 2-19.

feeding clock pulses to this register, it will step the number you wished to store (1010) right on out of the device into the next register, or the data is lost if there is not another digital circuit of some kind.

Now, the parallel inputs/outputs: data can be entered into all parallel inputs at the same time (say the binary number 1010 used in the serial input/output example just given) whenever the clear input (CLR) line is at a logic high level and any parallel input A through D is at a logic low level. For instance, if you place a logic high on parallel input A, a low on input B, high on C, and low on input D, it will turn FF1 off, FF2 on, FF3 off, and FF4 on. Then when all parallel inputs are again at a logic high, the register will be in memory mode, i.e., holding the binary word 1010. If you check the Q outputs (A, B, C, D), you should find QA = logic low, QB = logic high, QC = logic low, and QD = logic high. All this information is sent at the same instant to the next digital circuit and is, as you can see, much faster than a serial type operation. Resetting all flip-flop outputs to a logic low level simply requires that you place a logic low level pulse on the CLR input line. It should be pointed out that there are 4-bit bidirectional shift register IC's; for example, the 4-bit **CMOS MC14194B** and **TTL** 4-bit right-shift/left-shift register

OPERATING MODE	INPUTS ($\overline{\text{RESET}}$ = 1)					OUTPUTS (@ t_{n+1})			
	S1	S0	DSR	DSL	$D_{P0\ 3}$	Q0	Q1	Q2	Q3
HOLD	0	0	X	X	X	Q0	Q1	Q2	Q3
SHIFT LEFT	1 1	0 0	X X	0 1	X X	Q1 Q1	Q2 Q2	Q3 Q3	0 1
SHIFT RIGHT	0 0	1 1	0 1	X X	X X	0 1	Q0 Q0	Q1 Q1	Q2 Q2
PARALLEL	1 1	1 1	X X	X x	0 1	0 1	0 1	0 1	0 1

Table 2-4: Truth table for the 4-bit bidirectional shift register shown in Figure 2-20.

7495. Figure 2-20 is a layout of the MC14194B with its operation illustrated using the truth table shown in Table 2-4.

The MC14194B's synchronous reset input, when at a low logic level, overrides all other inputs, resets all stages, and forces all outputs (Q1, Q2, Q3 and Q4) to a logic low. When reset is at a logic high, the two mode control inputs S0 and S1 control the operating mode, as shown in Table 2-4. Both serial and parallel operation are triggered on the positive-going transition of the clock pulse. The parallel data (Dp), data shift (DSR, data shift-right, and DSL, data shift-left), and mode control (S0, S1) are all shown, with pin numbers for this **IC**, in Figure 2-20.

Referring to Figure 2-20 again, you will notice that numerous gates, inverters, and, of course, four flip-flops are combined to make the more complex **IC**'s such as this one. Basically, this is how logic gates and flip-flops are arranged to produce both registers and counters. Figure 2-21 is an example of a **TTL** family synchronous 4-bit counter. Note the similarity between this logic diagram and the one shown in Figure 2-20.

Synchronous operation is provided by having all flip-flops clocked simultaneously so that the outputs change coincidentally with each other, when so instructed by the count-enable inputs and internal gating. A buffered clock input triggers the four J-K master-slave flip-flops on the positive-going edge of the clock input waveform.

During a count operation, the flip-flop's J and K inputs are held at a logic high level by conditional **NAND** gates. What is happening is that the clock pulse cannot reach any one flip-flop unless the others are on. For example, QD will go to a high logic level when the next clock pulse occurs if QA, QB, and QC are at a logic high level. You'll also notice that there are other gates (other than those connected to the J and K inputs to the flip-flops). These are for controlling the count, parallel loading, and clearing operations. This is a binary-up counter. The only difference between a binary-down counter and a binary-up counter is that the count is reduced by one for each clock pulse using a down-counter. There are also up/down counters. The **TTL** family 74193 **IC** is an example of a 4-bit binary up/down counter. The counter **IC**'s, such as the ones we have discussed, have a carry output that is provided for arithmetic operation. In fact, for the same reason many also include a borrow output.

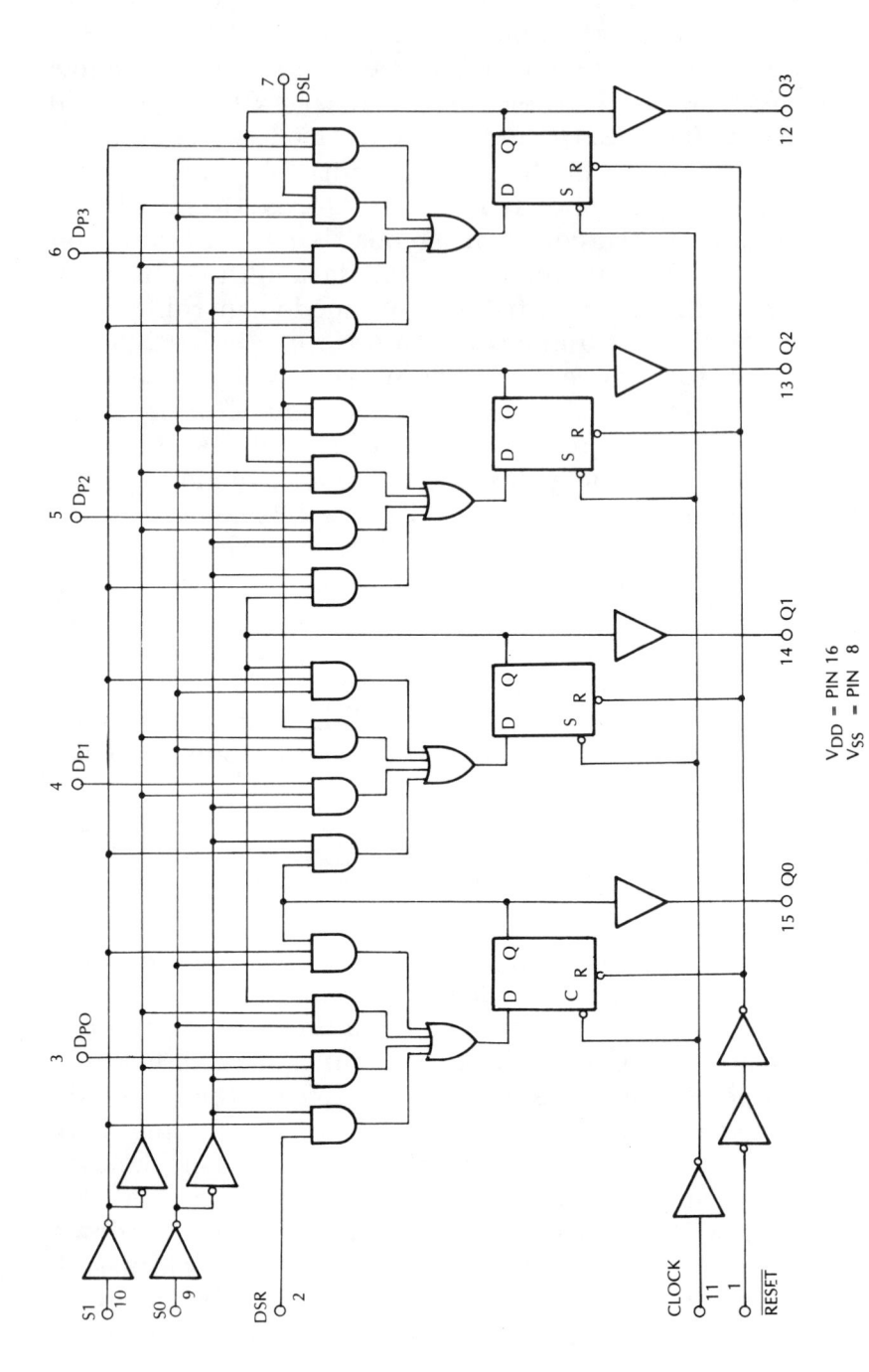

Figure 2-20: 4-bit bidirectional shift-right/shift-left register (MC14194B), capable of operating in the parallel load, *serial* shift-left, serial shift-right, or hold mode.

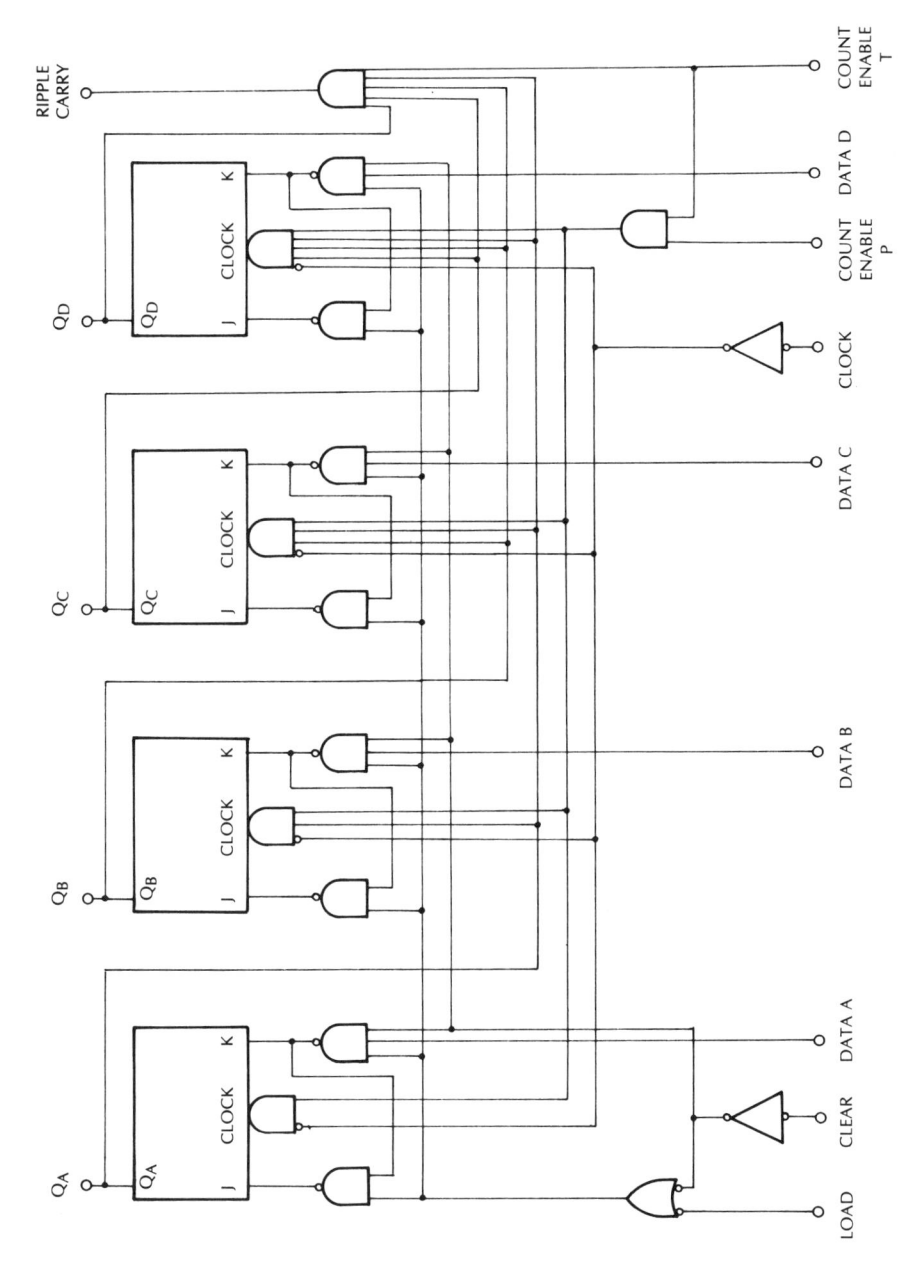

Figure 2-21: Logic diagram for a synchronous 4-bit binary (74163) counter.

Understanding Memory IC's

As seen in previous sections of this chapter, a flip-flop is a basic memory device. There are, of course, many other types of memory circuits used to store binary bits that represent data. In Chapters Three and Four you will learn that memory **IC**'s store binary data in the form of instructions for a control program used in computer systems.

Before the microcomputer system is studied in detail, let us first develop an understanding of the **IC**'s where the data may be stored. Basically, there are two: *Random Access Memory* (**RAM**) and *Read Only Memory* (**ROM**). A **RAM** contains data that may be electrically moved at any time new data is stored in its place. The other type—**ROM**—has data (computer programs, etc.) fixed in the memory and uses several different techniques to store the permanent pattern.

You will find that memory **IC**'s are manufactured using **TTL, CMOS,** and **ECL** techniques. However, in all cases they are built around a system of addressable flip-flops usually called *memory cells*. Each memory cell is only one of many identical devices in a memory **IC**. Now, it stands to reason that if a memory **IC** contains many flip-flop storage elements, there must be some provision that will let the computer, etc., select one particular memory cell at a time. Memory **IC**'s have a *memory cell select* and there are provisions for choosing either a writing (store) or a reading operation. For example, the MCM145101, a 256 × 4-bit static RAM, is a random access memory having 1,204 separate memory cells, that can accept, store, and read out 256 different 4-bit words. Figure 2-22 shows the block diagram for this **CMOS** memory **IC**.

Since this memory has 256 word-storage locations, it follows that it must have a cell array that adds up to 1,024. Refer to Figure 2-22 and you'll see that this **IC** does—32 rows, 32 columns. The address inputs (cell selection by row and column) are A0 through A7. These address inputs are processed in the row and column decoders which, in turn, select the correct cell for information storage (much more about this later).

There is a control terminal that determines whether data will be written into memory via the data inputs (DI) or read out via the data outputs (DO), plus other control operations. Writing a 4-bit word into memory is a matter of setting each of the **IC**'s

UNDERSTANDING TODAY'S SOLID STATE CIRCUITS

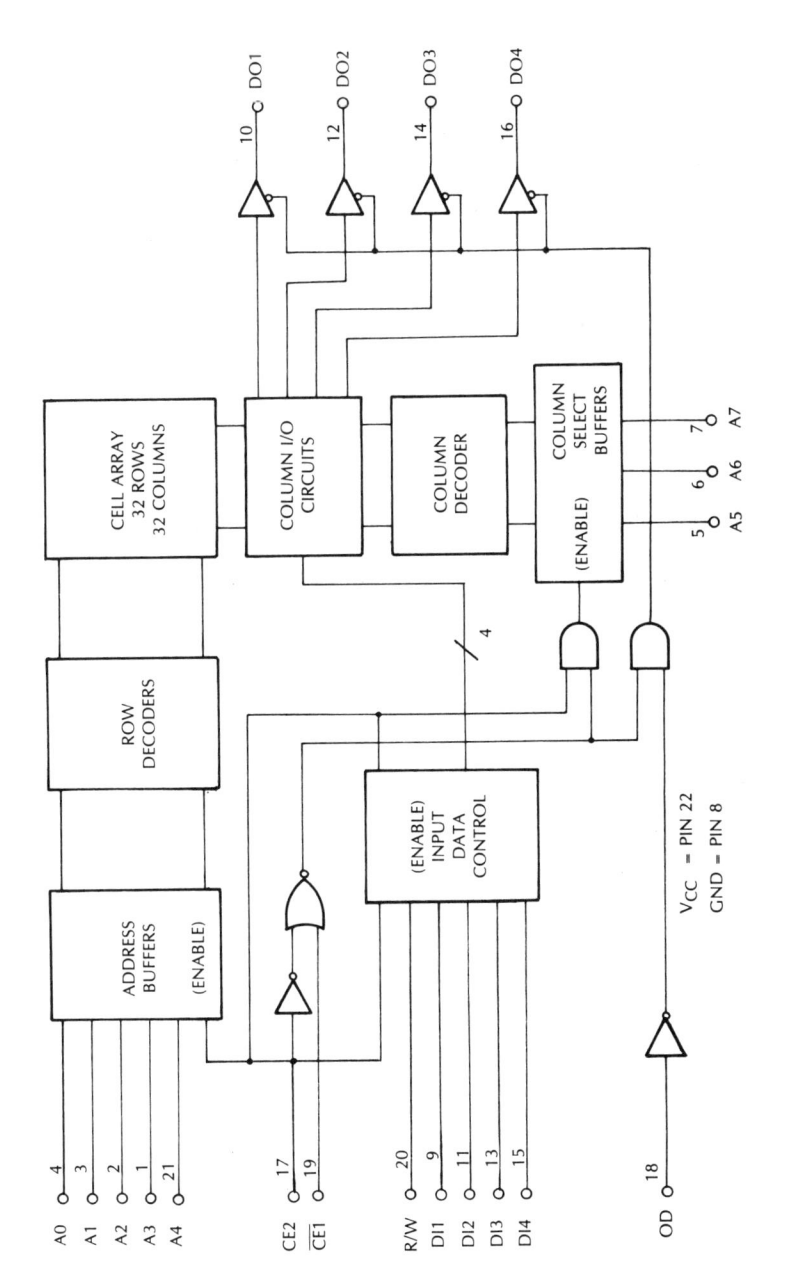

Figure 2-22: The MCM145101 256 × 4-bit memory IC block diagram.

pins to the proper logic level. See the truth table and pin assignment in Figure 2-23.

As we have shown, the primary feature of a **RAM IC** is that it is possible to read and write data with it. **A ROM,** however, is capable of performing only read-out operations. Forgetting erasable programmable read only memories (**EPROM**'s) for the moment, the programming of a **ROM IC** is internal and permanent. The MCM14524 device is a **CMOS** large-scale integration (**LSI**) **ROM** organized with a 256 × 4-bit pattern. It is, in other words, a 1024-bit **ROM.** The internal **ROM** programming is specified by the user, but it is the manufacturer's (Motorola, in this case) responsibility to implement the actual programming of the **IC** during the last phase of the manufacturing process. Programming of an **IC** such as this example is usually a custom job.

The advent of programmable read only memories (**PROM**'s) and erasable programmable read only memories has been a boon to electronics experimenters. To program a **PROM,** you blow selected fuses that are in each memory cell. A new **IC**

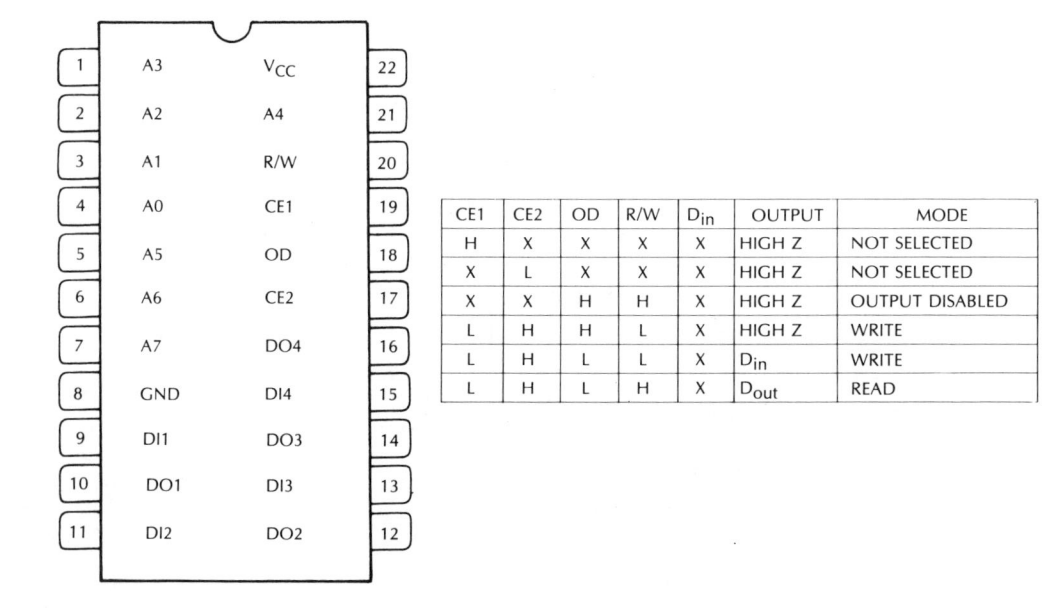

CE1	CE2	OD	R/W	D_{in}	OUTPUT	MODE
H	X	X	X	X	HIGH Z	NOT SELECTED
X	L	X	X	X	HIGH Z	NOT SELECTED
X	X	H	H	X	HIGH Z	OUTPUT DISABLED
L	H	H	L	X	HIGH Z	WRITE
L	H	L	L	X	D_{in}	WRITE
L	H	L	H	X	D_{out}	READ

Figure 2-23: Pin configuration and truth table for a MCM145101 256 × 4-bit static **RAM.**

of this type will have all its on-chip memory cell fuses intact and the outputs will be 1's or 0's, depending on how it was manufactured. You blow selected fuses electrically with external power supplies. Blowing the fuse of each memory cell may place that cell at a logic high, whereas in other **PROM**'s it could represent a permanently stored logic low.

The **PROM** is usually considered to have permanent data stored in its memory. **EPROM**'s, as the name suggests, are erasable. One type has a quartz window on top, and this window is transparent to short wave ultraviolet light. You erase the memory's contained data simply by exposing the window to an external source of ultraviolet light. As shown in Figure 2-24, the transparent lid is easily placed under the ultraviolet light. However, because this light is harmful to eyes, the light and **EPROM** are always placed inside a closed case. The erasure time is usually 15 to 20 minutes when using a lamp with 12,000 $\mu W/cm^2$ power rating. Another **IC** that contains an **EPROM** of this type is the microcomputer MC68701, which will be explained in Chapter Four.

A different type, an MOS device, is called a 16 × 16-bit electrically erasable programmable read only memory. This **EPROM,** the MCM2801, offers in-system erase and reprogram

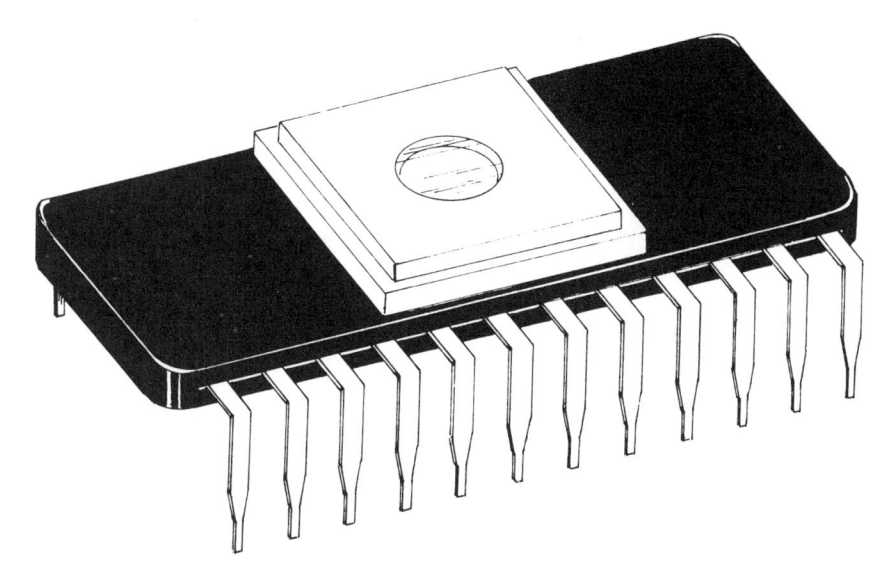

Figure 2-24: 1024 × 8 alterable (ultraviolet) ROM.

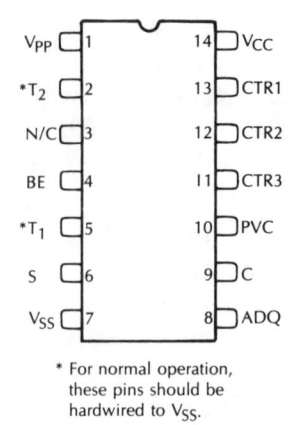

```
    Vpp  ⊏1   14⊐ Vcc
    *T₂  ⊏2   13⊐ CTR1
    N/C  ⊏3   12⊐ CTR2
    BE   ⊏4   11⊐ CTR3
    *T₁  ⊏5   10⊐ PVC
    S    ⊏6    9⊐ C
    Vss  ⊏7    8⊐ ADQ
```

* For normal operation, these pins should be hardwired to V_{SS}.

PIN NAMES	
ADQ	MULTIPLEXED ADDRESS/DATA-IN/DATA-OUT
C	CLOCK
PVC	PROGRAM VOLTAGE CONTROL
CTR 1, 2, 3	CONTROL
BE	BLOCK ERASE
S	CHIP SELECT
T₁, T₂	TEST PINS

Figure 2-25: 16 × 16-bit serial electrically erasable PROM.

capability. Figure 2-25 shows a pin assignment diagram and pin names for this **IC**.

Working with Decoders/Encoders

In the world of digital electronics, encoders are usually thought of as devices that convert other numbering systems (such as decimal or octal) into binary number systems, whereas decoders are devices that convert binary numbering systems back into other numbering systems. A **CMOS/BCD**-to-decimal decoder/binary-to-octal decoder **IC** (the MC14028B) block diagram and truth table are shown in Figure 2-26. Ordinarily, a decoder **IC** would be used at the output terminals of a digital system, but an encoder usually is used at the inputs.

The MC14028B is constructed so that an 8-4-2-1 **BCD** code on the four inputs (A, B, C, and D) provides a decimal (one-of-ten) decoded output, while a 3-bit binary input—A, B, and C (pins 10, 13, and 12)—provides a decoded octal (one-of-eight) code output with input D (pin 11) forced to a logic low level. This **IC** is useful for code conversion, address decoding in computer applications, or such applications as memory selection.

Frequently you will find that the **IC**'s designed as encoders are called *priority encoders*. As the name implies, this device was designed to give a priority to certain inputs. When more bits-per-output are required, the encoder may become very complex. Because of this, it is usual to find a **ROM** being

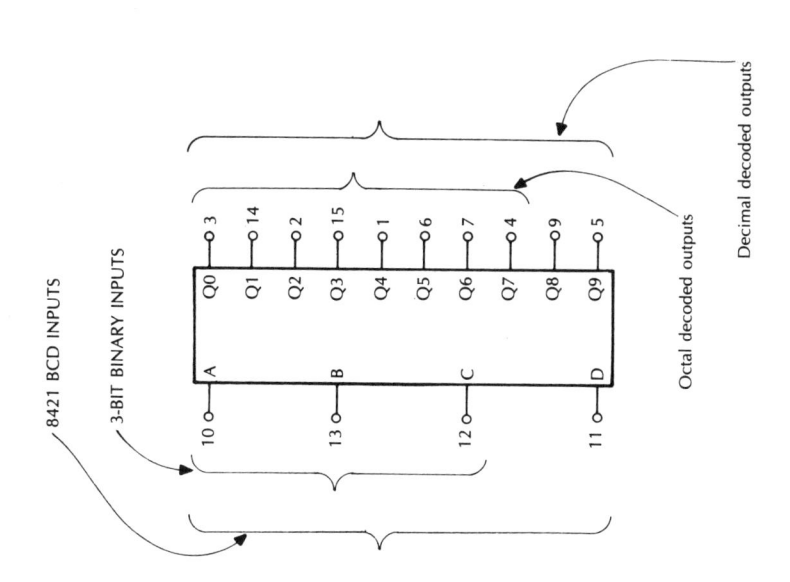

Figure 2-26: Block diagram and truth table for the BCD-to-decimal decoder/binary-to-octal decoder (MC14028B).

used as an encoder in computer systems. Two typical encoder IC's of the type we are discussing are the **TTL** 74147 10-line-to-4-line and the **CMOS** MC14532B 8-bit (eight data inputs, five data outputs) priority encoders. Figure 2-27 shows a truth table and pin configuration for the 74147, and Figure 2-28 is a block diagram and truth table for the MC14532B.

The 74147 encodes nine data lines to 4-line (8-4-2-1) BCD. The implied decimal zero condition needs no specific input condition because zero is encoded when all nine data lines are at a high logic level. The inputs (E1 and E) are for cascading purposes. They are provided to allow octal expansion without

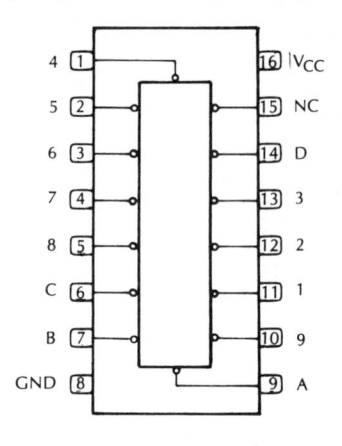

	INPUTS								OUTPUTS			
1	2	3	4	5	6	7	8	9	D	C	B	A
H	H	H	H	H	H	H	H	H	H	H	H	H
X	X	X	X	X	X	X	X	L	L	H	H	L
X	X	X	X	X	X	X	L	H	L	H	H	H
X	X	X	X	X	X	L	H	H	H	L	L	L
X	X	X	X	X	L	H	H	H	H	L	L	H
X	X	X	X	L	H	H	H	H	H	L	H	L
X	X	X	L	H	H	H	H	H	H	L	H	H
X	X	L	H	H	H	H	H	H	H	H	L	L
X	L	H	H	H	H	H	H	H	H	H	L	H
L	H	H	H	H	H	H	H	H	H	H	H	L

H = High logic level.   L = Low logic level.
X = Irrelevant.

Figure 2-27: Pin configuration and truth table for a 74147 10-line priority encoder.

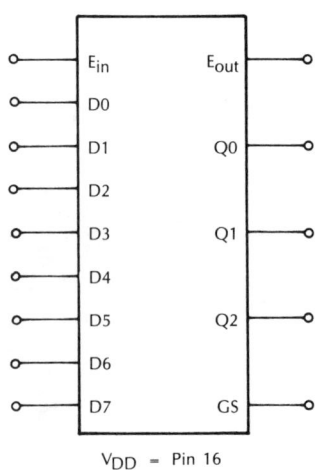

INPUT									OUTPUT				
E_{in}	D7	D6	D5	D4	D3	D2	D1	D0	GS	Q2	Q1	Q0	E_{out}
0	X	X	X	X	X	X	X	X	0	0	0	0	0
1	0	0	0	0	0	0	0	0	0	0	0	0	1
1	1	X	X	X	X	X	X	X	1	1	1	1	0
1	0	1	X	X	X	X	X	X	1	1	1	0	0
1	0	0	1	X	X	X	X	X	1	1	0	1	0
1	0	0	0	1	X	X	X	X	1	1	0	0	0
1	0	0	0	0	1	X	X	X	1	0	1	1	0
1	0	0	0	0	0	1	X	X	1	0	1	0	0
1	0	0	0	0	0	0	1	X	1	0	0	1	0
1	0	0	0	0	0	0	0	1	1	0	0	0	0

X = Don't care.

V_{DD} = Pin 16
V_{SS} = Pin 8

Figure 2-28: Block diagram and truth table for a MC14532B 8-bit priority encoder.

UNDERSTANDING TODAY'S SOLID STATE CIRCUITS   75

the need of external circuitry. In other words, you can connect one encoder **IC** to another to expand your operation, if necessary. For this **IC**, data inputs and outputs are active at the low logic level.

The MC14532B, like other encoders, has a primary function of providing a binary address for the active input with the highest priority. This **IC** has eight inputs (D0 through D7) and an enable input (E_{in}). You will also notice that five outputs are available; three are address outputs (Q0 through Q2), one group select (GS), and one enable output (E_{out}). Subjects such as addresses are covered in detail in the next two chapters.

Essentials of Display Decoders

Seven-segment displays (found on almost all digital readouts, digital watches, calculators, computers, etc.) usually need a **BCD**-to-7-segment decoder/driver to operate correctly. The standard **TTL** family (7446, 7, and 8) **BCD**-to-7-segment decoder/driver **IC**'s all have the same pin configuration. See Figure 2-29.

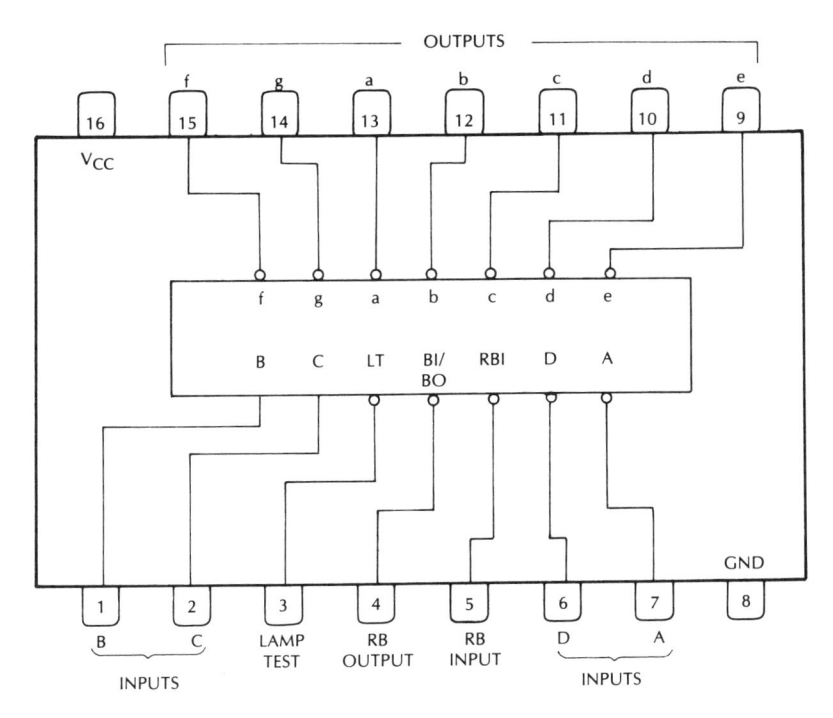

Figure 2-29: Pin configuration for standard TTL BCD-to-7-segment decoder/driver IC's.

A block diagram for the **CMOS BCD**-to-7-segment decoder MC14558B is shown in Figure 2-30. This **IC** decodes 4-bit binary coded decimal data, dependent upon the state of the auxiliary inputs enable (pin 3) and **RBI** (pin 5), and provides an active high 7-segment output for a display driver. Notice, the outputs are labeled a, b, c, d, e, f, and g (this is standard for all these decoders, both **TTL** and **CMOS**). Each of the lettered outputs corresponds to one segment in the actual display device. See the lettered figure 8 next to the block diagram in the illustration. A truth table for this **IC** is shown in Table 2-5.

Each input/output condition for a particular numerical display is shown, except for the auxiliary inputs RBI = ripple blocking input, and RBO = ripple blocking output. To perform a lamp test (check all seven outputs simultaneously), you would place a 0 on pin 3, 0 on pin 5, X (don't care) on **BCD** input code, and 0 on pin 4. Also, you can block all segments by placing a 0 on pin 3 and a 1 on pins 5 and 4, or by placing a high on pin 3 and a low on pin 5, **BCD** inputs, and pin 4. To perform all numerical displays requires, as explained, the setup shown in Table 2-5.

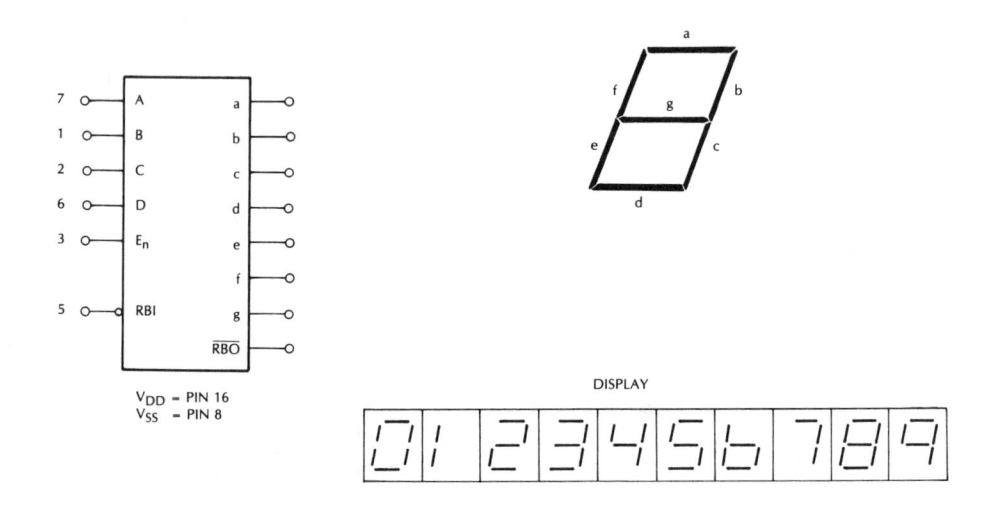

Figure 2-30: Block diagram for the MC14558B BCD-to-7-segment decoder.

UNDERSTANDING TODAY'S SOLID STATE CIRCUITS 77

INPUTS						OUTPUTS*							
E_{in}	$\overline{RBI}$	D	C	B	A	a	b	c	d	e	f	g	$\overline{RBO}$	DISPLAY
Pin 3	Pin 5	Pin 6	Pin 2	Pin 1	Pin 7	Pin 13	Pin 12	Pin 11	Pin 10	Pin 9	Pin 15	Pin 14	Pin 4	
1	1	0	0	0	0	1	1	1	1	1	1	0	1	0
1	X	0	0	0	1	0	0	0	0	1	1	0	1	1
1	X	0	0	1	0	1	1	0	1	1	0	1	1	2
1	X	0	0	1	1	1	1	1	1	0	0	1	1	3
1	X	0	1	0	0	0	1	1	0	0	1	1	1	4
1	X	0	1	0	1	1	0	1	1	0	1	1	1	5
1	X	0	1	1	0	0	0	1	1	1	1	1	1	6
1	X	0	1	1	1	1	1	1	0	0	0	0	1	7
1	X	1	0	0	0	1	1	1	1	1	1	1	1	8
1	X	1	0	0	1	1	1	1	0	0	1	1	1	9
1	0	0	0	0	0	0	0	0	0	0	0	0	0	BLANK
0	0	X	X	X	X	1	1	1	1	1	1	1	0	8
0	1	X	X	X	X	0	0	0	0	0	0	0	1	BLANK

* All non-valid BCD input codes produce a blank display.   X = Don't care.

Table 2-5: Truth table for the MC14558B BCD-to-7-segment decoder. This example is typical of the inputs required for these decoders, whether TTL or CMOS devices.

CHAPTER THREE

Simplified Digital Basics

The purpose of this chapter is to help you get a handle on the personal computer revolution. It will answer questions like

"Why the various numbering systems?"

"What makes a microcomputer tick?"

The heart of a computer is a microprocessor chip; however, it also must have such things as a keyboard (so you can enter data into the system) and, for example, random-access memory (**RAM**), internal read-only memory (**ROM**), and a separate video monitor display. You will learn about all of these units, plus much more, in this chapter.

Understanding Numbering Systems Used in Modern Computer Units

As you learned in Chapter Two, any data used by digital **IC**'s must be represented by binary notation, i.e., 1's and 0's. Also, as you probably have guessed by now, before a base 10 number can be used by a digital computer, it must be converted to high and low logic voltage levels. You will remember that this type of conversion was discussed in Chapter Two (see section titled "Working with Decoders/Encoders"), and octal conversion was also briefly mentioned in that section.

Let's now look a bit deeper into the various numbering systems used in computers (for example, base 2 . . . binary; base 8 . . . octal; and base 16 . . . hexadecimal) and see why one or the other is more convenient to use. First refer to Table 3-1, which illustrates the relationship of each binary number and each octal number to the equivalent base 10 number.

You'll be surprised at what the octal numbering system can do for you. Notice that each octal based number has a binary equivalent that has three digits (1's and 0's). The important point is that a binary number can be divided into groups of three digits each, starting from the right-hand side of the binary number, and each group of three binary digits may be represented by its octal equivalent. In fact, any binary number can be converted directly from an octal number, or any octal number can be written directly from a binary number. For example:

$$\begin{array}{ll} 010 & 101 \quad \text{binary} \\ 2 & 5 \quad \text{octal} \end{array}$$

Let's take another example of this technique and carry it a bit farther. Given a binary number containing many 1's and 0's, convert it to a decimal number using the octal system. For instance, what is the decimal equivalent of the binary number 1101010100?

$$\begin{array}{llll} 1 & 101 & 010 & 100 \quad \text{binary} \\ 1 & 5 & 2 & 4 \quad \text{octal} \end{array}$$

Reading right to left:

$$\begin{array}{rl} 4 \times 8^0 = & 4 \\ 2 \times 8^1 = & 16 \\ 5 \times 8^2 = & 320 \\ 1 \times 8^3 = & \underline{512} \\ & 852 \text{ decimal} \end{array}$$

DECIMAL	OCTAL	BINARY
0	0	000
1	1	001
2	2	010
3	3	011
4	4	100
5	5	101
6	6	110
7	7	111

Table 3-1: Relationship of binary based numbers to each octal based number and its equivalent base 10 number.

SIMPLIFIED DIGITAL BASICS 81

The conclusion to be drawn is that the octal numbering system can provide a shorthand method for bridging the gap between decimal and binary. Although personal computers rarely use octal directly, the octal numbering system can be used as an aid in programming.

Now that we have discussed the octal system, let's examine the hexadecimal (base 16). As you have seen, the octal number system can be used to simplify binary-to-decimal conversion. In this case, we have shown that three binary digits are represented by one octal digit. The hexadecimal system (from here on out we will call it simply the "hex system") makes conversion even easier because each hex digit represents four binary digits. What this means is that 16 binary digits can be represented with six base eight symbols or four hex symbols. Obviously, four hex symbols are easier to work with. However, as we know, the decimal system has only ten digits, or characters (0, 1, . . . , 9), and, of course, six additional characters are required for the hex system. In this system the letters A, B, C, D, E, and F are used, in addition to the numbers 0 through 9. Table 3-2 shows the relationship between decimal, binary, and hex.

To illustrate the usefulness of the hex system, we will convert a 16-digit binary number to its base 10 equivalent. For our example, we will use binary number 1101111101010100. The first step is to divide the binary number into groups of four each, starting from the right, and then write the hex equivalent of each group (use Table 3-2 to find the hex equivalents). Your work should look like this:

```
   1101      1111      0101      0100   binary
    D         F         5         4     hex
```

Now we know that DF54 is the hex equivalent of the binary number 1101111101010100. Our next step is to convert the hex DF54 to decimal, like this:

$$
\begin{array}{rcl}
\text{D} \quad \text{F} \quad 5 \quad 4 & & \\
4 \times 16^0 & = & 4 \\
5 \times 16^1 & = & 80 \\
(F = 15) \times 10^2 & = & 3840 \\
(D = 13) \times 10^3 & = & 53248 \\
\hline
& & 57172 \text{ decimal}
\end{array}
$$

DECIMAL	BINARY	HEX
0	0	0
1	1	1
2	10	2
3	11	3
4	100	4
5	101	5
6	110	6
7	111	7
8	1000	8
9	1001	9
10	1010	A
11	1011	B
12	1100	C
13	1101	D
14	1110	E
15	1111	F
16	10000	10
17	10001	11
18	10010	12
25	11001	19
26	11010	1A
27	11011	1B
32	100000	20
33	100001	21

Table 3-2: Relationship between decimal, binary, and hex numbers.

A term you will certainly run across sooner or later in your daily work is "USASCII," which stands for United States American Standard Code for Information Interchanges. You will probably see this written in most of today's literature as ASCII (usually pronounced *askey*). Therefore, let's examine the alphanumeric code system.

The ASCII Alphanumeric Code System

To begin, *machine language* (another term for the binary high and low logic 1's and 0's we previously discussed) is the *only* language a digital computer can work with, i.e., understands or recognizes. The ASCII standard code is used for interchange of information (whether input or output) between peripherals such as typewriters (keyboards) and line printers. For example,

SIMPLIFIED DIGITAL BASICS

some computer systems have an ASCII keyboard with a numeric keypad.

ASCII is an 8-bit code, and therefore is excellent for hex representation. Incidentally, you will find that only seven digits are actually required for ASCII. However, it is usually considered to be an 8-bit code. Now, since hex-to-binary conversion is relatively simple, as we have shown, and is used in integrated circuits, ASCII can be adapted to any personal computer system. Table 3-3 shows the ASCII codes.

Let's say that the computer is instructing a line printer to type the name "Louise." Using Table 3-3, we find that the

CHARACTER	ASCII CODE	CHARACTER	ASCII CODE
@	1000000	FORM FEED	0001100
A	1000001	CARRIAGE RETURN	0001101
B	1000010	RUBOUT	1111111
C	1000011	SPACE	0100000
D	1000100	!	0100001
E	1000101	"	0100010
F	1000110	#	0100011
G	1000111	$	0100100
H	1001000	%	0100101
I	1001001	&	0100110
J	1001010	'	0100111
K	1001011	(	0101000
L	1001100	)	0101001
M	1001101	*	0101010
N	1001110	+	0101011
O	1001111	`	0101100
P	1010000	−	0101101
Q	1010001	.	0101110
R	1010010	/	0101111
S	1010011	0	0110000
T	1010100	1	0110001
U	1010101	2	0110010
V	1010110	3	0110011
W	1010111	4	0110100
X	1011000	5	0110101
Y	1011001	6	0110110
Z	1011010	7	0110111
[	1011011	8	0111000
\	1011100	9	0111001
]	1011101	:	0111010
↑	1011110	;	0111011
NULL	0000000	<	0111100
HORIZ TAB	0001001	=	0111101
LINE FEED	0001010	>	0111110
VERT TAB	0001011	?	0111111

Table 3-3: ASCII codes.

computer must transmit the following ASCII code digits on the data lines to the typewriter:

Character		ASCII Code
L	=	1001100
O	=	1001111
U	=	1010101
I	=	1001001
S	=	1010011
E	=	1000101

Table 3-4 shows the conversion between ASCII and hex. To make a conversion, select a letter, symbol, or number from the ASCII character column shown. We will use the dollar sign ($) for our example. Next, locate the dollar sign on the ASCII-to-hex conversion table. Then read up the column (up to the number 2). This is your most significant hex digit (MSD). Now, move horizontally to the left and find the least significant hex digit (4). Thus, hex 24 equals the dollar sign in ASCII, or $ equals 24 in hex, whichever way you would like to look at it. This process can be reversed. Table 3-4 shows that hex 24 equals $, and Table 3-3 shows that ASCII code $ equals binary 0100100.

It is important to keep in mind that a microprocessor unit

		\multicolumn{8}{c}{MOST SIGNIFICANT HEX DIGIT}
		0	1	2	3	4	5	6	7
LEAST SIGNIFICANT HEX DIGIT	0	NUL	DLE	SP	0	Ø	P	\	p
	1	SOH	DC1	!	1	A	Q	a	q
	2	STX	DC2	"	2	B	R	b	r
	3	ETX	DC3	#	3	C	S	c	s
	4	EQT	DC4	$	4	D	T	d	t
	5	ENQ	NAK	%	5	E	U	e	u
	6	ACK	SYN	&	6	F	V	f	v
	7	BEL	ETB	/	7	G	W	g	w
	8	BS	CAN	(	8	H	X	h	x
	9	HT	EM	)	9	I	Y	i	y
	A	LF	SUB	*	:	J	Z	j	z
	B	VT	ESC	+	;	K	[	k	1
	C	FF	FS	,	<	L	\	l	:
	D	CR	GS	–	=	M	]	m	}
	E	SO	RS	.	>	N	↑	n	~
	F	SI	US	/	?	O	←	o	DEL

Note: Parity bit in most significant hex digit not included and characters in colummns o and 1 (as well as SP and DEL) are non-printing.

Table 3-4: ASCII-to-hex conversion table.

does not use ASCII. This code must be converted to binary for use in a microprocessor and converted back into ASCII for use at the keyboard, etc.

Introduction to Today's Computer-on-a-Chip

Computers-on-a-chip . . . what are they? A single *microprocessor* unit such as Motorola's MC6800 is an integrated circuit that performs many of the functions, such as arithmetic and control, usually found in a digital computer. However, the MC6800 by itself does not contain the memories (**RAM**'s and **ROM**'s) and input/output (I/O) functions of a computer. Therefore, an **IC** such as this one is called a *microprocessor* (**MPU**), not a *microcomputer* (**MCU**). But, add the necessary memories and I/O functions to a system and now a microcomputer unit is formed. There are some **IC**'s that contain all these functions. In effect, when an **IC** contains all the basic functions, it is a computer-on-a-chip.

To review briefly, in most computer systems there is a read only memory to store the computer instructions, a random access memory to store temporary data (the data to be manipulated by the computer program), a microprocessor, and an I/O integrated circuit to make the system interface with outside or peripheral equipment (intersystem communications). For example, video terminals, hard copy machines, keyboards, etc., are required. Include all these functions on a single chip and you have a computer-on-a-chip, or **MCU**.

One 8-bit microcomputer unit is Motorola's MC6805R2. This unit contains a microprocessor (referred to as a *Control Processor Unit*, or **CPU**, which is another name for microprocessor), on-chip clock, **ROM, RAM,** I/O, 4-channel 8-bit analog-to-digital (A/D) converter, and timer. This **IC** features 64 *bytes* of **RAM** and 2048 bytes of user **ROM**. *Note:* A byte is a unit composed of eight binary digits (8-bits). Figure 3-1 shows the pin assignment for the MC6805R2.

All the circuit elements of the **MCU** are in integrated circuit form except, of course, external wiring, readouts, keyboards, and the like. Thus, you do not have access to the circuit elements, nor can you change them in the way they func-

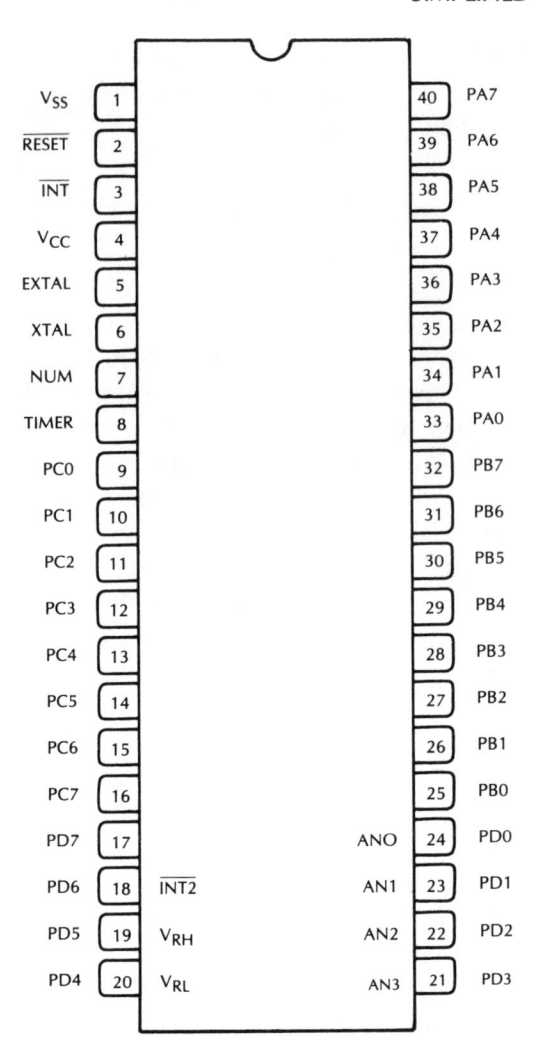

```
                    ┌───┐
      V_SS    │ 1        40 │  PA7
      RESET   │ 2        39 │  PA6
      INT     │ 3        38 │  PA5
      V_CC    │ 4        37 │  PA4
      EXTAL   │ 5        36 │  PA3
      XTAL    │ 6        35 │  PA2
      NUM     │ 7        34 │  PA1
      TIMER   │ 8        33 │  PA0
      PC0     │ 9        32 │  PB7
      PC1     │10        31 │  PB6
      PC2     │11        30 │  PB5
      PC3     │12        29 │  PB4
      PC4     │13        28 │  PB3
      PC5     │14        27 │  PB2
      PC6     │15        26 │  PB1
      PC7     │16        25 │  PB0
      PD7     │17   AN0  24 │  PD0
      PD6     │18   AN1  23 │  PD1
      PD5     │19   AN2  22 │  PD2
      PD4     │20   AN3  21 │  PD3
              INT2  V_RH V_RL
```

Figure 3-1: 8-bit microcomputer unit (MC6805R2) pin assignments. See Figure 1-18, Chapter One, for the actual IC package used for this MCU. Pin assignment signal descriptions will be covered in the following pages.

tion. However, it is possible to order a custom-made **MCU**. To initiate a **ROM** pattern for this **MCU**, one must contact a Motorola representative.

It should be pointed out that you need not understand every detail of the internal circuits of this **IC** (or any similar one) to effectively use and troubleshoot **MCU**'s. Nevertheless,

SIMPLIFIED DIGITAL BASICS

you do have to know what **MPU** registers and counters are and how (in basic terms) they operate.

Basic Microcomputer Address, Data, and Control Lines

By now you are somewhat familiar with the **RAM, ROM, I/O**, and **MPU IC**'s, but we are now interested in addressing and controlling these devices. Figure 3-2 is a basic microcomputer system.

It is important to understand that the **MPU** will do only what the program in the computer directs it to do. These instructions are sent over connecting wires (called *lines*) between the **MPU, RAM,** and **ROM**. Collectively, these lines are usually called *bus, address bus,* or *data bus.* See Figure 3-2.

Some **MPU**'s today (such as the MC6802, etc.), contain 16-bit memory addressing (A15 ..... A0). An address is generated using a binary code—logic highs and logic lows. As you will remember, a group of eight bits is one byte, which means that an address 16 bits wide would be equal to two bytes wide. The other lines (8-bit data lines) shown in Figure 3-2 are also typical in modern **MPU**'s. As the name "data" suggests, these lines

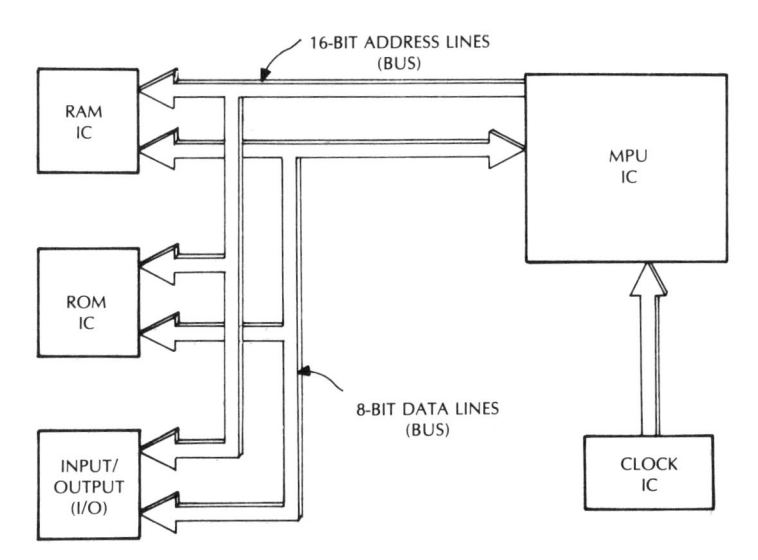

Figure 3-2: Basic microcomputer system.

(usually labeled D6 ..... D0) are used to send and receive data; therefore, data bus. Usually, data bus is bidirectional between memory and peripheral devices. Again, data is in the form of binary words, i.e., high logic and low logic.

There is a third type of input to an **MPU**. This is called a *control bus*. These lines are used to control the order of operation of the entire system. For example, it may be desired to write into memory or, on the other hand, it may be that the function is a read only operation. Some of the control bus lines on a **MPU** are READ/WRITE (R/W), bus available (BA), HALT, IRQ, RESET, NMI, RE, VMA, MR, and (E). There are two other connections found on the MC6802: the crystal connections (EXTAL and XTAL). See Figure 3-3 for the pin assignments on the MC6802 **MPU** with clock and 128 bytes of on-board **RAM**.

MPU Signal Descriptions

Just as the **MPU** must address the device that it wishes to communicate with (using address pins A0 ..... A15), it also must have some channel through which it can send data, or receive data, as has been explained. This is done by using DATA BUS pins (D0 ..... D7). The other pins shown in Figure 3-3 are basically control lines (except V_{cc} and V_{ss}, of course). A brief description of each of these various operations follows. See Figure 3-4 for a block diagram of a *typical* **MPU** using a MC6802.

Halt:

When this input (pin 2) is in the logic 0 state, all activity in the machine will halt.

Read/Write (R/W):

This output (pin 34) informs the peripherals and memory devices whether the **MPU** is in a read (1) or a write (0) state.

Valid Memory Address (VMA):

This output (pin 5) signals peripheral devices that there is a valid address on the address bus.

SIMPLIFIED DIGITAL BASICS 89

Bus Available (BA):

This signal (pin 7) will indicate when the microprocessor has stopped and the address bus is available, and when it is activated; i.e., goes to a logic 1 state from its normal logic 0 state.

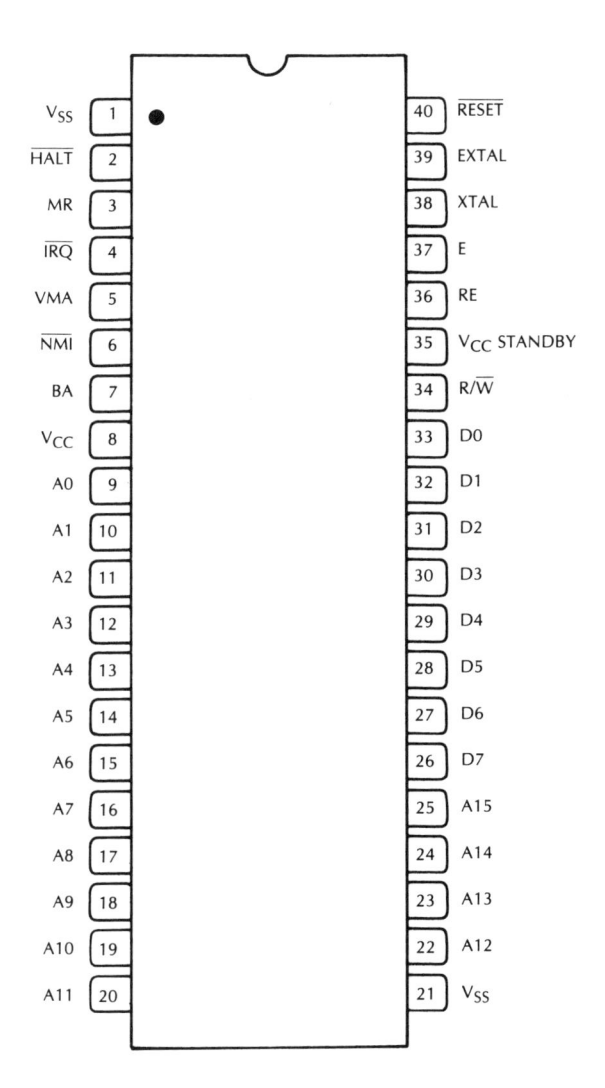

```
         ___                              _____
        V_SS  [ 1              40 ]  RESET
        HALT  [ 2              39 ]  EXTAL
         MR   [ 3              38 ]  XTAL
         ___
         IRQ  [ 4              37 ]  E
         VMA  [ 5              36 ]  RE
         ___
         NMI  [ 6              35 ]  V_CC STANDBY
                                          ___
         BA   [ 7              34 ]  R/W
         V_CC [ 8              33 ]  D0
         A0   [ 9              32 ]  D1
         A1   [10              31 ]  D2
         A2   [11              30 ]  D3
         A3   [12              29 ]  D4
         A4   [13              28 ]  D5
         A5   [14              27 ]  D6
         A6   [15              26 ]  D7
         A7   [16              25 ]  A15
         A8   [17              24 ]  A14
         A9   [18              23 ]  A13
         A10  [19              22 ]  A12
         A11  [20              21 ]  V_SS
```

Figure 3-3: Pin assignment for a MC6802 microprocessor. See text for MPU signal descriptions.

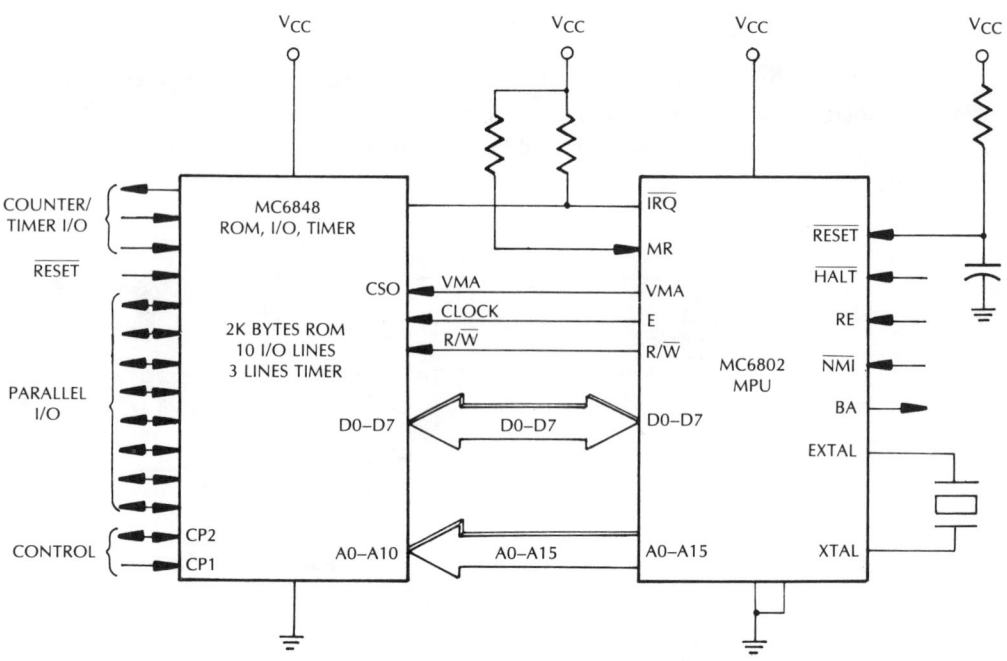

Figure 3-4: Basic microcomputer designed using an MC6802 MPU and a MC6848 ROM, I/O, timer. This minimum component system may be expanded with other parts of the M6800 microcomputer family.

Interrupt Request ($\overline{IRQ}$):

A logic 0 on this input (pin 4) requests that an interrupt sequence be generated within the machine.

Reset:

This input (pin 40) is used to reset and start the **MPU** when a power failure occurs, or during initial startup.

Non-Maskable Interrupt ($\overline{NMI}$):

A logic 0 on this pin (pin 6) requests that an interrupt sequence be generated within the **MPU**.

Programming Fundamentals

In the last section it was said that an **MPU** does only what the program in the **ROM** directs it to do (see Figure 3-4);

therefore the experimenter must work up a set of instructions written in sequential, logical, and simple steps. This series of instructions is then placed in memory to tell the **MPU** what you want it to do. A complete computer program is often referred to as *software*.

Early **MCU**'s used a set of front panel switches to represent the binary digits. These data switches were set to one binary word at a time until the program was completely entered into the computer. Even today, simple, single-board computers can be programmed this way. When directly programming, an experimenter writes a program to be placed into memory, using machine language (binary code), and enters each step. Chapter Two explained the basic operation of a **ROM** and **EPROM** in the section titled "Understanding Memory IC's." Figure 3-5 shows a simple programming circuit that could be used to enter a binary code into a **TTL PROM**.

The actual programming procedure for a **TTL** could be performed with these steps:

Step 1. Connect V_{CC} and V_{SS} to the **PROM IC**.

Step 2. Select the desired binary number (address) with the logic input switches.

Step 3. Set output switch (S1) to the lowest output (usually B0, S0, etc.).

Step 4. Depress switch S2 for about half a second and release. Watch the current monitor and do not let the current exceed the manufacturer's maximum value. This step will place a logic 0 or 1 in the **IC**'s internal circuit, depending on what type **ROM** you are working with.

Step 5. Permit the chip to cool down for a few seconds, then check with a logic probe to see if the proper logic has been entered. If it is incorrect, repeat Step 4 once again.

Step 6. Set switch S1 to the next output pin (usually S1, B1, etc.), and repeat Steps 4 and 5.

Step 7. Continue to select each binary number you want to place in memory and program the bits of each word one at a time until the programming of the PROM is completed.

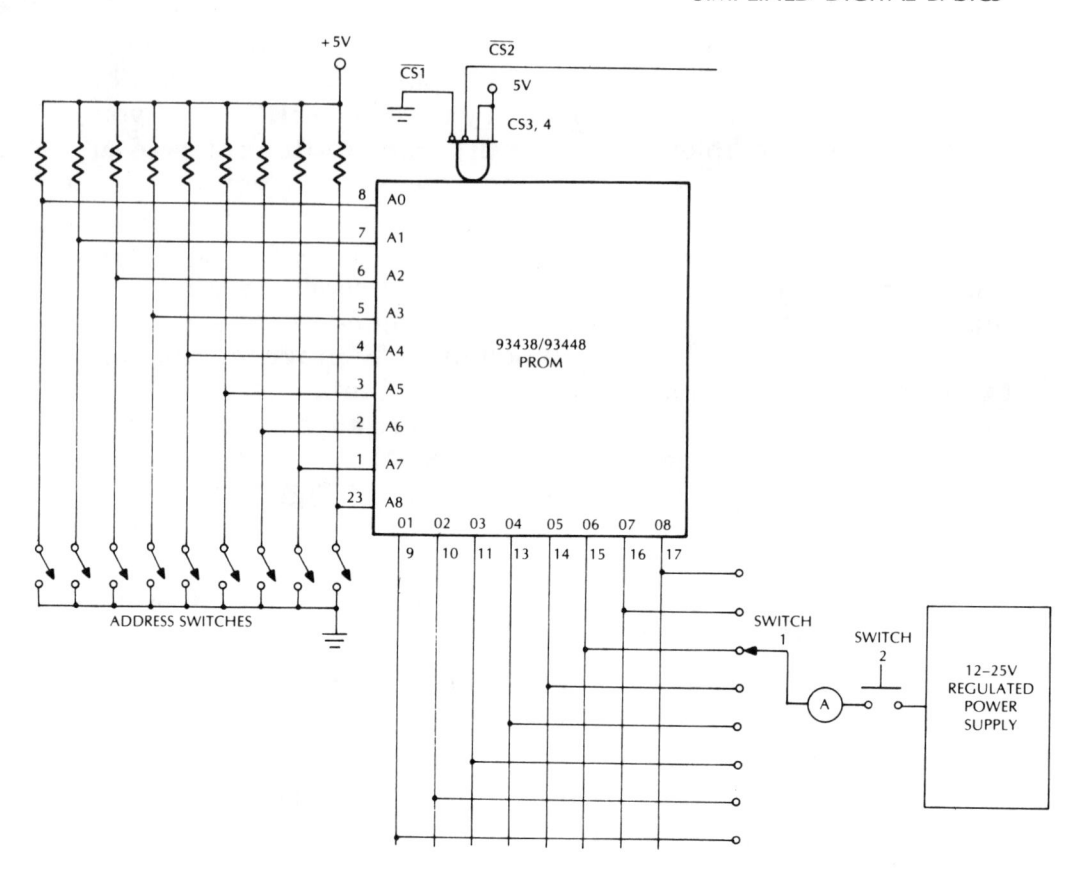

Figure 3-5: Programming a PROM.

It should be noted that if a mistake is made during the programming (even one bit in one setup of the address line switches), the entire **PROM** is made useless for that program. The **EPROM**'s discussed in Chapter Two are now quite popular because they overcome this problem.

If you want to write your own software for an entire computer system, your first step is to ask yourself, "Is the problem computer solvable?" Not all problems are. For instance, ask a computer whether you should get married or not and your chances of getting a correct answer would be better if you consulted a mystic. On the other hand, "How far will a spacecraft in free space fall in ten seconds?" is computable by using the equation d = gt^2/2, where d is the distance in feet, t is time in seconds, and g is the acceleration due to gravity (as any physics student knows, 32 ft/s for each second). Knowing which problems are computable and which are not makes the difference between whether you will succeed or fail at creative programming. Writing a program requires only paper and pencil; however, after you have written the program, it is frequently fixed in **ROM** by a manufacturer. There are several different methods used to place a program in a **ROM**. One of these is called *mask programmable* (a logical technique in which certain bits of a word are blanked out or inhibited).

The writing of the step-by-step procedure for solving a certain problem, whether it be writing letters, handling credit, doing a payroll, analyzing a business, or solving a scientific riddle, may be done by a programmer. Before someone actually starts writing a program, it is common practice to work up a graphical representation illustrating the logic steps required to solve the problem. This sequence of steps is known as a *flow chart*. A very simple flow chart is shown in Figure 3-6. This flow chart depicts a step-by-step procedure for reading two numbers into a computer, dividing the first by the second, and multiplying by 2π (6.28), then printing the result.

From the flow chart, you would write a program on a coding form. Remember, the **MCU** cannot use the flow chart as is. As we have shown, the **MPU** has its own language

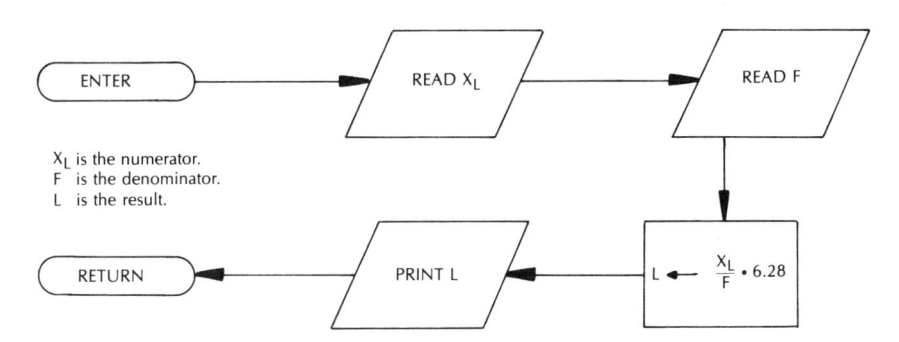

X_L is the numerator.
F is the denominator.
L is the result.

Figure 3-6: Basic flow chart for L = X_L/F (2π), where L equals inductance, X_L is the inductive reactance, F is frequency in Hertz, and 2π is 6.28.

(machine language) based on binary numbers and not alphabetic letters or words. In our example problem, the program tells the computer to use X_L, F, and L as three names to stand for the numerator, denominator, and the answer, to READ in two numbers, to divide the first (X_L) by the second (F), to multiply by 6.28, and to print out the number associated with the name L, then return to start.

In the real world, most manufacturers of **MCU**'s define a two-, three-, or four-letter code that describes the function of each instruction. This code, known as *mnemonic*, has an equivalent hex (or binary) number to represent the function. As an example, the problem for solving the value of L called for dividing and multiplying. An 8-bit code for divide might be 0011 1110 (hex, 3D). The hex code used for X_L and F in the problem is F and C. If the problem were presented to a computer, the answer L (9.42) would be printed.

We have shown that a **ROM** can be programmed by writing a sequence of instructions in binary code that the microprocessor can interpret directly (see Figure 3-5 for the programming setup). For instance, let's again use our imaginary instruction to divide (0011 0001) and multiply (0011 1101). If these two bytes are programmed into memory, either one will appear on a data bus (8-bit line) when the respective **MPU** data port is open. The **MPU** will either divide or multiply (probably the contents in two registers, X_L and F in our example problem).

To perform simple experiments using a **ROM** and **MPU** on a breadboard, writing a program in machine language is practical. However, at best it is a tedious job and subject to many errors. You can simplify the task and cut down on the errors by using hex instead of binary. A better, but more expensive, system is to use alphanumeric symbols to represent machine language. This system is usually called *Assembly Language*. The assembly program directs a computer to operate on a program in symbolic language to produce a program in machine language.

In summary, as we have stressed, all **MPU**, **ROM**, **EPROM**, etc., operate with binary numbers only. However, most manufacturers do define a mnemonic code (usually two or three letters of the alphabet) that describes the function of each instruction. Some mnemonic codes and operations used by

MNEMONIC	HEX	OPERATION
ADDA	9B	ADD
ABA	1B	ADD ACCUMULATOR
CLR	4F	CLEAR
CMPA	91	COMPARE
ASL		SHIFT LEVER, ARITHMETIC
LSR		SHIFT RIGHT, LOGIC

Table 3-5: Mnemonics.

Motorola are shown in Table 3-5. For example, ABA (mnemonic code) is the instruction for the **MPU** to add accumulator B to A. This instruction would be stored in **ROM** memory in binary form; i.e., 1B = 0001 1011 in memory.

How and Why Addressing Modes Are Used in a Microcomputer System

Figure 3-5 shows a program-loading circuit that could place instructions into memory in sequence, one by one. These instructions are executed in sequence, step by step, by an **MPU**, unless it is told to follow some other instruction in a different location in memory. As the **MPU** performs whatever instruction it has received, it is often necessary to have data. The needed data is generally located in memory, and sometimes is found in the next memory location.

Memory locations in **MCU**'s such as the MC6805R2 we previously studied are eight binary digits wide. Incidentally, there are many 8-bit chips; for instance, the **EPROM** microcomputer unit MC68705U3, the microcomputer unit MC6805P4, and a host of other **MPU**'s, memory, etc. However, there are 4-bit single-chip **MPU**'s such as the **CMOS** MC141000.

As you would guess, instructions in an 8-bit **MCU** memory are eight bits wide (one byte) and the instructions in a 4-bit system are four bits wide. It should be pointed out that, in many cases, although each instruction is one byte, it can require the use of more than one memory location. When an instruction requires the use of a certain memory location and one or two following locations, it is referred to as a 2-byte, 3-byte, etc., instruction.

Memories are divided into locations called *addresses,* as has been explained, and the contents of these addresses can be instruction, data, or a combination of both. There are several methods for arranging the data bytes in memory. The **MCU** we have been talking about (the 8-bit MC6805R2) has ten addressing modes available for use by the programmer. But **MPU**'s (the heart of an **MCU**) vary in design, with each design programmable only via its own set of instructions. The device that will be covered in this and the next chapter is the MC6800 family, manufactured by Motorola.

In general, each family of processors (8080, 8040, Z80, and 8085, or the M6800, MC6809E, and all 6800 family **MCU**'s) require a certain internal programming language. Nevertheless, most information (*but not all*) will apply to the entire family of the particular manufacturer. Instructions not used by Motorola's 6800 family will not, however, be covered.

Before we discuss addressing modes, you will need to understand the on-chip registers and accumulators available in the **MPU**'s. If you have forgotten registers, it is suggested that you review the section titled "Principles of Registers and Counter IC's" in Chapter Two, before proceeding farther. Note: In most instances, *accumulator* is simply another word for *register*.

There are several generation additions to the M6800 family that have architectural improvements (arrangement of counters, registers, arithmetic logic units, and so forth, within the **MPU**) which include additional registers, instructions and addressing modes, but, to make it as simple as possible for the moment, we shall first consider only the M6800 **MPU** registers and accumulators:

A and B Accumulator:

The **MPU** contains two 8-bit accumulators which are used to hold operands and results from an arithmetic logic unit (**ALU**).

Index Register (IR):

This register contains two bytes, i.e., a 2-byte register. It is used to store data or a 16-bit memory address for the index mode (to be explained later) of memory addressing.

Program Counter:

This counter is a 2-byte (16-bit) register that points to the current program address. In other words, it contains the address of the next instruction to be fetched from memory. Early in the execution phase of each instruction, the program counter is automatically advanced to the address of the next sequential instruction in memory.

Stack Pointer (SP):

Most processors (for example, the 8080 family or M6800 family) have a special method of handling subroutines to insure an orderly return to the main program. This is needed when there is a program within a program. The stack (a 16-bit register) holds the memory address of the instruction to be executed after the subroutine is completed. The stack is normally a Read/Write memory that may have any location (address) that is convenient. It should be pointed out that various **MPU**'s have different methods of maintaining their stack. *A stack pointer is an internal register of the CPU.*

When an **MPU** has a reserved area for this use—stack pointer—it contains the address of the most recent stack entry. In other words, the stack pointer always "points" to the top of the stack. This method is a last-in-first-out (LIFO) memory and permits a virtually unlimited subroutine stacking.

Condition Code Register (CC):

The condition code register indicates the results of an arithmetic logic unit operation. It is a 1-byte register that is used to test the results of certain instructions (this will be explained in detail later).

Now that you have a basic understanding of the registers and accumulators in an **MPU**, you will also need a brief description of the various addressing modes. This is one of the other basic features you must comprehend before working with **MPU** problems. "Addressing modes" refer to the manner in which you cause the **MPU** to obtain its instructions and data. And, as we have shown in previous pages, you must have a method for addressing the **MPU**'s on-chip registers and all external memory locations. The fundamental addressing modes for the M6800 family follow:

Inherent Addressing Mode:

The instructions in this mode are always in 8-bit form and do not require an address in memory. An example of an inherent instruction is the mnemonic instruction "ABA." This instruction tells the **MPU** to "add the contents of accumulators A and B together and place the result in accumulator A."

Accumulator Addressing Mode:

Like the inherent mode, the accumulator mode is also an 8-bit instruction. This instruction is directed to either the A or the B accumulator. For example, the mnemonic instruction 1NCB causes the contents of accumulator B to be increased by one. This is an increment operation and the Boolean/arithmetic operation is written as $B = 1 \rightarrow B$. Another operation is complement 1's, mnemonic either COMA or COMB. Example: COMB (complement the B accumulator) instruction; after execution of the COMB instruction, each of the eight bits in the B register will have been inverted, i.e., all 1's become 0's and all 0's become 1's (Boolean/arithmetic operation $\overline{B} \rightarrow B$).

Immediate Addressing Mode:

This addressing mode is used for operations such as "Load accumulator A," mnemonic LDAA (Boolean $M \rightarrow A$). For example:

operator	operand	comment
LDAA	#20	load 20 into ACCA

This instruction causes the **MPU** to immediately load accumulator A with the value 20. Notice, the *operand* is the value (number 20, in our example) to be operated on.

Direct Addressing Mode:

Direct addressing generates a single 8-bit operand, and hence can address only memory locations 0 through 155 in our example chip, the M6800 (this chip has extended address, which permits use of more locations, but we will come to that in the next mode). In this mode, the *address* where the data is located is in the next memory location.

The **MPU**, after encountering the operation code (opcode) for instruction Load Accumulator A (LDAA, direct) at a certain memory location, looks in the next location for the address of the operand. It then sets the program counter equal to the value found there and fetches the operand (in our example, a value to be loaded into accumulator A) from that location. Because the address of the data must be specified in 8 bits, the lowest direct address is 0000 0000, and the highest is 1111 1111 (hex-FF).

Extended Addressing Mode:

The direct and extended addressing modes differ only in the range of memory locations to which they can direct the **MPU**. As was pointed out in the direct mode, only address 0 through 155 can be used with the M6800. A 2-byte operand is generated using this **MPU** for extended addressing, enabling the M6800 to reach the remaining locations 256 through 65535. The instructions of the direct addressing mode are two bytes wide and the extender mode instructions are three bytes wide. One memory location is used for instruction and the next two for data location, in the extended mode.

Extended address is a method of reaching any place in memory. However, direct addressing is a faster method of processing data because, in most applications, it uses fewer bytes of control code.

Relative Address Mode:

In the last modes, we have discussed direct and extended. The address obtained by the M6800 is an absolute numerical address located immediately following memory location and, in extended, the next two memory locations.

In the relative addressing mode, the next instruction to be executed by the **MPU** is located at some memory address other than the one following. This mode of operation is implemented for the **MPU**'s *branch* instructions. These instructions specify a memory location relative to the program counter's current location. Branch instructions generate two bytes of machine code: one for the "instructions opcode" and one for the "relative" address.

The M6800 is capable of branching both forward and backward. The next location in memory after the branch instruction contains the information that tells the **MPU** how to branch (branch ahead or branch back from its present location) for its next instruction. The **MPU** will then continue executing instructions from the new location in memory.

Indexed Addressing Mode:

With this mode, the numerical address is variable and is dependent upon the real time contents of the index register. This mode of addressing uses a 2-byte instruction, in which an instruction followed by a number that is added to the contents of the index register to form an address, or source statement such as

operator	*operand*	*comment*
STAA (store accumulator)	X	Put A in indexed location

causes the **MPU** to store the contents of accumulator A in the memory location specified by the contents of the index register. It should be mentioned that the label X is reserved to designate the index register, when working with the MC6800 **MPU**. For example, the TSX instruction for this **MPU** causes the index register to be loaded with the address of the last data byte put onto the "stack."

Summary

An **MPU** signal description (pin by pin) was given in a previous section of this chapter. The last sections described the various accumulators and registers, and the different addressing modes used to load them. The most important point presented in the last section (addressing) is that if, while you are making up a program, the actual data is in the next one or two bytes, you should use the instruction code for the immediate mode. Or, if the data is one byte wide, use the code for the direct mode. On the other hand, if the data is located at an address that requires two bytes to describe, you should then use the instruction code for the extended mode. That is to say that the instruction to load A or B register can require any of several different modes, and the hex code for each mode is different, hence the binary code will be different. Referring to Table 3-6, you will see that it is absolutely critical that you consult the manufacturer's instruction sheet before programming a specific **MPU**.

MODE	INSTRUCTION CODE
LOAD A IMMEDIATE	86 1000 0110 ← BINARY
	8 6 ← HEX
LOAD A DIRECT	96 1001 0110 ← BINARY
	9 6 ← HEX
LOAD A EXTENDED	B6 1011 0110 ← BINARY
	B 6 ← HEX
LOAD A INDEXED	A6 1010 10110 ← BINARY
	A 6 ← HEX

Table 3-6: Addressing modes to load A register, showing mnemonic instruction code (hex) and binary code.

CHAPTER FOUR

Practical Guide to State-of-the-Art On-Chip Microcomputers

Whatever your prime interest in a microcomputer is, be it system building, maintenance and service, programming, or experimenting with the various chips, you should familiarize yourself with modern on-chip microcomputer units. Microcomputers vary in design, with each design programmable only through its own set of instructions. Some of the **MCU**'s and **MPU**'s we will cover in this chapter are the MC6805 **MCU** family and the MC68000 **MPU** family . . . certainly the state-of-the-art in the **MPU** technology.

This chapter is primarily devoted to **MCU** operation, and it is assumed that the reader has read Chapter Three and gained a knowledge of **MPU**'s. If you have studied **MPU** basics thoroughly, you should have no difficulty in understanding the diagrams and descriptions presented in this chapter.

All examples used in the following pages were obtained from actual **MCU** literature provided for the author by the Motorola Corporation Semiconductor Division, located at 3501 Ed Bluestein Boulevard, Austin, Texas 78721. However, the basic approaches to using **MCU** chips presented here can be applied to almost any microprocessor/stored-program system now being developed (much of the information in this chapter is advance information), to those that will be manufactured in the future, as well as to all the **MPU**-based **IC**'s now in use. *Note:* This statement does not apply to some of the new computer language such as Ada, artificial intelligence, Forth, and the like.

About Today's Microcomputer Units

As we indicated in Chapter Three, the **MPU** is the heart of the system. But in order to compose a computer system, support circuitry is required. As you will remember, in addition to **ROM** (instruction memory) there is also a need for working data storage (**RAM**), and the **MPU** must provide the required timing, plus control signals. Of course, I/O data instructions and address are also necessary. One example of today's **MCU** units is the MC6805P2. This computer-on-a-chip contains a **CPU** (central processing unit), on-board clock, **ROM, RAM,** I/O, and timer. Figure 4-1 shows a block diagram of this 8-bit **MCU** unit. This **MCU** comes in a 28-pin package that is illustrated in Figure 4-2, with pin assignments labeled.

An important point is that you, as a technician, realize that once you understand the input and output signals of an **MCU** family, you'll find one unit is very similar to another. For instance, refer to Figure 3-1 in Chapter Three (pin assignments for the MC6805R2) and then back to Figure 4-2 (pin assignments for the MC6805P2), and it should be apparent there is not much input signal difference except that the MC6805R2 has more user **ROM** (2048 bytes versus 1100 bytes), more I/O lines (24 versus 20), etc., and, of course, 40 pins, whereas the MC6805P2 is a 28-pin **IC**.

Because the MC6805P2 is a simpler device, let's examine its block diagram (see Figure 4-1) and see just what its input and output signals are.

V_{CC} and V_{SS}:

Notice that pins 1 and 4 are V_{SS} and V_{CC} respectively. This is not true of all **MPU**'s and **MCU**'s. For example, the MC68701 uses pins 1 and 7 for V_{SS} and V_{CC}, the MC6805P4 uses pins 1 and 3, but the MC6800, MC6805R2, and MC68705R3 all use pin 1 (V_{SS}) and pin 4 (V_{CC}), as does the MC6805P2 we are discussing.

STATE-OF-THE-ART ON-CHIP MICROCOMPUTERS

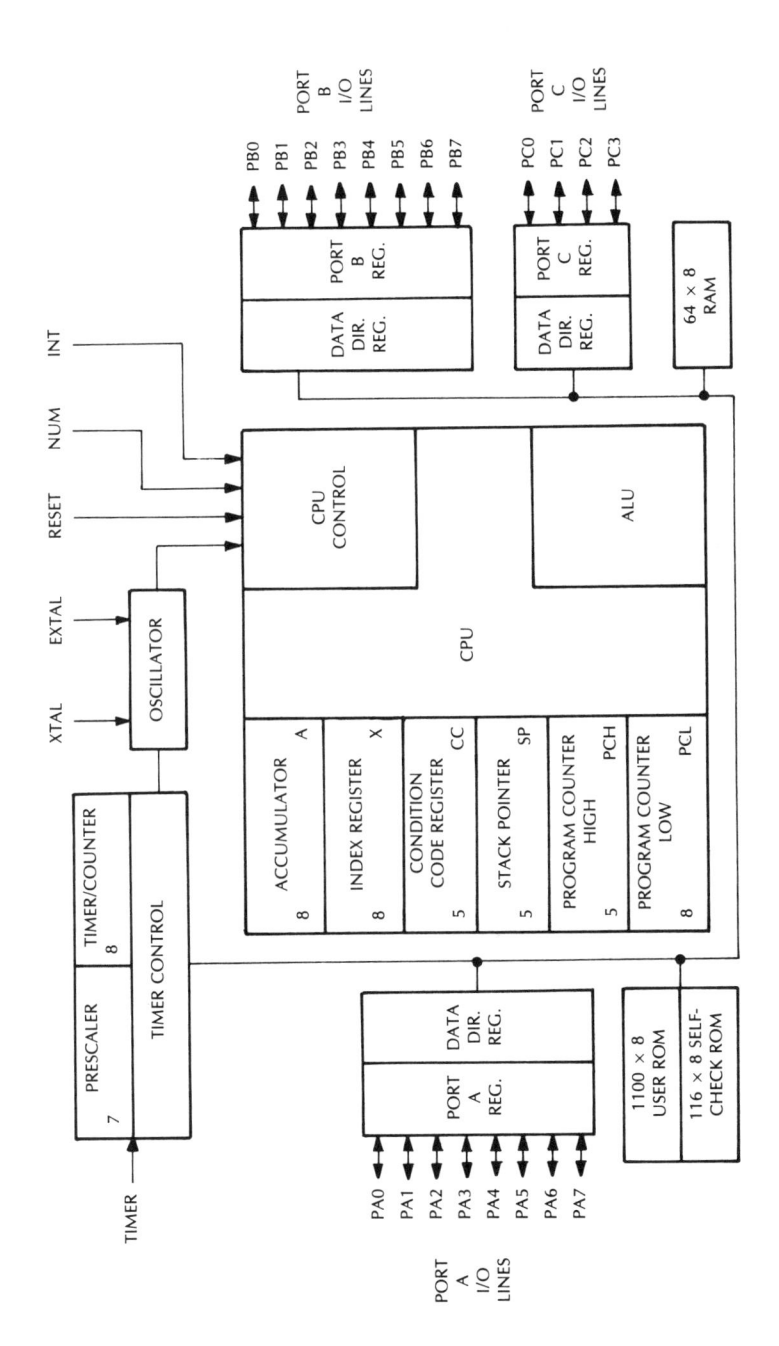

Figure 4-1: A state-of-the-art MCU block diagram (MC6805).

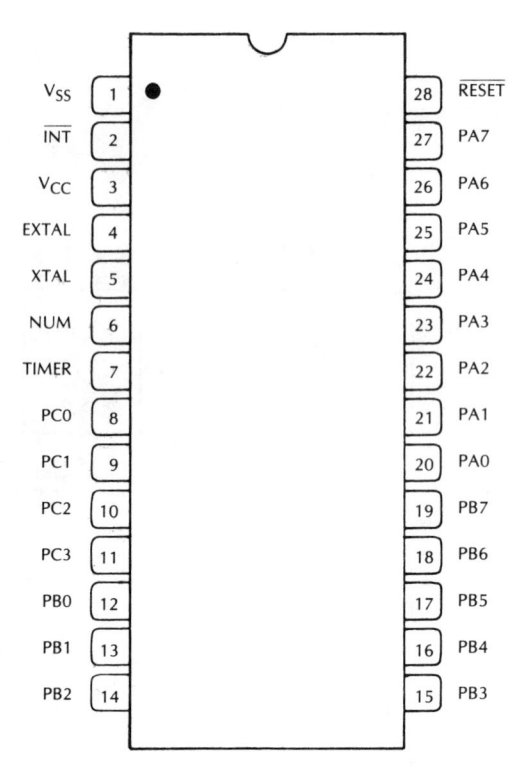

```
        _____
Vss  1 |●                28| RESET
INT  2 |                 27| PA7
Vcc  3 |                 26| PA6
EXTAL 4|                 25| PA5
XTAL 5 |                 24| PA4
NUM  6 |                 23| PA3
TIMER 7|                 22| PA2
PC0  8 |                 21| PA1
PC1  9 |                 20| PA0
PC2 10 |                 19| PB7
PC3 11 |                 18| PB6
PB0 12 |                 17| PB5
PB1 13 |                 16| PB4
PB2 14 |                 15| PB3
```

Figure 4-2: Package drawing and pin assignments for the 8-bit MCU MC6805P2.

$\overline{Int}$:

This pin feeds the **CPU** control (see block diagram, Figure 4-1) and provides a way of applying an external interrupt to the **MCU**. Although different **MCU**'s use different pins for the input, they all perform the same basic task. However, **MCU**'s have three different ways in which they can be interrupted (as has the MC6805P2), and some have four ways—the MC6805R2, for example. Regardless of what method is used for interrupt, when any interrupt occurs, processing is suspended.

XTAL and EXTAL:

The block diagram (Figure 4-1) shows these pins (Figure 4-2) connected to the on-board oscillator (clock circuit). If you are working with the MC6805P2, a crystal (usually 4.0 MHz), a

STATE-OF-THE-ART ON-CHIP MICROCOMPUTERS

resistor, a jumper wire, or an external signal may be used to generate a system clock. With other **MCU**'s, sometimes you can use only a crystal or resistor. For instance, the MC6805R2 requires an "AT" cut crystal at a maximum frequency of 4.0 MHz.

Timer:

This input (pin 7, Figures 4-1 and 4-2) allows you to use an external input to decrement the internal timer circuitry. More about this function will be presented later in the chapter.

Reset:

This pin is tied to the **CPU** control, as shown in Figure 4-1. Basically, you can reset the **MCU** at times other than those automatically used with the circuits built into the **MCU**.

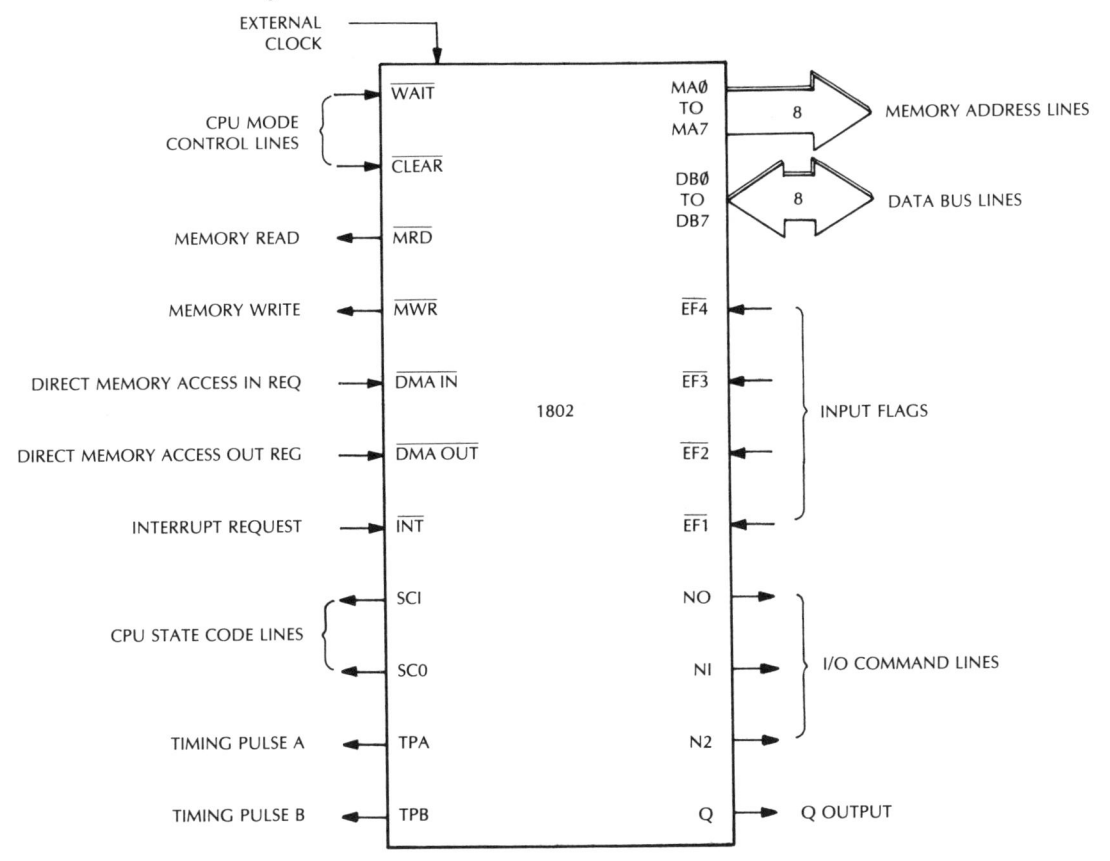

Figure 4-3: Inputs and outputs for the 1802 (A) and 8080 (B).

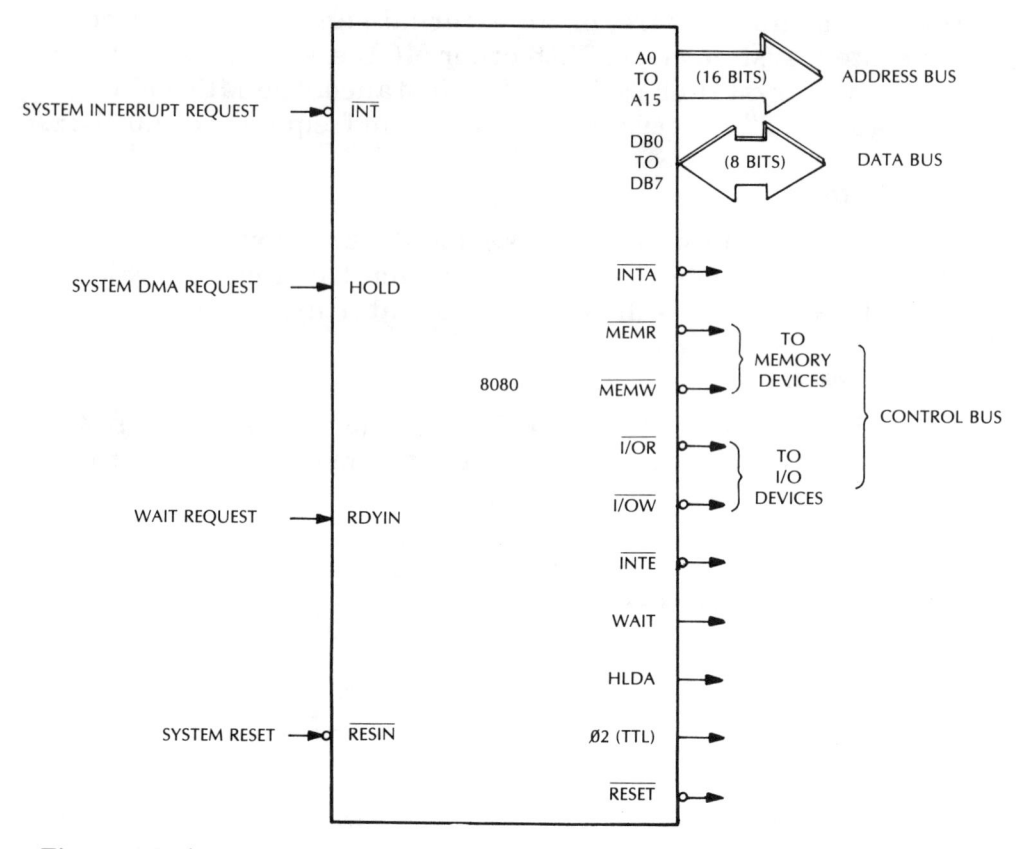

Figure 4-3: (continued)

NUM:

You, as an **MCU** user, do not use this pin (pin 6, Figure 4-2). It must be connected to V_{SS} (pin 1, Figure 4-2). All other lines such as PA0 . . . PA7, and PC0 . . . PC3 (Figure 4-1) are I/O ports, as shown.

Other **MCU**'s may have more or fewer I/O lines, accumulators, and registers (this one has five registers available to the programmer), and more or fewer **RAM** (64 bytes, in this case), **ROM** (1100 × 8 user **ROM** and 116 × 8 self-check **ROM**, shown in Figure 4-1) but, as we have explained, in general all the **MCU**'s in a certain family will follow the same number-bit and architecture, use basically the same control operations, and, in the case of the M6805 family, have software compatibility similar to that of the M6800 8-bit microprocessing unit. In fact, this is usually true of all **MCU**'s. Whatever the basic **MPU** is—for instance, the 8080—the entire **MCU** family will be developed using similar hardware and software.

As an example of the similarity between inputs and outputs, let's compare the MC6805P2 with the 1802 and 8080. We have explained all the inputs and outputs of the MC6805P2 in the last few pages, so all we need now is the other inputs and outputs. See Figure 4-3.

Notice that there are differences in signal descriptions, but not so many that you can't quickly grasp what is going on once you have a thorough understanding of one of the various families of **MPU**'s. Nevertheless, it should be apparent that you must read the manufacturer's instruction sheet because **MPU**'s and **MCU**'s vary greatly in terms of their capabilities and specific language or instruction sets. However, as we are trying to point out, they are similar in many functions.

The M6805 Family of MCU's

To use an **MCU** for any of a host of applications, you need only become familiar with

1. The microcomputer signal descriptions.
2. The program language.
3. The manufacturer's mnemonics and equivalent hex (or binary).

This requires learning about one hundred new terms and abbreviations (you know most of them), such as the names of the various addressing modes, interrupts, and I/O, **MPU**, **MCU**, **ROM**, **RAM**, etc.

Four of Motorola's 6805 family are the MC6805P2 (which we have discussed, see Figure 4-1), the MC6805P4, the MC6805R2 (Figure 3-1 shows the pin assignments), and the MC146805G2. We will now examine this family in order to increase your knowledge of **MCU**'s.

First, you already have a fair understanding of the MC6805P2, so let's start off with the MC6805P4, which we have not examined. This **MCU** is an 8-bit unit that contains a **CPU**, on-chip clock, **ROM**, **RAM**, I/O, and timer. Table 4-1 shows a

HARDWARE FEATURES	
MP6805P4	MP6805P2
8-BIT ARCHITECTURE	SAME
112 BYTES OF STANDBY RAM	64 BYTES
STANDBY RAM POWER PIN	NONE
MEMORY MAPPED I/O	SAME
1100 BYTES USER ROM	SAME
20 I/O LINES	SAME
ON-CHIP CLOCK	SAME
SELF-CHECK MODE	SAME
MASTER RESET	SAME
5V SINGLE SUPPLY V_{CC}	SAME

Table 4-1: Comparison of the hardware features of two MC6805 family MCU's.

comparison of the hardware features of the MC6805P4 and MC6805P2.

As you can see from Table 4-1, both these **MCU**'s are very similar in both hardware and software features. At this point, you may be wondering if there is any great difference between the 6800 family of **MCU**'s. To answer this, notice that the MC6805P4 has 112 bytes of **RAM**, whereas the MC6805P2 has only 64 bytes of **RAM**. As was explained earlier, a RAM is an area where data can be stored. This data may be erased at any time and new data stored in its place—obviously very important to the user.

Another **MCU**, the MC6805R2, has a greater amount of user **ROM**—2048 bytes. Referring to Table 4-1, you'll find these two **MCU**'s have only 1100 bytes of user **ROM**. You will remember that the program that tells the **MPU** what to do is stored in sequential locations in **ROM**, but the user cannot change the contents of **ROM** locations, as he can **RAM**. Therefore, if your need is for a greater amount of **MPU** instructions than either the MC6805P2 or P4 can store, one possible solution to the problem is to use the MC6805R2.

There are several other differences between the MC6805R2 and the two other **MCU**'s we have been discussing. For example, this **MCU** unit has 24 **TTL/CMOS** compatible I/O lines. *Note:* All three devices' I/O lines are **TTL/CMOS** compatible, with eight of these lines being LED compatible. Also, this **IC** can be interrupted four different ways, whereas the other two have only three (more about interrupts later); in addition, this unit has an internal A/D converter. To increase your understanding of these and other **MCU**'s, let's now re-examine each of their internal components in greater detail.

How the RAM's Operate

For our example of how **RAM**'s operate, we will assume that there are 128 bytes of **RAM** and that the **MPU** has the capability to work with any bit in **RAM**. To begin, a byte in our example is 8-bits; hence, we are talking about a total of 1,024 memory cells. Because an illustration of that many memory cells would be difficult (and serve no real purpose), we will use a simplified drawing of a 16-bit memory array. This is shown in Figure 4-4.

A computer memory is formed from a large array of semiconductor elements, each capable of storing a single binary 1 or 0, usually organized into groups of bits (in most of our examples, 8-bits, or a byte). In the example shown in Figure 4-4, each memory cell has a row selector line (A0 ... A7) connected to it. By selecting one of the A0-A3 lines and one of the A4-A7 lines, the desired memory cell is accessed by an **MPU** and conditioned to either write in (store) a bit or read out a bit. For instance, if Row Select line A2 and Column Select line A5 go to a logic high, memory cell 10 is selected. If the **MPU** reads only the contents of cell 10, its output will be either 1 or 0, depending on what's stored in the memory cell. The **MPU** can write a 1 or 0 into memory cell 10 if it places the proper control data input on the Data In bus.

With this simple example, the **MPU** needs only eight lines (A0 ... A7) to select any of the 16 memory cells. However, this basic memory cell arrangement restricts the **MPU** to reading in or reading out only one bit at a time. Figure 4-5 shows the function diagram of a practical modern **RAM** that is compatible with Motorola's family of 6800's.

The eight **RAM** data bus pins D0 through D7 are tied to **MPU** data bus pins D0 through D7. Address pins A0 through A6 are tied to **MPU** address pins A0 through A6. These seven address pins are used by the **MPU** to select an 8-bit word on a particular chip addressed.

If you want to include more than one **RAM** in a discrete component computer system, all A0's would be tied to the A0

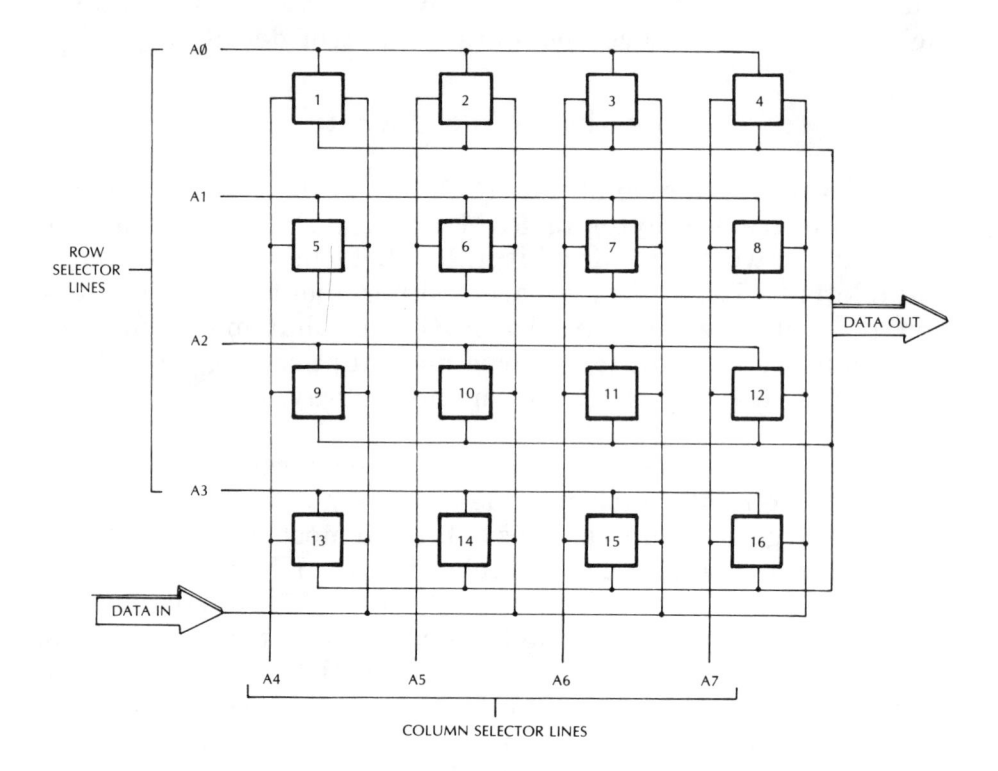

Figure 4-4: Simplified illustration of a 16-bit memory that requires eight address lines to select any of the memory cells.

address line of the **MPU**, all A1's would be tied to the A1 address line from the **MPU**, and so on. Incidentally, using this same line of thinking (adding more memory), it should be apparent that the basic memory cell array shown in Figure 4-4 can be extended to include many more cells: i.e., add additional row selector lines, column selector lines, and memory cells.

When using more than one **RAM** in a system, there must be some provision to select a certain **RAM**. You will find that many **RAM**'s have chip select pins. For example, six might be available. Whatever the number, the chip select pins are tied to the address bus from the **MPU** in such a way that only one **RAM** chip is addressed at a time.

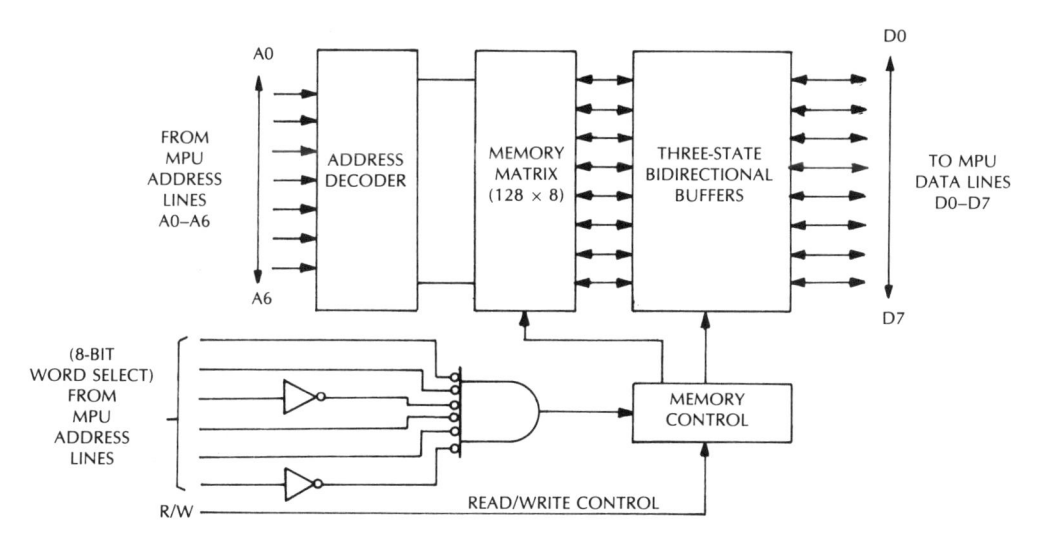

Figure 4-5: Function block diagram of a RAM that can store 128 bytes, i.e., 128 8-bit words.

To illustrate what would happen if you tied three 128-byte **RAM**'s to the same **MPU**:

128 8-bit words × 3 = 384 8-bit words, or 3,072 individual memory locations.

But, in practical applications, your actual storage is only 384 locations because read-out is in 8-bit words.

Understanding ROM's

In Chapter Two, we explained that the basic **ROM** may be programmed by the manufacturer and, in some instances, by the user. In fact, for most systems, numerous **ROM** programmers are available. You also learned that a field-programmable read-only memory (**FPROM**) is a memory that you can program for yourself. An example of such a device is Signetic's 8223, 256-bit bipolar **PROM** (organized 32 × 8-bit word). Still better, the erasable programmable read-only memory (**EPROM**) is also available to the experimenter. National Semiconductor Corporation has a 2048-bit (MM5203) that is a 16-pin **IC** with a quartz window that is transparent to ultraviolet light (see Figure 2-24).

If your needs call for a higher performance system, faster than **TTL** memories, Motorola Emitter-Coupled Logic (MECL) circuits could very well be the answer. They offer a 32 × 8 **FPROM** (15ns), the MCM10549L ceramic DIP, and a 256 × 4 **FPROM** (24ns), the MCM10549L. See Figure 2-8 in Chapter Two for an illustration of a MECL basic gate.

In the **MCU** category, we find 1-, 4-, 8-, 12-, and 16-bit processors. Of course this means that a **ROM** for each **MPU** must be constructed using the same architecture, i.e., an 8-bit **MPU** requires an 8-bit **ROM**, etc. In a microprogrammed **MPU**, an instruction (from main memory **RAM**) causes the program memory (**ROM**) to generate a bit pattern which, in turn, initiates the binary signal (control) that activates the circuit for execution of the instruction.

The type of memory utilized to store these bit patterns may be any of the **ROM**'s we have discussed up to this point. You will find that this group of bit patterns (instructions) that inform the **MPU** what operation to perform (often called the "instruction set") are usually produced by **MPU** manufacturers such as Fairchild, Intel, Intersil, MOS, Motorola, National Semiconductor, RCA, Rockwell, Signetics, Texas Instruments, and Zilog.

As an example of a *mask-programmable* (the chip selects, etc., are manufactured in the device) **ROM,** we will use the MCM6830 that is compatible with **TTL**, contains 1024 bytes, and is a 24-pin **IC**. This **ROM** is much like the **RAM** circuit shown in Figure 4-4 (see Figure 4-6). But since data cannot be stored in a **ROM**, a **ROM** must be addressed solely to obtain data. Notice, Figure 4-5 shows no read/write control line.

This custom-programmed **ROM** is a 3-state chip. Three-state control is a technique that permits an **MPU** to see a certain device (in this case, a **ROM**) as having three states—1, 0, and an open circuit (high impedance). When the **ROM** is not addressed by the **MPU**, it goes into the third state; thus it is possible to shut the **ROM** off from the system by use of the control inputs to the device. Notice that the user can define whether the chip-select inputs are active high or active low. Also note that the user must define the binary word to be stored at each address. This is true, of course, regardless of whether you use a **FPROM**, an **EPROM**, or a **PROM**.

There are two obvious reasons why an experimenter would usually want to use an **EPROM** instead of a mask-

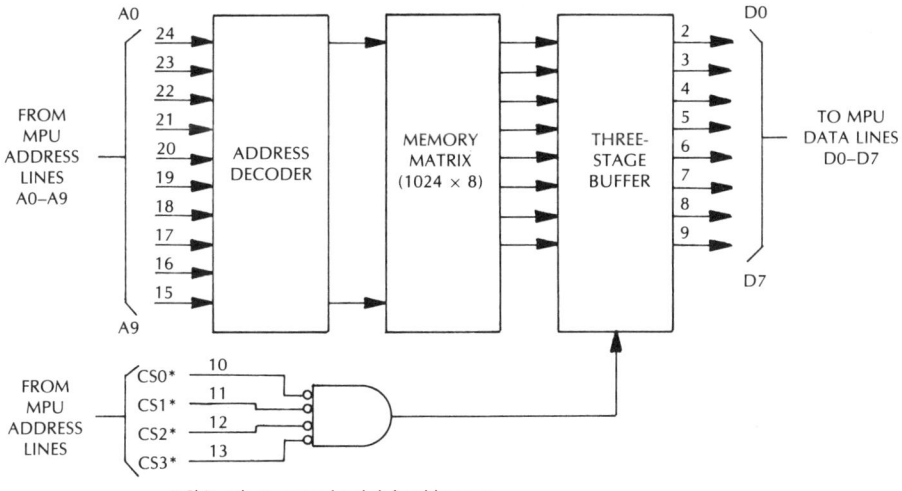

Figure 4-6: Functional block diagram of Motorola's MCM6830 mask-programmable (certain signals defined for the manufacturer, by the user) ROM.

programmable **ROM**. First is the cost of having a manufacturer do the programming, and second, and perhaps more important, the experimenter can work up his program and then try it. If it isn't correct, the **ROM** is erasable.

Whatever type **ROM** you use, it must match the **MPU** interface lines. As you will remember, if the **ROM** address lines are A0 . . . A9, then they must be connected to the **MPU**'s A0 . . . A9 lines. Also, assuming that the **ROM** has data bus lines, they must be connected to the **MPU**'s D0 . . . D7 lines. Furthermore, the chip-enable inputs have to be defined—active high or active low—for the different type **MPU**'s. As a general rule, you will find more address lines are used for addressing internal locations in **ROM** in a system than are used for internal address of **RAM**; for instance, seven lines for **RAM** and, perhaps, ten lines for **ROM**.

Input/Output Signals

We have shown that a microprocessor and memories can be manufactured to form a simple and almost complete microcomputer (see Figure 4-1). Incidentally, in many instances, you will find the **ROM**'s and **RAM**'s are separate **IC**'s connected to an **MPU** and, in other cases where more memory is desired, connected within an **MCU** system. Notice, we said "almost complete." The missing element is your ability to communicate with the microcomputer hardware. The **MCU** usually has an internal I/O integrated circuit to make the system interface with the outside world, as explained in Chapter Three. Without some I/O hardware, it would be impossible for you to enter data into or out of an **MCU** system.

Input/Output Ports

By referring to Figure 4-1, you will find that the MC6805P2 **MCU** has 20 input/output lines (A0...A7, B0...B7, C0...C3). Figure 4-7 is an illustration showing typical port I/O circuitry found in **MCU**'s such as this one.

All pins on the MC6805P1 (ports A, B, and C, see Figures 4-1 and 4-2) are programmable as either inputs or outputs under software control of the corresponding Data Direction Register (see Figure 4-7). In this case, you would program the corresponding bit in the Port Data Direction Register (DDR) to a logic 1 for output, or a logic 0 for input. On reset, all DDR's are initialized (to set addresses to their starting value or some prescribed points in the routine) to a logic 0 state to put the ports in the input modes. On the other hand, the port output registers are not initialized on reset but may have signals applied before setting the DDR bits, to avoid bits that do not fall into the active high or active low logic levels. *Example:* Logic high could be a minimum 3.5 volts, logic low, a maximum 0.4 volts, and other levels may be referred to as "undefined." To put it another way, the circuit will not respond at these levels.

When programmed as outputs, the latched output data is readable as input data, regardless of the logic levels at the output pin, due to output loading (see Figure 4-7). In case you are not familiar with *latch*, it is a term used to indicate data storage such as in a flip-flop. In fact, very often a D type flip-flop is called a *latch* or *latching circuit*. Or we could say that a certain buffer has a latch function that permits each bit in the buffer to be latched to 1 or 0 by the data pulses. In this case, the latched byte (stored 8-bits) is held in the buffer until the register is ready to accept new data.

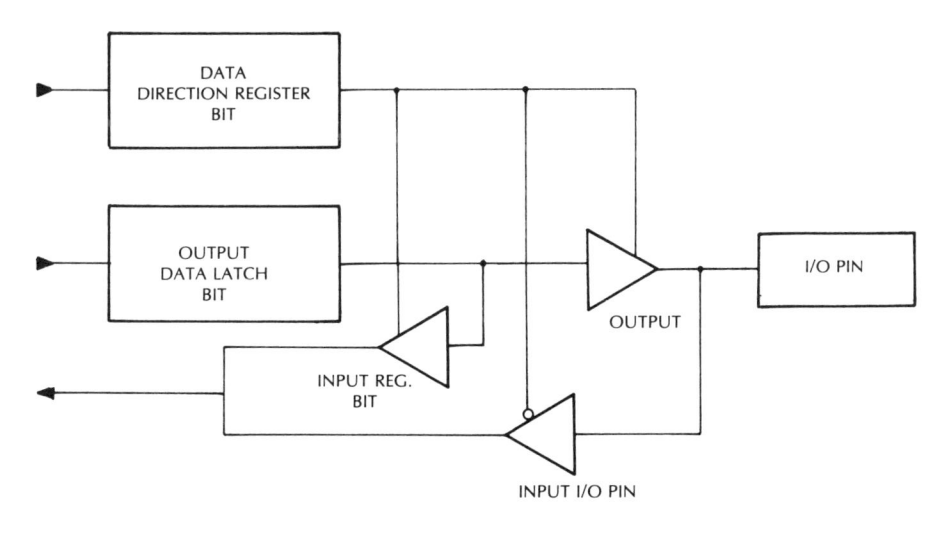

DATA DIRECTION REGISTER BIT	OUTPUT DATA BIT	OUTPUT STATE	INPUT TO MCU
1	0	0	0
1	1	1	1
0	X	3-STATE	PIN

Figure 4-7: Typical port I/O circuitry used in the MC6805 family of microcomputers.

There are always cautions when using I/O ports on any MCU. For instance, the corresponding DDR's for ports A, B, and C (see Figure 4-1) are write-only (binary bit inputs) registers (registers $004, $005, and $006 in the MCU M6805P2). You cannot read these registers; that is, a read operation on these registers is undefined. Also, the latched output data bit (see Figure 4-7) may always be written. What this all boils down to is that any write to a port writes all of its data bits, even though the port DDR is set to input. You can use this to initialize the data registers and avoid undefined outputs. Nevertheless, you must take care when using the read/modify/write instructions on the MC6805 family of MCU's, because the data read corresponds to the pin level of the DDR (is an input at 0), and corresponds to the latched output when the DDR is an output (1).

When working with microprocessors, etc., it is important that you realize that data bytes can be transmitted in both parallel and serial form from the peripheral to the **MPU**. The parallel method is the fastest and requires the simplest I/O circuit in the **MPU**. But when using some peripherals, a *serial data bit stream* is the conventional way to transmit data. Figure 4-8 shows the relationship between a peripheral typewriter (using a digital in/out signal) and an **IC** called *Asynchronous Communications Interface Adapter* (ACIA). The ACIA permits data to be transmitted in serial format with only one line instead of the eight required for parallel transmission. An **IC** such as the MC6805 (an ACIA chip) can be used to convert from serial to parallel or vice versa. *Note:* **MCU**'s such as the MC68701 have on-chip serial communications interface and parallel I/O.

A considerable amount of today's computer equipment is being connected to telephone lines. In this case, you will sometimes find that another **IC** is required. This **IC**—for example, the MC6860, a modulator/demodulator (Modem)—is placed in series between the ACIA and telephone input/output line. A block diagram of such a hookup is shown in Figure 4-9.

Motorola's extensive line of M6800 **MPU** peripherals interfacing **IC**'s are directly compatible with their newer MC68000 family (16-bit **MPU**'s). A few of these devices that may be useful to you are listed in Table 4-2. Again, **MCU**'s such as

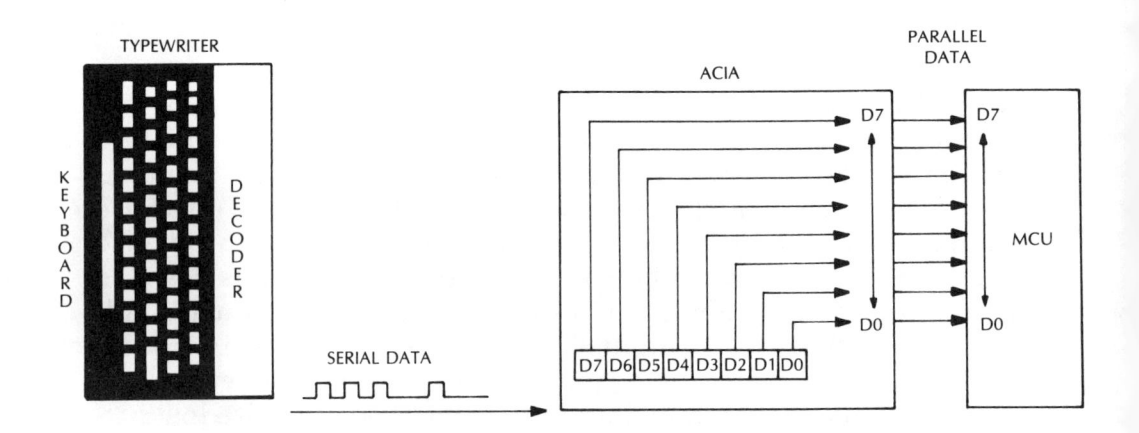

Figure 4-8: Input relationship between a digital serial output peripheral and a parallel input MCU, using an ACIA chip such as the MC6805, etc.

STATE-OF-THE-ART ON-CHIP MICROCOMPUTERS 119

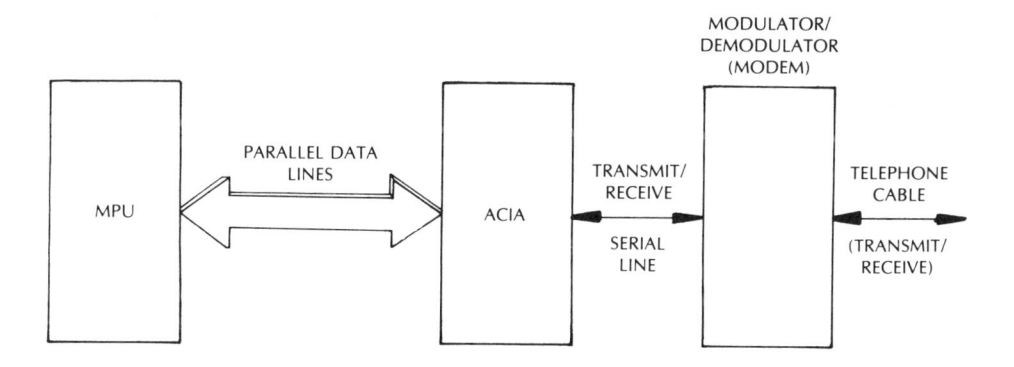

Figure 4-9: Modulator/demodulator, ACIA, and MPU. The input/output signal to the Modem from the telephone cable is a sinusoidal signal. All other lines, both serial and parallel, carry data in digital form. The purpose of the Modem is to make the conversion between sinusoidal and digital signals.

PRODUCT NUMBER	NAME
MC6821	PERIPHERAL INTERFACE ADAPTER
MC6840	PROGRAMMABLE TIMER MODULE
MC6843	FLOPPY DISK CONTROLLER
MC6845	CRT CONTROLLER
MC6850	ASYNCHRONOUS COMMUNICATION INTERFACE ADAPTER
MC6852	SYNCHRONOUS SERAL DATA ADAPTER SERIAL
MC6854	ADVANCED DATA LINK CONTROLLER
MC68488	GENERAL PURPOSE INTERFACE ADAPTER

Table 4-2: Adapter and controller IC's for peripheral interface to M6800 (8-bit) and MC68000 (16-bit) MPU's.

the MC68701 have both serial and parallel interface and, incidentally, this **MCU** also features on-board **EPROM** rather than maskable **ROM**.

Registers

Adapter and control **IC**'s such as the ACIA MC6850 (see Table 4-2) contain various registers, just as do the **MCU** units. The MC6805 has four 8-bit registers that may be addressed:

1. Status Register (SR).

2. Receive Data Register (RDR). These are read-only registers (the **MPU** cannot write into them).
3. Transmit Data Register (TDR).
4. Control Register (CR). These last two registers (items 3 and 4) are addressable but not readable, i.e., the **MPU** cannot read data out of them, although it can instruct them.

In addition to these registers, there are control lines. Some of these are 3-chip select lines (CS0, CS1, $\overline{CS2}$), one Register Select line (RS), one Interrupt Request line ($\overline{IRQ}$), one Enable line (E), one Read/Write line (R/W), and seven Data Control lines. Many of these lines (pins 2, 5, 6, 23, and 24) are used to send and receive data when connected to peripheral equipment, or when the system includes a modem such as the MC6860 listed in Table 4-2. Figure 4-10 shows the pin configuration for the MC6850 ACIA.

The Peripheral Interface Adapter (PIA), product No. MC6821, also has several registers: two peripheral data registers (A and B), two data direction registers (A and B), and two control registers (A and B). It also has two separate 8-bit bidirectional peripheral data buses that can be used for interfacing with external equipment. See Figure 4-11.

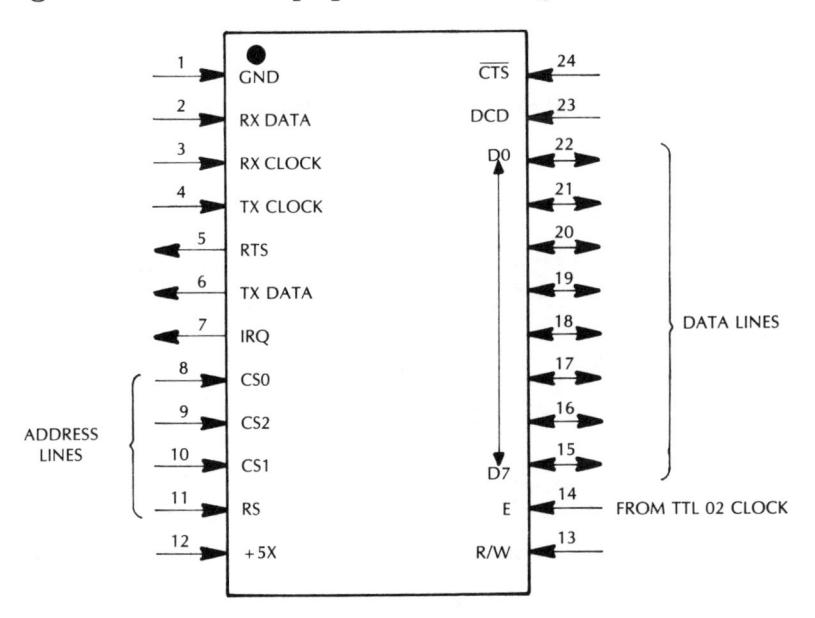

Figure 4-10: Pin configuration for the MC6850, an ACIA unit.

STATE-OF-THE-ART ON-CHIP MICROCOMPUTERS 121

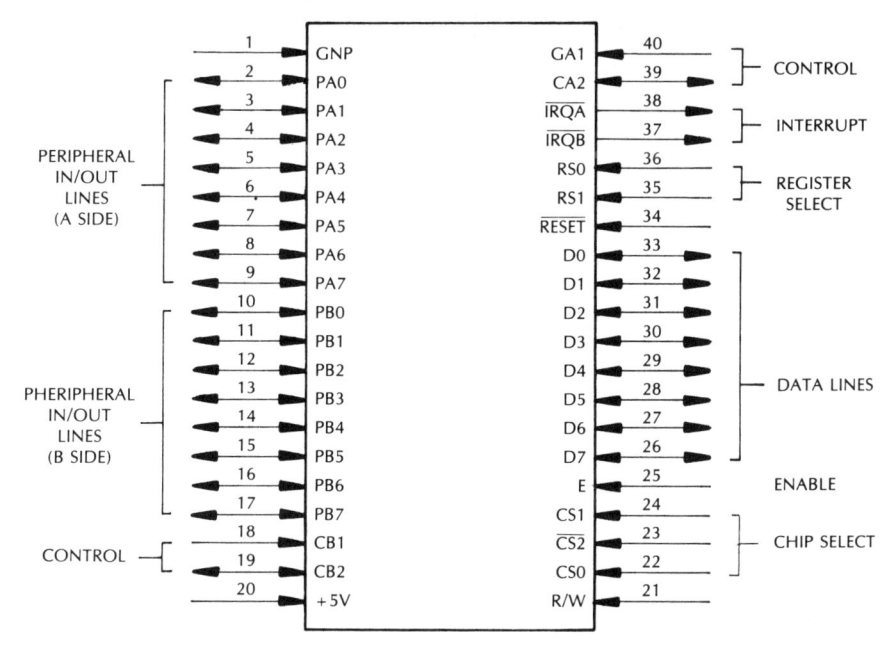

Figure 4-11: Pin configuration for the MC6821 listed in Table 4-2.

You will notice that the software features for these **IC**'s (the MC6805 and MC6821) are similar to the port I/O's we discussed in the preceding section titled "Input/Output Signals," where we used the MC6805 **MCU** family as our example. Basically, the only difference is that the I/O circuitry is located on the **MCU** chip, but usually not on **MPU** chips. In other words, in most cases you would not need an ACIA with an **MCU** but would when connecting to a **MPU**.

The **MCU** has the necessary registers manufactured into the chip. There are 20 input/output pins on the microcomputer chip MC6805P2. All I/O pins (ports A, B, and C, see Figure 4-1) are connected to ports A, B, and C register. See Figure 4-7, which shows a typical port I/O circuit.

Timer Circuitry and How It Works

The electrical pulses used by the **MPU**, **ROM** and **RAM** (whether they be internal in an **MCU**, or external when using a separate **MPU**) are generated by a *clock* or *timer*. It should be pointed out that these components, **ROM**'s, **RAM**'s, **MPU**'s, etc., do not actually generate the binary pulses we have been discussing in previous pages, but produce their bytes on the address and data lines, when they receive clock pulses. Also, as we know, all digital systems must have a clock signal. Microcomputers such as the MC6805 family are no different. However, the hardware features of the **MCU**'s may be different. For example, the on-chip clock generator found on the MC6805 units uses an 8-bit timer, where others such as the MC68701 **MCU** feature a 16-bit 3-function *programable* timer.

Inside an **MCU** such as the MC68701 is a timer (see Figure 4-12) that can be used to perform input waveform

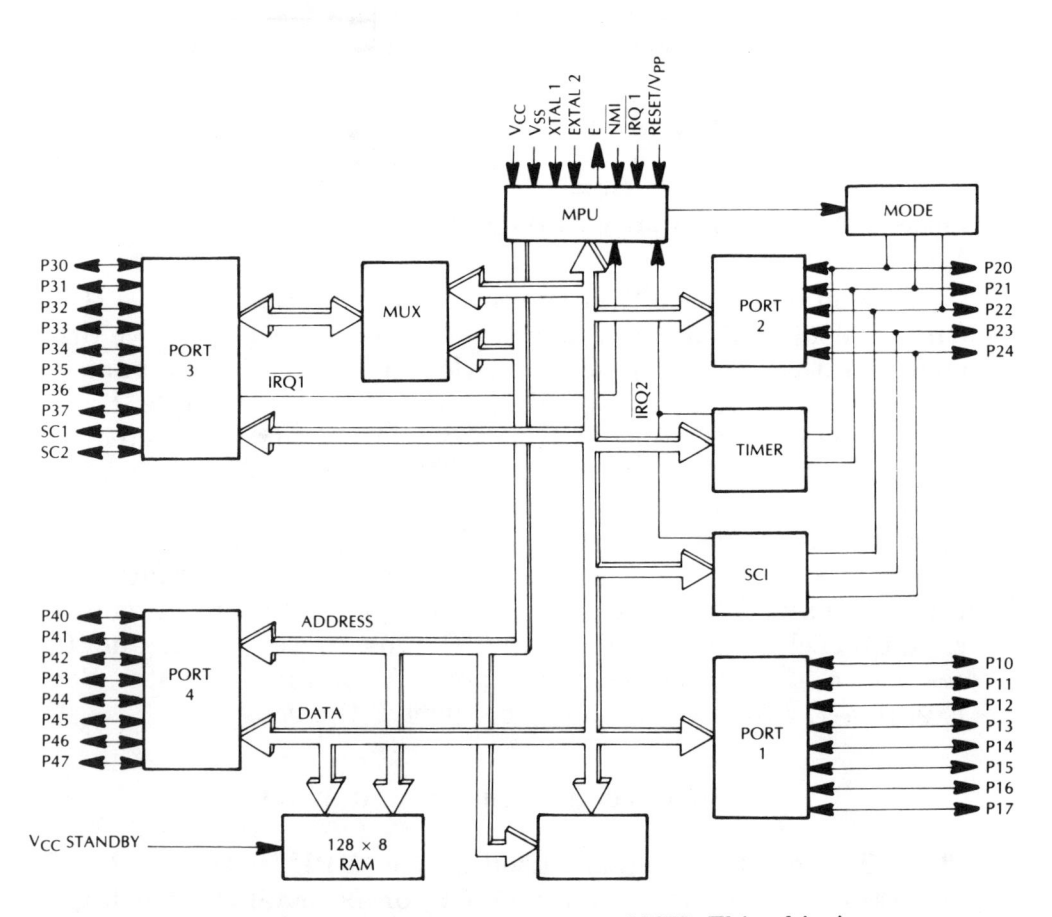

Figure 4-12: Block diagram for the MC68701 MCU. This chip is contained in a 40-pin package that uses ultraviolet erasure. Figure 2-24 shows a similar type package with an ultraviolet window.

measurements while independently generating an output waveform. Pulse widths, in this case, can vary from several microseconds to many seconds.

What makes this **MCU** timer special is that it can be programmed, as we mentioned. The key timer element in this system is a 16-bit free-running counter. See Figure 4-13 for a block diagram of the timer. You will also notice that there are two 16-bit registers, the Output Compare Register and the Input Capture Register. The other register, the Timer Control and Status Register (TSCR), is an 8-bit register.

To illustrate the role of each block in the system, let us first see what they are and what they do. The mnemonics included with each heading applies only to the MC68701 family of **MPU** units. When a dollar sign is shown, it indicates the number is in hex.

Counter ($09:0A)

The heart of the timer is the free-running 16-bit counter. This counter is incremented (caused to change state) by E (Enable, see Figure 4-12). That happens to be pin 40 on this chip. Timer Overflow Flag (TOF, see small center blocks in Timer Control and Status Register, Figure 4-13), is set whenever the counter contains all 1's. Let's digress for a moment and say a few words about "flag." Generally, a counter will send a *flag*, or signal, to other circuits, indicating a full count. This flag can be used for any number of functions. For instance, the flag can be used to halt or reset operation of the **MPU**, or it might be used to pulse another counter, etc.

Output Compare Register ($0B:0C)

This register is a read/write circuit used to control an output signal waveform, or provide a time-out flag, based on one's preference. It is compared with the free-running counter (shown in Figure 4-12) on each cycle that is an input to the **MPU** in the same illustration. When a match occurs, Output Compare Flag (see OCF block in Figure 4-13) is set and the Output Level (OLVL) is clocked to an output control register that is shown in the block diagram. If port 2 (see Figures 4-12 and 13), bit 1, is configured as an output, OLVL will appear at

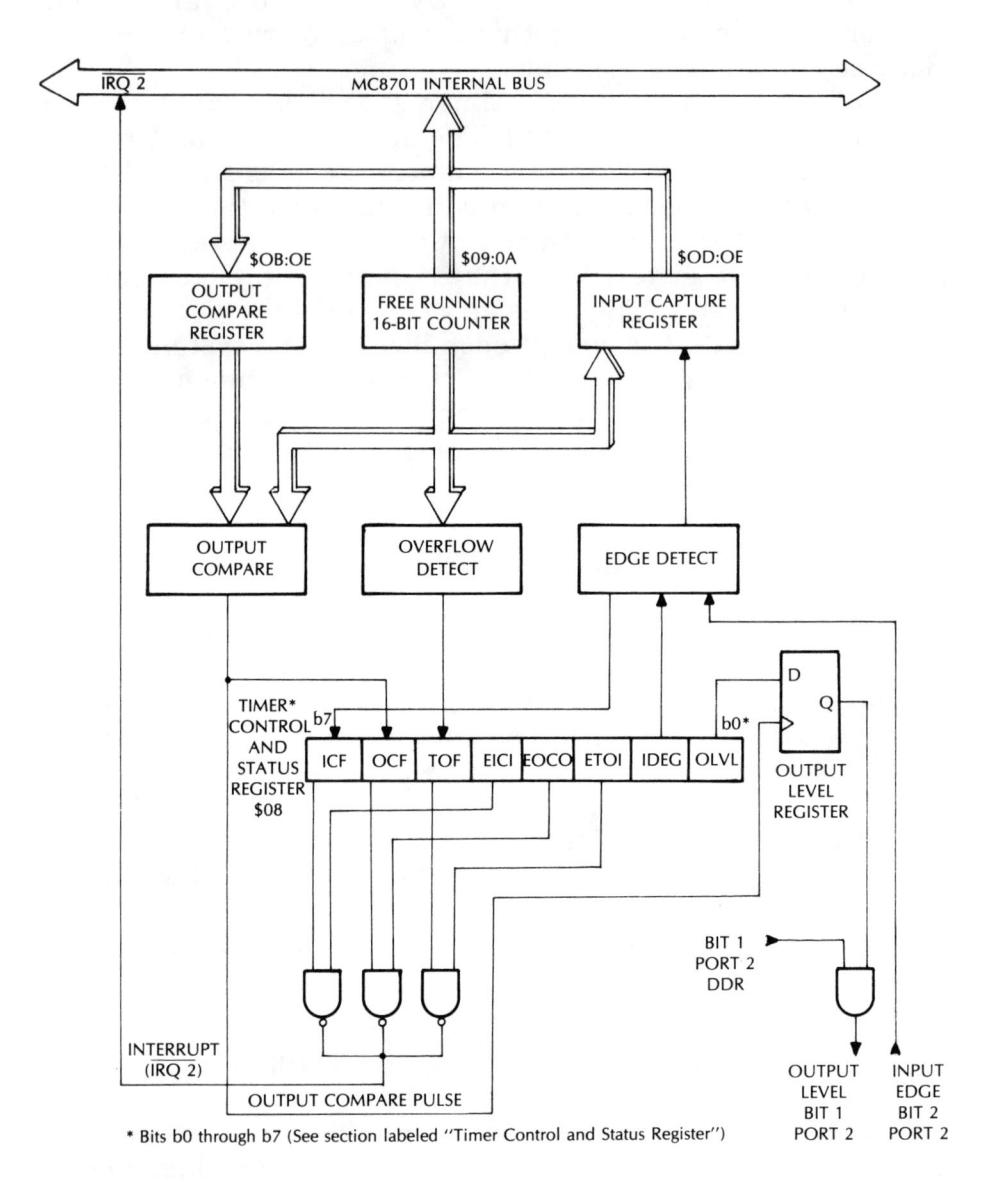

Figure 4-13: Block diagram of an on-chip programmable timer (MC68701 family).

P21 (see Figure 4-12) and Output Compare Register, and OLVL can then be changed for the next compare.

Input Capture Register ($0D:0E)

This register is a read-only circuit used to store the free-running counter when a correct input signal change (transition) occurs. The IEDG presents the proper input for this operation (see block diagram, Figure 4-13). Port 2, bit 0 should be configured as an input; however, the Edge Detect circuit (shown just below the Input Compare Register in the block diagram) always senses P20 (see Figure 4-12), even when it is used as an output. But an input capture can occur independently of ICF (shown in the Timer Control and Status Register): the register always contains the most current value. Also, the input pulse width must be at least two E cycles, if it is desired to capture under all conditions.

Timer Control and Status Register ($08)

We have repeatedly referred to this register, therefore let's examine its operation in more detail. First, refer to Figure 4-14, which is an extract from the center of the block diagram shown in Figure 4-13, and which shows a digital signal (bits b0 through b7).

This register, the Timer Control and Status Register (TCSR), is an 8-bit device with all bits readable, while bits 0 ... 4 can be written. The three most important bits provide the timer status and indicate the following factors:

1. If a correct bit level transition has been detected.
2. Whether a match has occurred between the free-running counter and the output compare register.

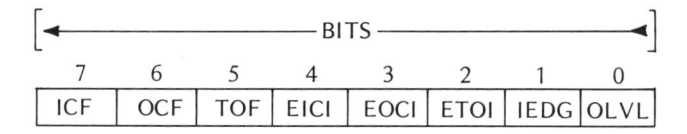

7	6	5	4	3	2	1	0
ICF	OCF	TOF	EICI	EOCI	ETOI	IEDG	OLVL

Figure 4-14: Digital bits b0 through b7 in the Timer Control and Status Register. See Table 4-3 for function description.

3. Whether the free-running counter is in an overflow condition.

Any one of these conditions can generate an IRQ2 (interrupt) signal and is controlled by an individual enable bit in the TCSR (see block diagram in Figure 4-13). Table 4-3 lists each function, starting with bit 0 (OLVL) and ending with bit 7 (ICF), shown in Figures 4-13 and 14.

Addressing Modes

In Chapter Three, under the heading "How and Why Addressing Modes Are Used in a Microcomputer System," we discussed the fundamental address modes for the M6800 family. As you might expect, the more advanced model, the MC68701 MCU, is very similar. It is an 8-bit single-chip MCU unit that has 6 addressing modes that can be used to reference memory:

1. *Inherent* (the operand): Registers, and no memory reference is required. These are 8-bit instructions.

TCSR 8-BITS	WHAT EACH BIT DOES
BIT 0 OLVL	Output level. OLVL is clocked to the output level register by a successful output compare and will appear at P21 if Bit 1 of the Port 2 Data Direction Register is set. It is cleared during reset.
BIT 1 IEDG	Input Edge. IEDG is cleared during reset and controls which level transition will trigger a counter transfer to the Input Capture Register.
BIT 2 ETOI	Enable Timer Overflow Interrupt. When set, an $\overline{IRQ2}$ interrupt is enabled for a timer overflow; when clear, the interrupt is inhibited. It is cleared during reset.
BIT 3 EOCI	Enable Output Compare Interrupt. When set, an $\overline{IRQ2}$ interrupt is enabled for an output compare; when clear, the interrupt is inhibited. It is cleared during reset.
BIT 4 EICI	Enable Output Capture Interrupt. When set, an $\overline{IRQ2}$ interrupt is enabled for an input capture; when clear, the interrupt is inhibited. It is cleared during reset.
BIT 5 TOF	Timer Overflow Flag. TOF is set when the counter contains all 1's. It is cleared by reading the TCSR (with TOF set), then reading the counter high byte ($09), or by RESET.
BIT 6 OCF	Output Compare Flag. OCF is set when the Output Compare Register matches the free-running counter. It is cleared by reading the TCSR (with OCF set) and then writing the to the Output Compare Register($0B or #0C), or by RESET.
BIT 7 ICF	Input Capture Flag. ICF is set to indicate a proper level transition; it is cleared by reading the TCSR (with ICF set) and then the Input Capture Register High Byte (%0D), or by RESET.

Table 4-3: Timer control and status register event table for the programmable timer manufactured in the MC68701 microcomputer chip.

2. *Immediate:* As explained previously, the operand or immediate byte(s) of the instruction where the number of bytes matches the size of the register. Also, these are 2- or 3-byte instructions.
3. *Direct:* This is a 2-byte instruction, as you may remember. The least significant byte of the operand address is contained in the second byte of the instruction and the most significant byte is assumed to be $00.
4. *Extended:* As explained in Chapter Three, in this addressing mode the next instruction to be executed by the **MPU** is located at some other address than the one following.
5. *Indexed:* The indexed mode of addressing uses a 2-byte instruction. The number contained in the second byte (next location after the instruction) is often referred to as an "offset." The unsigned offset contained in the second byte of the instruction is added with Carry to the index register and used to reference memory without changing the index register. This new address, as in the extended mode, contains the data.
6. *Relative:* As explained in Chapter Three, in this addressing mode the next instruction to be executed by the MPU is located at some address other than the one following.

In other words, this mode is used only for branch instructions. These are 2-byte instructions and, if the branch condition is true, the program counter is overwritten with the sum of a signal single byte displacement in the second byte of the instruction and current program counter. What this does is to add many addresses—branch range; for example, relative addressing mode for the MC68701 **MCU** provides a branch range of -126 to 129 bytes from the first byte of the instruction.

Software

Although a single instruction can have several modes, it is important to realize that each addressing mode (Inherent, Immediate, Direct, etc.) is just a different way of telling the **MCU** where to reference memory. For example, the Motorola data sheet for their MC6800 lists 72 instructions—variable length; but, as we now know, each instruction can have more than one addressing mode. In the case of the MC6800 **MPU**, it adds up to 197 valid machine codes. Newer units such as the MC68701 **MCU** have more—82 instructions—and, in this case, there are 220 valid machine codes. Table 4-4 lists seven of the possible accumulator and memory instructions for the MC68701 **MCU**.

Don't forget, the mnemonic (MNE) code is assigned by the manufacturer. For instance, ABA (Add accumulator B to A) is represented by the hex code 1B (see column titled Inherent Mode, under Operation Code OP). This would actually appear in **ROM** memory as 00011011 (see Table 3-2; the extra three 0's are needed in an 8-bit system). Also, the symbol ~ shown in Table 4-4 stands for the number of cycles and the other symbol, #, is read as the number of program bytes.

In the previous pages it was said that when a dollar sign is shown, it indicates the given number is in hex. However, there are other symbols that are used to tell you (not the **MCU**) certain functions, etc. For instance, when a # sign is shown, the instruction is in the immediate mode, and the number following this sign is located in the next byte of memory. As another example, the index addressing mode may be shown with a hex number, say 19 (00011001 binary, 25 decimal), followed by a comma and X; i.e., it could be $19,X. In this case, the number

ACCUMULATOR AND MEMORY OPERATIONS	MNE	IMMED OP	~	#	DIRECT OP	~	#	INDEX OP	~	#	EXTEND OP	~	#	INHER OP	~	#
ADD ACMLTRS	ABA													1B	2	1
ADD B TO X	ABX													3A	3	1
ADD WITH CARRY	ADCA	89	2	2	99	3	2	A9	4	2	B9	4	3			
	ADCB	C9	2	2	D9	3	2	E9	4	2	F9	43	3			
ADD	ADDA	8B	2	2	9B	3	2	AB	4	2	BB	4	3			
	ADDB	CB	2	2	DB	3	2	EB	4	2	FB	4	3			
ADD DOUBLE	ADDD	C3	4	3	D3	5	2	E3	6	2	F3	6	3			
AND	ANDA	84	2	2	94	3	2	A4	4	2	B4	4	3			
	ANDB	C4	2	2	D4	3	2	E4	4	2	F4	4	3			
SHIFT LEFT ARITHMETIC	ASL							68	6	2	78	6	3			
	ASLA													48	2	1
	ASLB													58	2	1

Table 4-4: Seven example accumulator and memory instructions for the MC68701 MCU. This chip has a total of 82 separate instructions and, because of the separate addressing modes, has 220 valid machine codes.

STATE-OF-THE-ART ON-CHIP MICROCOMPUTERS   **129**

following the $ is added to the contents of the index register to form a new effectual address. Of course, as has been pointed out, inside the chip the octal number added to the index register is 00011001.

Introduction to Programming Basics

Although you should now know that a program for a computer or processor consists of a sequence of operational instructions stored in memory, it is important to you, as a user, to know the **MPU**, **MCU** language and the manner in which the device operates. In short, you must know the *Instruction Set*. The set of all instructions common to a given **MPU**, **MCU**, or **CPU** is referred to as its Instruction Set. Each instruction in the Instruction Set (the writing of the Instruction Set is the task of the programmer) enables a single elementary operation such as the movement of a data byte, an arithmetic or logical operation on a data byte, or a change in instruction execution sequence.

A point that bears repeating, and one you may have missed, is that the size of the Instruction Set is a measure of the capabilities of the **MPU**'s, the **MCU**'s, or the **CPU**'s. Another such measure you may have noticed is the length of binary words the device can work with (usually 4-bit, 8-bit, or the new 16- and 32-bit). Generally speaking, the larger the Instruction Set or word size, the more powerful the **MPU**, **MCU**, etc. The MC68000 family of **MPU**'s, 16-bit **MPU** with over 1000 useful instructions (this includes the Instruction Set and all variations of instructions) is thus more powerful than the MC141000, a 4-bit **MCU** with 43 standard instructions.

After the programmer has written the program, it is stored in memory (**EPROM, ROM,** or **RAM**) as a sequence of bytes that represent the instruction. Then to review a bit, the program execution proceeds sequentially (i.e., memory location 8001 is executed after location 8000, etc.), unless a transfer-of-control, or branch instruction (branch-forward, branch-back, or jump, for example) is executed, which causes the **PCU** to set to a specified memory address. *Note:* The instruction *jump* causes the **MPU** to branch to another program (usually called a *subroutine*—a program within a program). Sets of rules or processes for solving a problem in a certain number of steps—

example, arithmetic procedure—are often programmed as subroutines.

The subroutine type of jump requires the **MPU** to store the contents of the program counter at the time the jump occurs. This enables the processor to resume execution of the main program after the last instruction of the subroutine has been executed (called *Return from Subroutine*, or *RTS*).

The *Stack Pointer*, a 2-byte register that was mentioned in Chapter Three, contains a beginning address, normally in **RAM,** where the status of the **MPU**'s register may be stored during a branch to a subroutine. A subroutine may call up another subroutine. This is called *nesting subroutines*. If the **MCU**, etc., being used has a stack for storing return addresses, the maximum depth of nesting subroutines is determined solely by the depth of the stack itself.

The MC8705R3, an 8-bit **EPROM MCU** with analog-to-digital conversion, is capable of subroutines and interrupts that may be nested down to an on-chip location $061 (31 bytes maximum), which allows the programmer to use up to 15 levels of subroutine calls (fewer, if interrupts are allowed). *Note:* Interrupts, as you would imagine, refer to a method in which the **MCU** is interrupted from doing its primary duties so that it may perform a certain task more important; for example, some type of emergency.

In summary, if you want to write a program in machine language, it is quite a task. For example, the instruction-load accumulator A in the immediate mode of addressing, with the decimal number 26:

Machine Language		*Mnemonics*
(Binary)	(Hex)	
10000110	86	LDAA (IMM)
00011010	1A	Data to be put in A

Of course, there is an easier way. Write the program in source language and let an assembler convert the source language for you. For instance, source language might read:

                LD   A $ 1A

The purpose of the dollar sign is to let the assembler know what base system your number is in. In this example, $ 1A indicates hex 1A (decimal 26, binary 00011010). If you had used

STATE-OF-THE-ART ON-CHIP MICROCOMPUTERS

SIGN	INSTRUCTION
#	Indicates the immediate addressing mode.
$	Indicates that the number following is in hex.
@	Indicates that the number following is in octal.
%	Indicates that the number following is in binary.
'	(apostrophe) Indicates an ASCII literal character.

Table 4-5: Example signs for address mode and number system used while programming.

%00011010, it would have indicated that your number is already in binary. The number you want to include in the program can be listed in decimal, hex, octal, or binary, *provided* the assembler is informed which numbering system and other required symbols (such as #, shown in Table 4-5) you are using.

At this point, you might ask, "What is an assembler?" The answer to this question is that an assembler is a special program designed to convert a source program (which you make up using a mnemonic code supplied by the manufacturer) into the machine language program. You will find that almost any company has an assembler (in all cases, a separate computer system) that can do this job.

Last, but very important, there are many rules that you *must* follow when writing a program to be used by an assembler. Therefore, you should contact the manufacturer of the particular **MPU** or **MCU** you want to program (before using mnemonics). Table 4-5 shows a few rules for M6800 software.

In this and the preceding chapter we have discussed only machine language and mnemonics, which are the two most basic computer languages. Remember, machine language is the only language the **MPU** can understand. However, you, as a complete computer system user, can use many different computer languages to program a computer system. For example, you can use BASIC (the computer based on the 8080, Z80, 6502, and 6800, can be programmed using BASIC, as can some of the 16-bit chips). There are about a dozen or so computer languages being used today, but it must be realized that many personal computer systems will handle only a very few of them. *It's very important to find out whether your computer system will handle a certain computer language before attempting to program it.*

CHAPTER FIVE

Complete Guidelines for the Circuit Builder, Using Modern Solid State Devices

This chapter is written for the circuit builder who is searching for fast, practical, and economical ways to breadboard, construct PC boards, and troubleshoot electronic projects. Every section stresses inexpensive, simplified techniques that you can use in each project you undertake. For instance, some electronic components (especially semiconductors) are extremely sensitive to too much heat. Even passive elements (resistors and capacitors, for example), which are normally thought of as insensitive to temperature variations, can undergo parameter changes that can cause malfunctions. This chapter will tell you exactly how to eliminate problems such as these before they happen.

All procedures and simplified techniques deal with troubleshooting solid state circuits, and each step is explained in detail. No workbench would be complete without the troubleshooting information pertaining to both analog and digital circuits found in the following pages. You will find answers to such questions as "How do I troubleshoot linear and analog IC's?" and "What is the best approach to troubleshooting digital IC's?" Every section in this chapter provides you with practical techniques that are essential to your success at the workbench.

Practical Techniques for Working With Resistors, Capacitors, and Inductors

Quick and Easy Component Replacement:

The best solution to our electronics circuit problems is often very basic and very simple. This section contains some time-saving ideas that I have found really work. For example, a trick fast-service TV technicians often use when removing and replacing resistors, capacitors, diodes, and the like, is shown in Figure 5-1.

Although sometimes I have found it hard to do, the procedure for removing and replacing (soldering into the circuit) an electronic component is as follows:

1. Use side-cutting pliers (diagonals) or long-nose pliers, and crush the component. You may have to cut the component out of the circuit but, either way, be sure to leave the original leads as long as possible and connected to the PC board; i.e., don't break the original solder job. *Note:* The author does not recommend crushing glass components such as diodes.
2. Clean all parts of the component off the PC board, then straighten, clean, and tin the old lead ends.
3. Make a mechanical connection between the new component leads and then solder, as shown in Step 3, Figure 5-1.

Passive components, such as resistors, are normally insensitive to temperature variations. But in today's close-tolerance circuits, when a soldering iron is applied to the leads these components can undergo parameter changes which are sometimes sufficient to influence circuit behavior. Therefore, after you have soldered a component to a PC board, it's best to check to see if it has changed value. For example, check diode front-to-back resistance. You should find a low resistance in the reverse direction. Capacitors should be checked for a short.

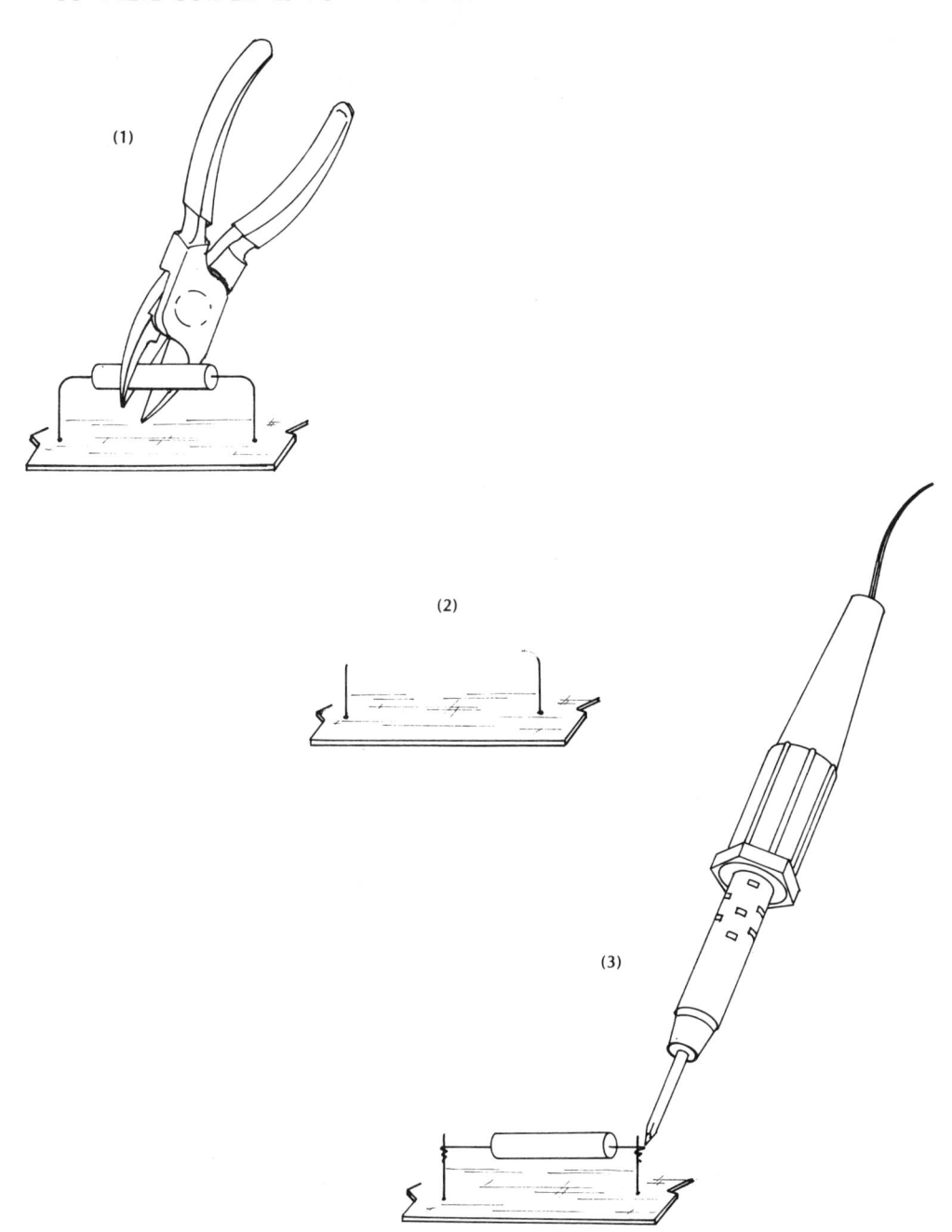

COMPLETE GUIDELINES FOR THE CIRCUIT BUILDER 135

Figure 5-1: Removing and replacing a small solid state component or resistor, etc., mounted on a PC board, breadboard, or chassis.

A capacitor should read a high resistance both ways; a low resistance reading is an indication of a shorted capacitor.

Resistance Measurements:

Here are a few pointers you should keep in mind when using the resistance function on your multimeter:

1. *First, remove power from the circuit under test.* Do not make any resistance measurements with power applied to the circuit you are working on.
2. Next, to avoid erroneous readings, short the test leads together and zero adjust *every time you change resistance scales.*
3. Do not let your fingers come in contact with the metal ends of the test probes, or touch the wire leads while you are making a measurement. Your body effect (resistance, capacitance, etc.) can cause errors in your readings. This is especially true on high resistance meter settings.
4. In all measurements, the most accurate readings are made when the meter pointer points to a value near the center of the meter scale.
5. When measuring a small resistance, keep your range switch in the lowest range possible (usually R × 1). When measuring a large value resistance, increase the range switch only as far upward as necessary to get a valid reading.
6. When using most low-cost multimeters, you will find that the only way you can make an accurate resistance measurement is to disconnect one of the resistor leads from the PC board or circuit.

All of us who use an electric soldering iron to solder resistors, capacitors, coils and the like to PC boards, have run into the problem of waiting for it to heat when there are long periods between soldering jobs. The problem is that leaving a soldering iron on for long periods will, due to overheating, eventually render the tip "inoperative." The solution is very simple. All you need is a diode and switch rated for a fairly high current and voltage (around 3 amps or better, at 200 volts or more).

COMPLETE GUIDELINES FOR THE CIRCUIT BUILDER 137

Soldering Iron Heat Control Circuit and Holder:

The circuit shown in Figure 5-2 can be used to change almost any electric soldering iron into a two-setting device: switch open, iron in the warm operating mode; switch closed, iron in the normal (hot) operating mode.

Next, you will need a soldering iron holder to mount your switch and diode circuit on. One possible solution is to use the soldering iron holder/cleaner from Radio Shack, No. 64-2078. This holder is constructed with an open coil that holds the soldering iron (to permit fast heat dissipation) and contains a sponge pad to keep your soldering iron tip clean.

If you like to do it yourself and save money, you might try using an old metal desk calendar stand for the base, and a piece of stiff wire bent into whatever shape is practical, and then you may want to improve the setup by using a lever switch (Radio Shack 275-016) for S1. If you do this, it will automatically set the circuit to the warm operating mode whenever you place the soldering iron in its holder.

Like resistors, capacitors can also undergo parameter changes due to elevated temperatures brought about by solder-

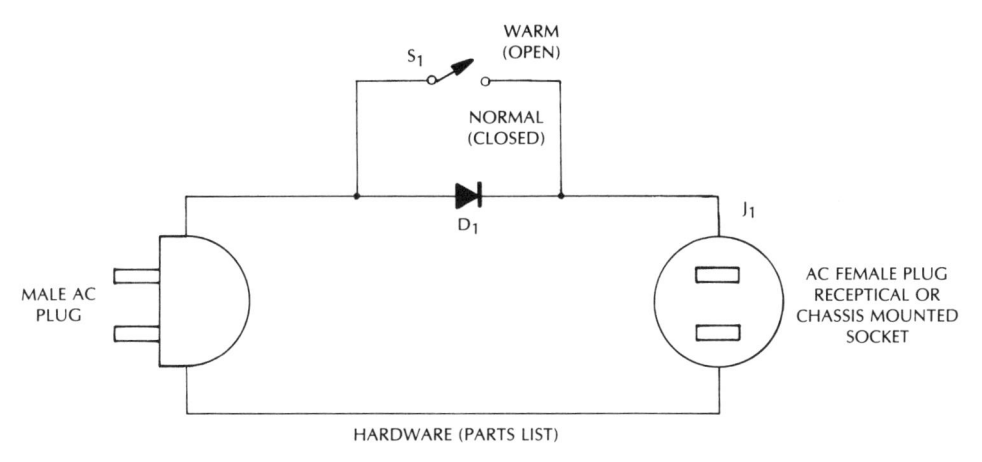

D_1 — Radio Shack 3A barrel type diode, 1N5400 series. 200V (# 276–1143) or 400V (# 276–1144), catalog # 328, or equivalent.

S_1 — SPST switch. Radio Shack subminiature 275–612, heavy duty 275–651, or any switch that is equivalent (such as a heavy duty slide switch, etc.).

J_1 — A chassis-mount AC socket, Radio Shack 270–642, or equivalent.

Figure 5-2: Two-setting soldering iron tip saver.

ing. In Chapter One, it was pointed out that in some cases (for example, in filter circuits), the exact value of a capacitor is very important. However, when we use the word *exact* in the science of electronics, we must ask, "How exact?" In other words, what is the tolerance, or how precise should we be? Let's say that you must have a 0.001 μF ± 1% capacitor. In the world of capacitors, this is very precise. In fact, most technicians would call it a *precision capacitor*. Buying a precision capacitor can be both expensive and frustrating. On the other hand, there is another solution that could be inexpensive and easy, assuming you already own one of the many digital capacitance meters now on the market.

Suppose that you have several 0.001 μF ceramics in your spare parts box. Now, it's very possible that they were rated, by the manufacturer, as having −20%/+80% when they were originally offered for sale. A little bit of arithmetic shows that this actually means each of the capacitors could be anywhere between 0.0008 and 0.0018 μF. But you are looking for 0.001 μF ±1%. In this case, you should measure no lower than 0.00099 and no greater a value than 0.00101.

How do you calculate your tolerance? First, set your limits (in our example, it was ±1%; however, you can use any percentage such as 2%, 5%, etc.), and then use this formula:

your limits = (cap value desired) × (1 ± percentage/100).

Example:

Using 2% tolerance and a 0.01 μF required capacitance,

your upper reading = 0.001 μF × 1 + 2/100
= (0.001)(1.02) = 0.00102 μF
your lower reading = (0.001)(1 − 2/100)
= (0.001)(0.98) = 0.00098 μF.

If you would like to calculate what the actual percentage (i.e., whether it is a 1%, 2%, etc.) of a certain capacitor is, use this formula:

$$\text{tolerance (\%)} = \frac{\text{what you measured} - \text{marked value}}{\text{marked value}} \times 100.$$

Measuring Coax Cable Capacitance:

When working on digital equipment, it is usually better if you *do not* use coax for direct connections to your oscilloscope.

COMPLETE GUIDELINES FOR THE CIRCUIT BUILDER 139

In general, always use a probe because coax can introduce excessive capacitance into your test setup. The capacitance of a coax cable is, as you may know, an integral part of its specifications. Sometimes you will find it expressed as capacitance per meter, per foot, per 1000 feet, or even per mile. But, in any case, it's usually in capacitance-per-unit-length. However, even if you don't know what it's supposed to be per-unit-length, you can still determine the value, which, as we have said, can be extremely important in some test setups—especially digital. Also, due to rough handling, etc., the capacitance value of any coax can change drastically. All you need is a capacitance meter and this formula:

$$\text{cap per-unit-length} = \frac{\text{cap value read off meter}}{\text{number inches, feet, etc.}}.$$

Simple Method for Measuring Capacitance and Inductance:

If accuracy isn't important, and many times it is not, you can use the setup shown in Figure 5-3 to measure the reactance of many capacitive and inductive devices. Then, with a small

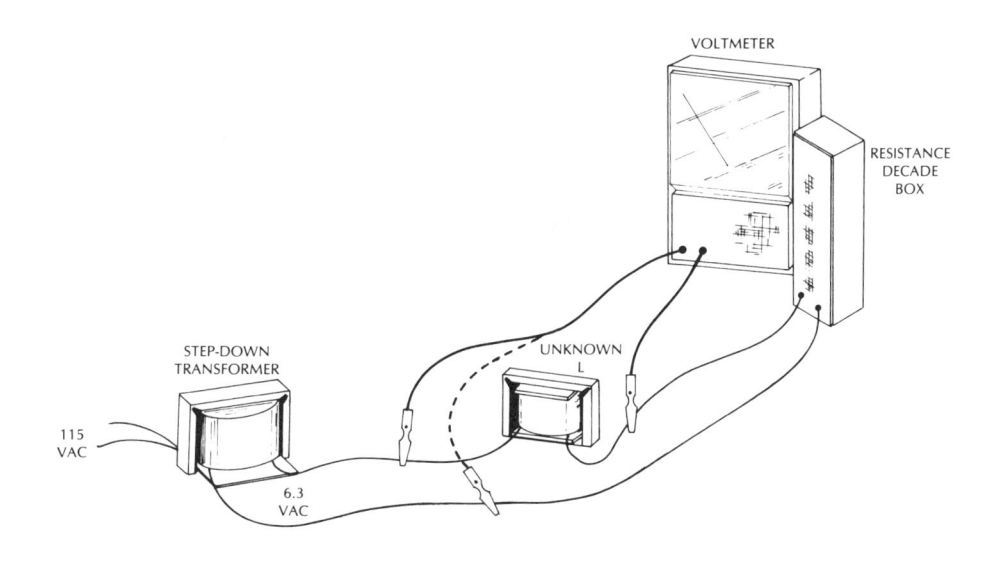

Figure 5-3: Place your inductor or capacitor as shown. See text for procedure.

amount of arithmetic, find the inductance or capacitor's value.

This is the procedure for using measuring setup shown in Figure 5-3:

Step 1. Place resistor decade box (*set at maximum resistance*) in series with the inductor (or capacitor), as shown.

Step 2. With no power applied, connect one lead of the decade box to the output leads (6.3V) of a power transformer (Radio Shack 273-1384). Connect your voltmeter leads as shown.

Step 3. Apply power to the primary (120V, 60Hz) and adjust the resistance until you measure the same voltage across both the resistor and the inductor (or capacitor). First, move one of the solid line leads as indicated by the dashed line.

Step 4. Read the value of the resistance off the decade box. The number is equal to the reactance of the device under test.

Step 5. The final step is to calculate the value of the inductor or capacitor, using the formula

$$L = X_L/376.8 \quad \text{or} \quad C = 0.00265/X_C.$$

For example, using a resistance value of 10 ohms (read off the decade box), and a hand-held calculator, our work, when finding the inductance, is

$$L = X_L/376.8 = 10/0.0265392, \text{ or about } 26.5 \text{ millihenrys}.$$

How to Select and Use Linear Multitesters

For those readers who may not already know, there are two common types of meter scales most of us have to use sooner or later. These are the linear and square-law types. Figure 5-4 points out the difference in calibration points. A square-law scale meter is a meter in which the deflection is proportional to the square of the applied energies. It is often used for radio frequency work.

In general, all linear scale multitesters (common multimeters, usually some sort of ohm, volt, milliampere scale

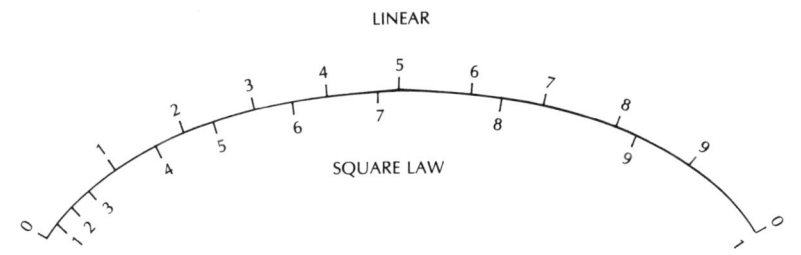

COMPLETE GUIDELINES FOR THE CIRCUIT BUILDER   141

Figure 5-4: A comparison of the linear and square-law meter scales.

combination) are built around the familiar D'Arsonval meter movement and have a set of linear scales, as shown in Figure 5-5.

Look in any electronics catalog (such as Radio Shack's) that includes test instruments and you will probably find multitesters listed as having a certain ohm-per-volt rating (for example, 10,000 ohms/volt, 20,000 ohms/volt, etc.). As was explained in Chapter One, ohms-per-volt is the rating of a voltmeter and, mathematically, it is the reciprocal of its full-scale current in amperes.

By referring to Figure 5-5, near the bottom of the meter faceplate, you will see that this meter has two ohms-per-volt ratings: one dc, at 20,000 ohms/volt, and one ac at 5,000 ohms/volt. Now, as you will remember, the higher the ohms-per-volt rating, the lower the loading effect on any circuit you connect your voltmeter to. In other words, the more accurate your reading will be. For this reason, we usually think that the

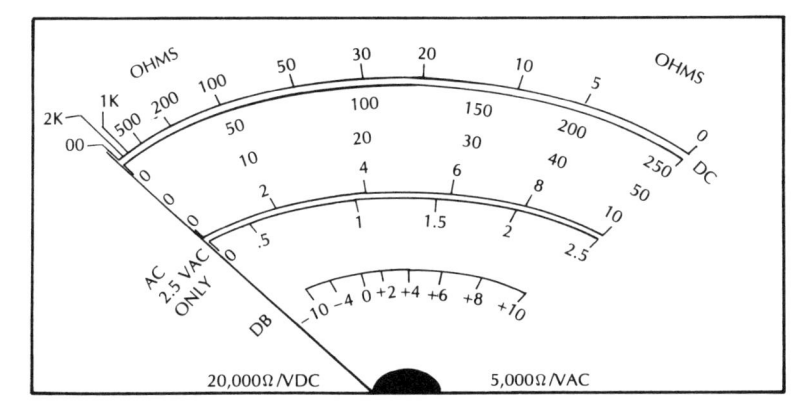

Figure 5-5: Scale plate of a typical modern multitester.

higher ohms-per-volt meters are the best. Although this is true, many catalogs give only the dc rating (because it is higher and looks better), or no such reading is listed. However, in almost every case you will find both the dc and ac ohm-per-volt rating listed somewhere on the meter faceplate. Look before you buy, and remember, the higher, the better.

At this point, you might ask, "Suppose I have a multi-tester that I think loads a circuit during a certain test. Can I check it?" Yes, you can. To make this test, you will need a resistor (preferably ±1%) with a value equal to the input resistance of the instrument being checked. A better quality multi-tester will usually have something like a 10-megohm input, and circuit loading should not be evident during the check.

Test Procedure:

Step 1. Connect your voltmeter to an operating circuit. Record the voltage reading.

Step 2. Place the resistor in series with the voltmeter plus lead and again measure the voltage of the operating circuit. Take both measurements (Steps one and two) at the same place on the circuit; that is, across the same two points.

Step 3. The voltage reading for Step 2 should be one-half (or *very* near) the voltage reading you had in Step 1. If your reading in Step 2 is greater than one-half, the voltmeter is loading the circuit and your direct measurements are not valid.

Figure 5-6 shows another type (different from that shown in Figure 5-3) of resistance scale found on several multitesters sold by Radio Shack and other retailers. Notice the crowded high-resistance end of the scale.

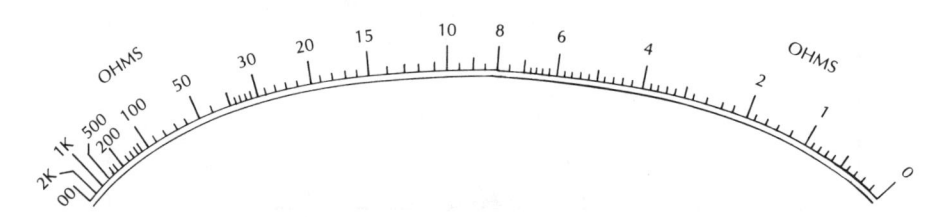

Figure 5-6: Resistance scale showing upper-end crowding. This is called a *10-ohms center scale* type.

A great many modern multitester accuracy ratings are listed as 4% for ac volts and 3% for all other ranges (dc, ohms, etc.). But this accuracy of 3% on the ohms scale actually means ± 3% of the *arc length* of the scale. Now, since the resistance scale is crowded toward the high value end of the scale (Figures 5-5 and 5-6), readings will be more inaccurate in this region. Or, for greater accuracy, resistance reading should be made in the lower value half of the resistance scale. The closer you can stay to zero, the more accurate your readings will be—although something like 5 or 6% is about the best you can do when using a ± 3% (10-ohms center-scale) scale length specification.

Breadboarding Solid State Linear Circuits

In checking out a breadboarded circuit, generally the solid state components are not one of the first items to suspect, provided you have read and followed the **CMOS** Handling Precautions given in Chapter Two. As a general rule, you will find all semiconductors (both discrete components and **IC**'s) to be both mechanically rugged and long lived. However, there are several elusive problems that can plague your circuits. In most cases, you will find the fault wasn't the design of the **IC**. Rather, it's how you wired or laid out the components on your breadboard. As an example, suppose you want to breadboard a dual wideband operational amplifier as an audio circuit. One such circuit, using a bipolar power supply, is shown in Figure 5-7.

In this circuit, the V_{EE} (negative) dc line doesn't share the same line as the output return (see Figure 5-7). This system uses a dual output (a negative line and a positive line) to feed the **OP AMP**. In fact, this points out one of the advantages of using two-pole (bipolar, if you like) power supplies. *They isolate your circuits from the ac power lines.*

Each audio input signal will appear across its respective resistor (R_2 or R_5, see Figure 5-7), and enter one of the **OP AMPS** (A or B) through the noninverting (+) input. In **OP AMPS**, the feedback loop must supply (at points A3 and B3) a voltage equal to that at points A1 and B1, to insure that the differential sum of the voltages between all these points is zero; i.e., A3 to A1 equals zero and B3 to B1 equals zero.

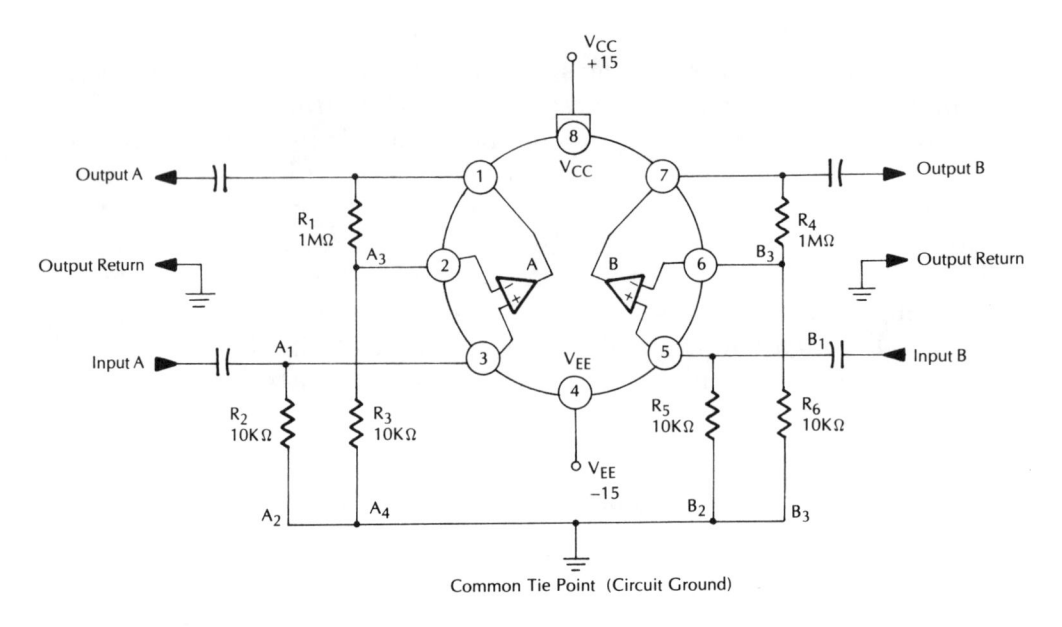

Figure 5-7: A dual wideband operational amplifier (MC4558 family).

Now, if during soldering, wiring, or plugging in the breadboard components, you should cause a resistance to appear between points A2 and A4, or between points B2 and B3, the **OP AMP** will see the voltage as part of the input signal. Of course, this will cause a voltage equal to the voltage drop developed by the ground loop resistance to appear across either R_3 and R_6—or both. To be sure that you don't create this problem, points A2, A4, B2, B3 must all be combined into a single tie point. This is circuit ground, not V_{EE} (the dc negative voltage output).

When you are working with shielded audio cables, there is another ground loop trouble that could cause hum problems. For instance, let's assume that you decide to use a MC3306P ¼-watt audio amplifier as an output power amplifier in a battery powered audio system. Figure 5-8 shows a schematic diagram of how this might be done.

The input to the preamp (transistor Q_1) is a shielded cable, and good grounding practice (for a circuit such as this) calls for the shield to be grounded at only one place (for example, point A). If you encounter a hum problem with a coaxial cable

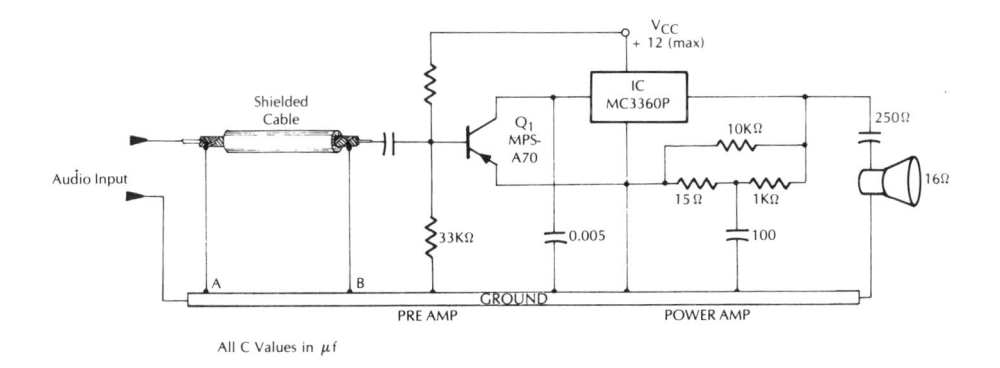

COMPLETE GUIDELINES FOR THE CIRCUIT BUILDER 145

Figure 5-8: This illustration shows shielded audio cable grounded at both ends. This may produce hum in your audio projects. *Shielded cable should be grounded at one point only.*

hookup such as the one shown, it's more than likely the result of your having grounded the cable at both ends. If you do this, there is a good chance that, due to the ground loop formed between points A and B, a 60 Hz signal (ac hum) will be induced into the desired audio signal path.

Test Equipment for Digital Systems

Digital Probes (Standard and ECL):

Testing digital integrated circuits today may require you to add additional test equipment to your shop. For instance, in the past you could check all logic families (**DTL, TTL, HETL, and CMOS**) with a logic probe. However, with the coming of the high speed and narrow threshold differentials of emitter-coupled logic (**ECL**) circuits, we must now have a high speed **ECL** probe.

The standard logic probes have a minimum detectable pulse width of 300 nano seconds, and as short as 1 nano second, with a maximum input signal (frequency) of 1.5 MHz and 50 MHz respectively. On the other hand, an **ECL** probe's specs will be input pulse width, 3 nsec min., and input pulse rate, 100 MHz min. As you can see, a low-cost, hand-held digital logic probe such as Radio Shack's model 22-300 would not be suitable for testing the **MECL** logic group we discussed in Chapter Two (particularly the **MECL III** family).

To use a typical logic probe, you connect the two wires that are provided for an external connection, to the power supply (usually, V_{CC} and ground of the circuit under test). Then simply touch a PC card run or the pin of an **IC**, and observe the indicators (or indicator, on some instruments).

The LED indicators will tell you if the signal level is logic high or low. Then you must know what the output of the digital circuit under test should be for a given set of inputs. With this information (such as Truth Tables, **IC** spec sheets, etc.), you usually can find the trouble quickly. In the case of digital **IC**'s, the trouble will more than likely be an input/output short to ground, or V_{CC}, input/output open circuit, or a short between **IC** pins. It's also possible, but not probable, that the **IC** may have an internal defect.

Digital Pulser Probe:

Applications for a digital pulser probe include the following:

1. Injection of substitute logic pulses to aid in digital circuit *debugging*.
2. Quick evaluation of digital breadboard changes or circuit modification.
3. Aid in development of **MPU** or other digital systems.
4. Isolate faulty **IC**'s or components such as transistors, capacitors, resistors, etc., in digital circuits.
5. Digital circuit (**IC**'s, etc.) stimulation. For instance, as an instructional aid to help you better understand a logic family or logic circuit operations.

Each one of these digital pulser probe applications is self-explanatory, except application number 1—debugging. You must keep in mind that all the **MPU**'s and **MCU**'s we have discussed in Chapters Three and Four respond to the *arrangement* of bits (8-bit bytes, 16-bit bytes, etc.). It makes no difference where the pulses originate: digital pulser probe (as we have suggested), or **MPU** memory. If there are false instructions in memory, the program is said to have a *bug* or *bugs*. The process you use in finding the undesired data (bit or bits), removing the instruction and replacing it with the correct information, is called *debugging*.

To use a logic pulser, first connect the power leads and then place the tip of the pulser to the desired input line. Signal injection usually is by means of a pushbutton switch, often located near the probe tip. When the button is depressed, a single high-going or low-going pulse (usually 1 or 2 μsec wide) is delivered to the circuit junction (called *node*) under test. In quite a few of today's logic pulsers, pulse polarity is automatic; high nodes are pulsed low and low nodes are pulsed high. Holding the trigger button down should deliver a series of pulses (for example, 20 pulses per second) to the circuit under test. *Note:* Pulses per second usually are referred to as a *baud rate*.

Logic Monitor:

As with the logic probe and logic pulser, an inexpensive logic monitor may be circuit-under-test powered. However, there are logic monitors on the market that contain a fully isolated power supply. Each has its own advantages. The circuit-powered design eliminates hunting for power leads and worries about accidental shorts or grounding. But an isolated power supply type eliminates loading of the circuit under test.

Some of these logic monitors examine up to 16 nodes at the same time, using 16-LED displays mounted on the instrument. For example, the connector/display unit clips over any DIP IC up to 16 pins. Basically, the monitor is nothing but 16 logic probes all connected and functioning together. However, this type does not have an isolated power supply. These are circuit powered.

Troubleshooting Digital Circuits

The procedures for troubleshooting and testing digital IC's and external components are similar to those used for analog circuits, but, in many cases, much easier. In troubleshooting digital circuits, unlike analog circuits, it is essential that you know what the output of a certain IC (be it a **TTL**, **CMOS**, **MOS** or **MPU** device) will be for a given set of inputs. If you will just keep in mind that you have to apply the manufacturer's specified set of inputs (in the form of digital words), you will find that locating a faulty component or making a test is fairly easy.

Generally, you will use one of the digital test instruments we have described in the previous pages to detect the presence or absence of a pulse, a train of pulses, and/or the static logic condition of the digital circuit. All of your standard electronic test gear such as a multitester (DMM, etc.), oscilloscope, and, of course, a pulse generator, can be used for checking electronic failures in digital systems. Nevertheless, when it comes to troubleshooting, you will find digital test instruments such as the logic probe and logic monitor are easier and faster to use on most projects.

Chapters One through Four have shown us that integrated circuits come in all sizes, shapes, and classifications, and are among the most widely used solid state components in the electronics field. Testing an IC is a fairly simple and straightforward matter. You first must know if the device is to be operated in a linear or a digital mode. Testing the linear operation of IC's is the same as testing any other linear device (transistor audio amplifiers and the like, see Chapter Six). Any signal modification such as amplification must be accomplished without amplitude, frequency, or other distortion. Furthermore, in most cases, your test instrument (signal injector) must be sinusoidal, although squarewave testing of linear circuits is frequently done.

As in the case of the linear IC, it is not possible to get to the internal circuits of a digital IC. Therefore, you must determine if all the IC circuits are operating properly or improperly, without ever seeing the internal circuits. In general, point-to-point signal tracing is the best method to use when troubleshooting or testing a complete system, whether it is a digital or a linear application. In this procedure, you introduce a pulse train (digital) or a sine wave (linear) at the circuit input, while monitoring various points throughout your circuit with an oscilloscope, logic pulser, etc.

When testing digital IC's, take care. For example, only if two shorted nodes are common to one IC can you begin to suspect the internal circuits of the IC. *Always* examine the IC's external circuits first; then, as a last resort, check the IC.

For the average experimenter, the logic probe provides the least expensive and quickest way to locate open signal paths in digital systems. The logic probe will not only show that a problem exists; it will also physically locate the particular IC or external component that is causing the malfunction.

DC Power Supplies for Linear Circuits

Unfortunately, when it comes to selecting a power supply circuit for experimental purposes, it sometimes seems desirable to save a few more pennies. Deeming one power supply about as good as another can, and has, caused quite a few experimenters considerable trouble. A perfect battery would have a zero voltage drop and zero internal resistance when connected equipment is turned on; so would a dc power supply. But, as we all know, all batteries and power supplies will have some internal resistance and a slight voltage drop during the time they are feeding energy to an operating circuit. What we are looking for, then, is a very stable dc voltage and low impedance over the connected equipment signal operating range. From the viewpoint of your breadboarded circuit, a voltage-regulated dc power supply will best provide the stabilization and low internal impedance needed.

As an example of what a well-regulated dc power supply will do for you (Chapter Seven will cover dc power supplies in detail), look at Figure 5-9. This graph shows the low frequency end of a test run on a high fidelity (hi-fi) audio amplifier.

Not only will you have a wider frequency range when working with hi-fi audio amplifiers, but a regulated dc supply will also produce less total harmonic distortion (THD) at a higher output power and less THD in the lower frequency range of the amplifier. The secret is in the low ac impedance of the voltage-regulated dc power supply used in the test run. In summary, when you purchase or build a voltage-regulated supply, you would like to have

1. Negligible output impedance from dc up to the highest frequency possible.
2. A constant dc output voltage (zero regulation) for all possible ac line voltage changes and overall possible load conditions—increased load (more current drain), decreased load, etc.
3. No power dissipated in the power supply circuits.
4. Overload protection.
5. Variable dc outputs.

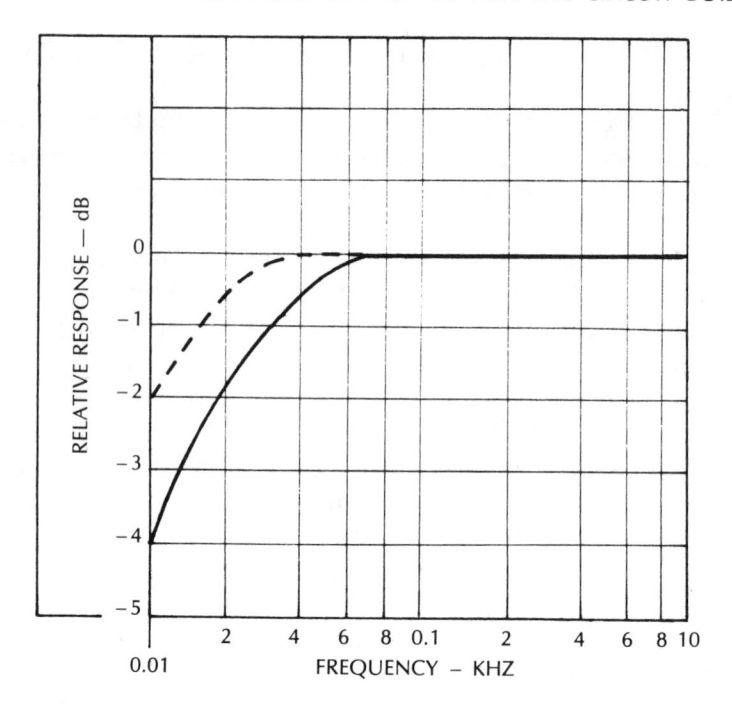

Figure 5-9: Performance curve from a test run on a hi-fi audio amplifier. Regulated power supply shown by dashed line; unregulated shown by solid line. 0dB on this graph equals 10 watts.

DC Power Supply Requirements for Digital Circuits

Not all electrical voltages applied to a **MPU, MCU, ROM, RAM,** and other digital **IC**'s are in binary form. For example, referring to Figure 3-1, note that there are two lines into this **MCU** (labeled V_{CC} and V_{SS}). These are the power supply lines connected to an external supply voltage of $+5.25$ Vdc ± 0.5 V for V_{CC} and ground for V_{SS}.

The power supply in this type of application is perhaps one of the most critical of "home-brew" computer components. This is because of the strict requirement just mentioned ($+5.25$ Vdc ± 0.5 V). To begin, your power supply must have a step-down power transformer, rectifier, and simple filter circuit, if it is to operate using common 120 V, 60 Hz ac line voltage as a source of energy. Both circuits shown in Figure 5-10 are full-

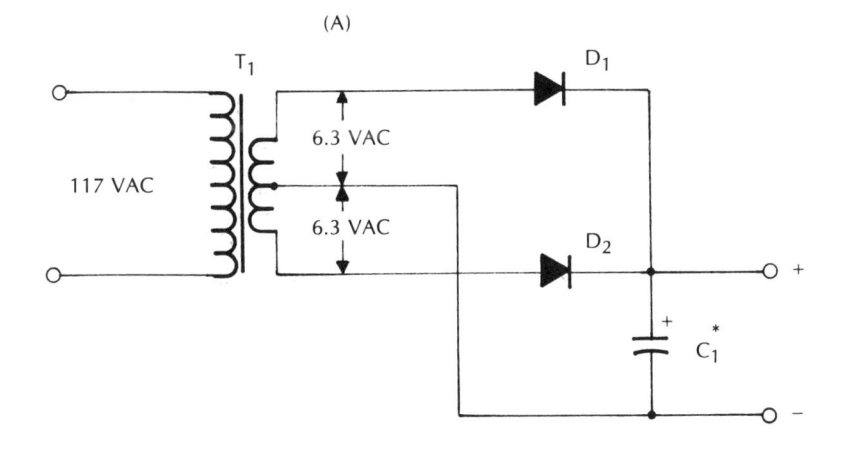

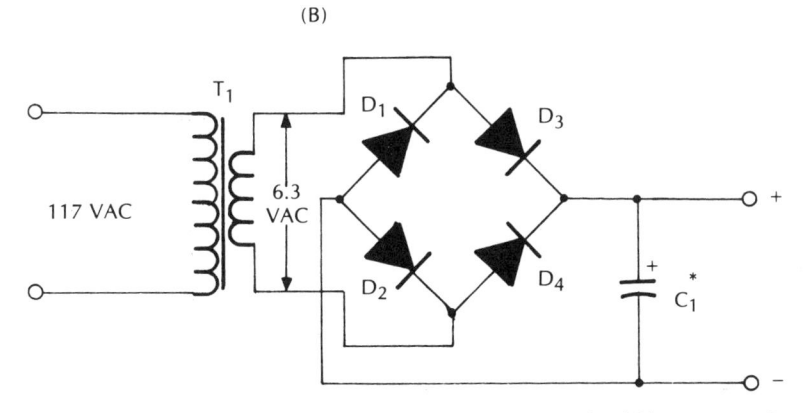

COMPLETE GUIDELINES FOR THE CIRCUIT BUILDER 151

*C₁ is 2000 μF per ampere. If used with Figure 5–11, should be 20,000 μF.

Figure 5-10: (A) basic full-wave rectifier, (B) full-wave bridge rectifier.

wave rectifiers, meaning that they make use of both alternations of the ac power line input to the power supply.

Using a full-wave rectifier not only makes it easier to filter; it will also produce a higher average output voltage than a simple half-wave. In addition, it will place less of a load on the transformer primary during operation. The full-wave rectifier will produce the *unregulated* dc voltage you need, but it will not provide the + 5.25 Vdc ± 0.5 V regulated voltage requirement. To do this, you must use some type of regulator following the rectifier. Figure 5-11 shows a 10-amp regulator using Motorola's MC1469R monolithic voltage regulator.

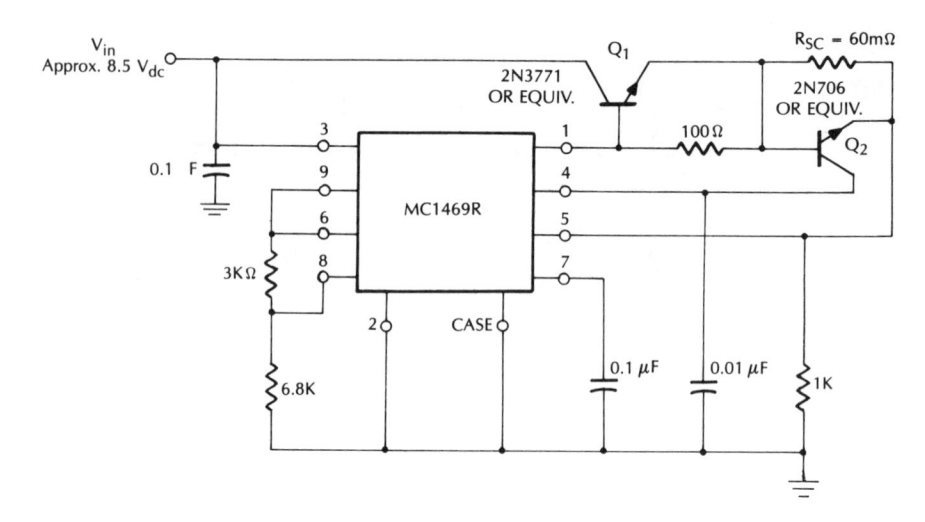

Figure 5-11: Regulator circuit constructed using Motorola's MC1469R regulator IC. See Chapter Seven for more about this regulator circuit.

One of your first questions may be, "Why suggest a 10-amp regulator?" While any given **TTL** gate may be required to sink only 1.6 mA of current, seldom will you have a project operating with only one gate or **IC** on. For example, a typical system using digital components may contain 30 or more gates, **IC**'s, etc., connected to a common ground bus. Therefore, with all, or a large portion, of these digital components enabled simultaneously, a large current demand may be placed on the power supply. *Note:* Current levels above about 5 amperes require special cooling (blower or fan) in addition to regular heat sinks.

A very important feature of this regulator is its current-limiting circuit, transistor Q_2 and R_{sc}. The maximum output current is set by resistor R_{sc} and is equal to $0.6/R_{sc}$. Using a controlled current regulator may very well save an expensive component during experiments. Notice, by no means is this a standard off-the-shelf resistor. You can make one by constructing a simple balance bridge, as described in the author's book "Electronic Troubleshooter's Handbook," sold in many book stores and by the publisher of this book. Or use ohms/foot data given in practical electronic reference data books. For example, bare wire, 46 gauge (AWG) is listed as 0.2377 ohms-per-foot.

Another refinement that should be included and connected to the output of the regulator circuit is a crowbar voltage protector. In other words, a separate circuit that monitors the output terminals of the regulator whenever a voltage you preset is exceeded. See Chapter Twelve, Project 1, for such a circuit. Also, much more will be said about dc power supplies in Chapter Seven.

CHAPTER SIX

How to Work with Practical Audio Circuits Using Solid State Devices

Look at almost any retailer's catalog—Radio Shack, Sears, Heathkit, ad infinitum—and you will quickly realize that audio technology is one of the most rapidly growing areas of electronics. The high-fidelity field, in particular, has evolved greatly within recent years. For example, many of today's receivers are all-digital, quartz locked, and include 16-station memory. They have 50 watts (rms) minimum per channel into eight ohms, from 20 to 20,000 Hz, with no more than 0.02% total harmonic distortion. Analyzing trouble symptoms in this type of receiver circuit is just one of the procedures that we will demonstrate in this chapter.

The following pages not only explain audio circuit troubleshooting in easily understood terms, but also include simplified testing and measurements procedures you will need when designing or modifying audio circuits.

Introduction to Audio Systems Using IC's

In Chapter Two, you were introduced to the 8-watt audio power amplifier TDA2002 (see Figure 2-4). Another low power audio device that comes in **IC** form is the MC1306P ½-watt complementary preamplifier and power amplifier designed in a single 8-pin dual-in-line package. Figure 6-1 is an application suggested by Motorola: a phonograph amplifier using a ceramic cartridge on the input.

Sometimes the difficult part of building a circuit such as that shown in Figure 6-1 is that you do not have an 8-ohm speaker on hand, or that your dc power supply will not produce 9 Vdc. Now, we know that it is risky to substitute a lower ohms value speaker for the one called for (8 ohms, in this case), but we can also ask, "What if we use a higher ohms value speaker"? Furthermore, we can ask, "How about using a lower dc supply voltage"? Figure 6-2 shows both an 8-ohm load and a 16-ohm load (speakers), and how they, and various power supply voltages, will affect the performance of the amplifier.

Let's say that we decide to place another 8-ohm speaker in series with the one shown in Figure 6-1. Or, basically the same thing, we decide to use a 16-ohm speaker in place of the 8-ohm one shown. In referring to Figure 6-2, it's apparent that the first thing we will probably notice is a decrease in output power regardless of where we set the supply voltage. As you can see, using an 8-ohm speaker at a total harmonic distortion (THD) of 1%, and a supply voltage of 10 to 12 Vdc, is about the best we can do. Of course, we can drive the amplifier harder but we will pay for it with a higher THD if we do.

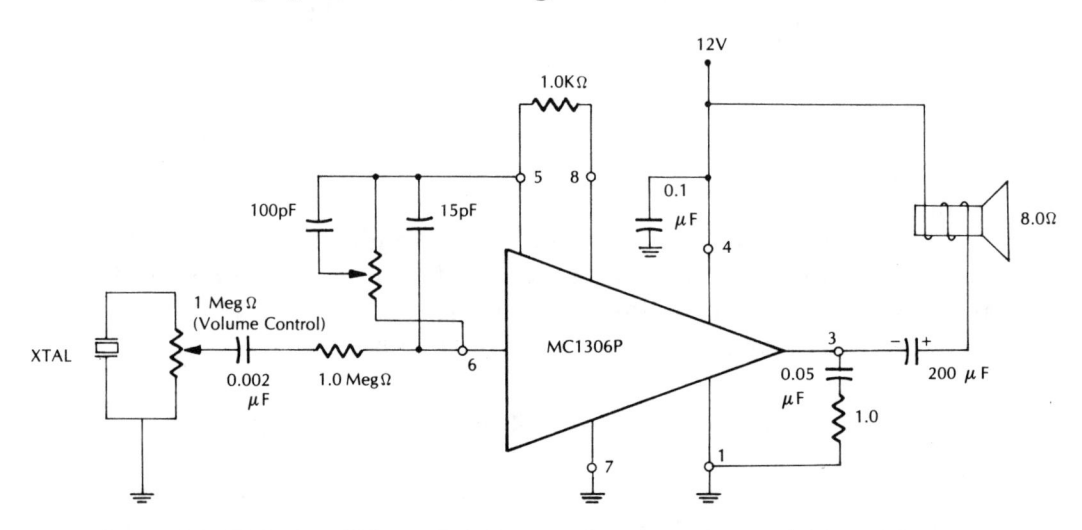

Figure 6-1: Using a MC1306P IC as a phonograph amplifier with a ceramic cartridge.

HOW TO WORK WITH PRACTICAL AUDIO CIRCUITS

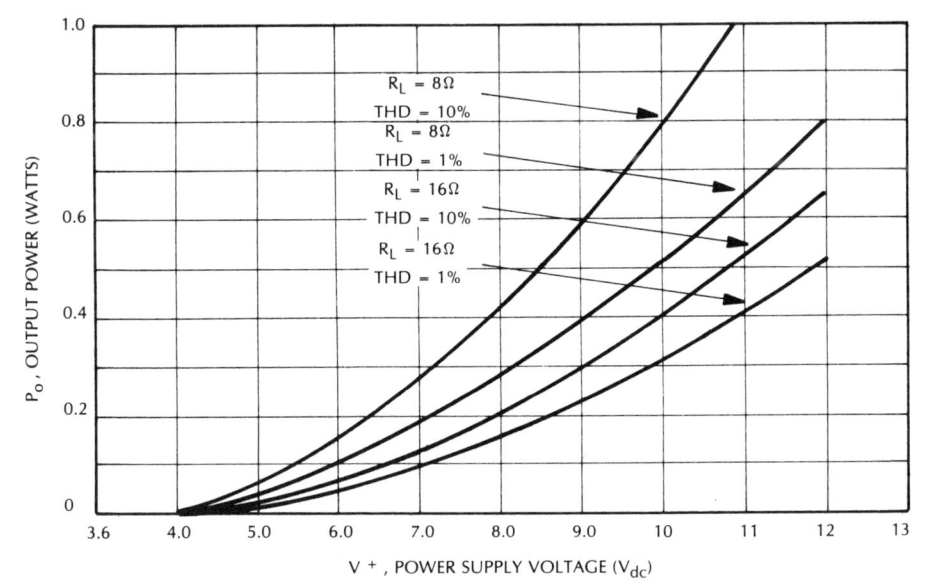

Figure 6-2: Output power versus power supply voltages using either 8- or 16-ohm speakers. Total harmonic distortion is also shown for both loads at various supply voltages and output powers.

Could we parallel two 16-ohm speakers? Sure can. In fact, we could also series two 4-ohm speakers. It all adds up to this: use any combination that results in a load of 8 ohms and don't overdrive the amplifier. Also, keep your supply voltage fairly near the recommended operating voltage—in this case, not more than 12 Vdc and not below about 8 or 9 Vdc. This particular device (the MC1306P) is rated by Motorola as a ½-watt amplifier. Figure 6-3 contains a distortion curve that illustrates what will happen to the THD content if you try to drive it to higher power outputs.

The basic test procedures for audio amplifiers designed around IC's are essentially the same as for transistor audio amplifiers. In other words, you must determine which stage fails to process its signal correctly. Once you have isolated the stage that messes up or blocks the signal, you're home free and clear because finding a malfunctioning IC is easy. However, it's almost as simple if it's a bad component in the amplifier circuit. In most cases, start by checking the battery (if you are using one as a dc supply), and then the capacitors and resistors, *in that order*.

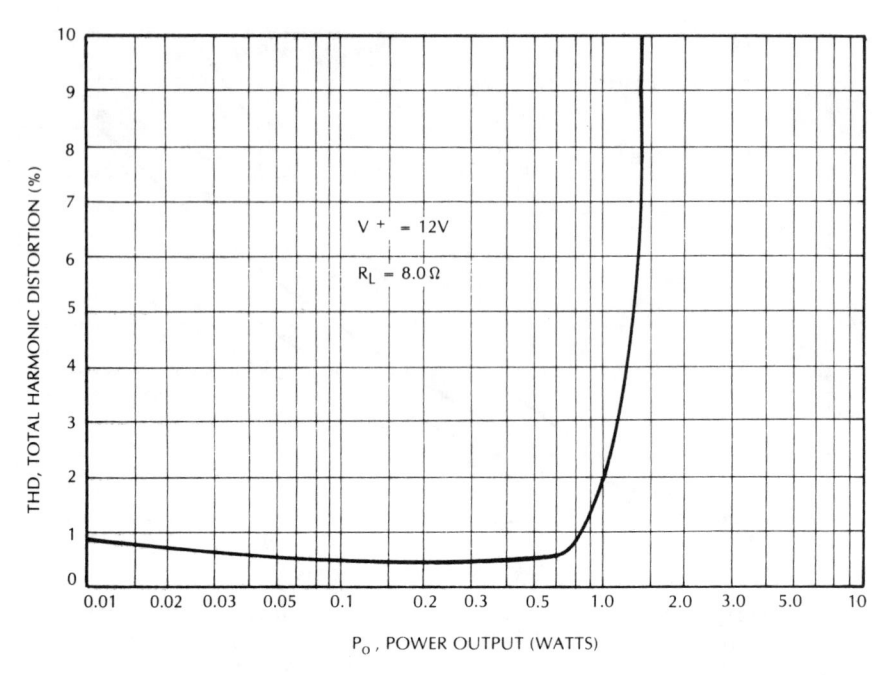

Figure 6-3: Distortion characteristics of the MC1306P ½-watt audio amplifier.

Discrete Power Transistors

Two individual audio circuit components (discrete devices) preferred by many circuit designers are a set of complementary power transistors; for example, the 2N6111 (PNP) and 2N6288 (NPN). Figure 6-4 shows the package used for both of these transistors.

A set of power transistors such as these two usually uses one of the basic output-transformerless (OTL) types of audio output configurations. A stripped-down circuit diagram for a complementary amplifier is shown in Figure 6-5.

Connecting two transistors in this manner causes each transistor to conduct over one-half of the input cycle, because Q_1 (2N6111) is a PNP type whereas Q_2 (2N6288) is an NPN type. To put it another way, this is a Class B power amplifier. The maximum transistor power dissipation occurs at the time when signal power output is 40% of its maximum value. At this time, the power dissipated by each transistor is 20% of the maximum power output (40 watts for these transistors). Maximum rat-

HOW TO WORK WITH PRACTICAL AUDIO CIRCUITS   159

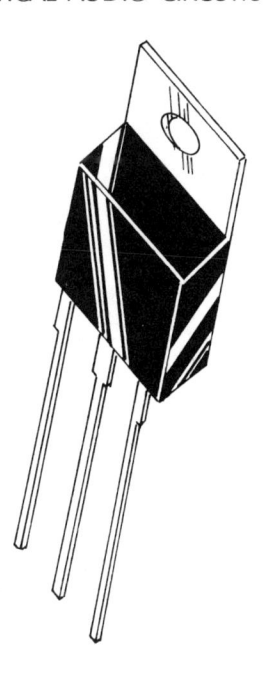

Figure 6-4: Type of case used for the 2N6111 and 2N6288 complementary power transistors. See Figure 6-6 for mounting hardware required.

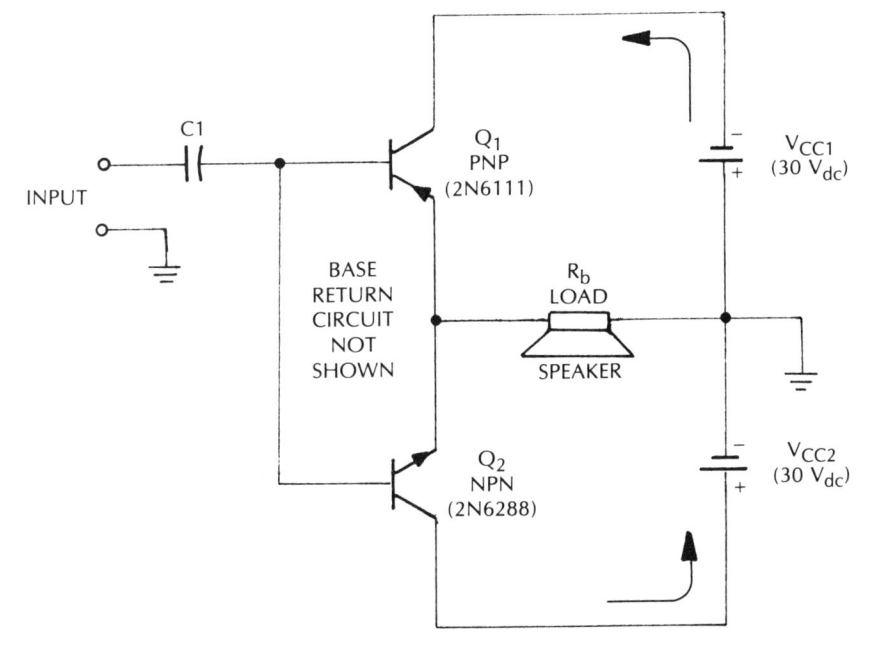

Figure 6-5: Simplified diagram of a complementary symmetry amplifier.

HEX HEAD SCREW

(1) RECTANGULAR STEEL WASHER

TRANSISTOR CASE (SEE FIGURE 6–4)

RECTANGULAR MICA INSULATOR

HEAT SINK

NYLON BUSHING

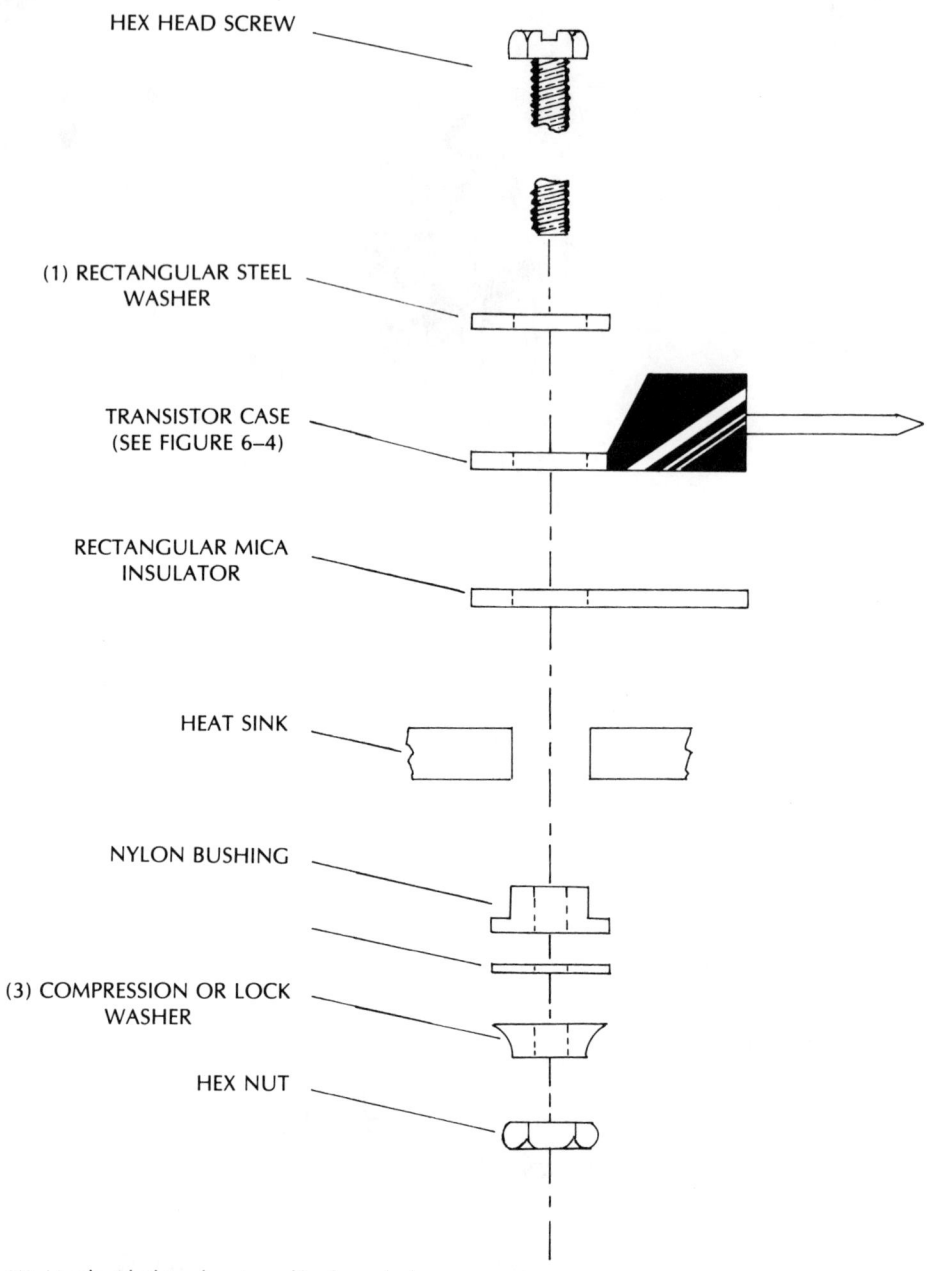

(3) COMPRESSION OR LOCK WASHER

HEX NUT

(1) Used with thin chassis and/or large hole.
(2) Required when nylon bushing and lock washer are used.
(3) Compression washer preferred when plastic insulating material is used.

Figure 6-6: Mounting hardware for 2N6111 and 2N6288 power transistors discussed in text.

HOW TO WORK WITH PRACTICAL AUDIO CIRCUITS

ings are V_{cc} = 30 Vdc, V_{cb} = 40 Vdc, I_c = 7 Adc continuous, and I_b = 3 Adc. Forty watts is quite a bit of power; therefore these transistors, like any medium or high power transistor, *must be mounted on a heat sink before making any kind of operational checks.* Figure 6-6 shows the mounting hardware for these devices. *Note:* Transistors with collector current of 5 amperes and above usually require additional cooling by use of a small fan or blower.

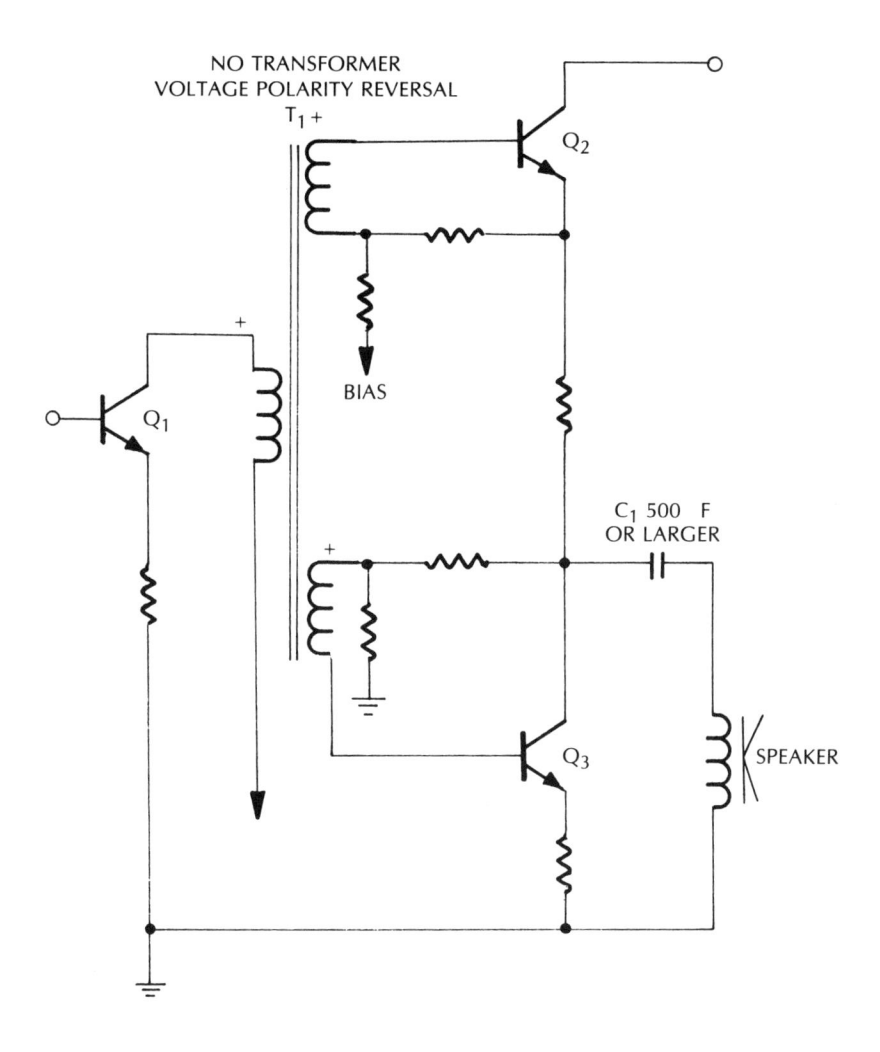

Figure 6-7: A popular OTL amplifier (a stacked circuit), using two NPN transistors and a single dc power supply.

There are several OTL circuits in which one can use discrete transistors. For example, a popular audio output amplifier using individual transistors is the so-called *stacked circuit* (see Figure 6-7).

An example of a Darlington pair (also called *Darlington amplifier, double-emitter amplifier follower,* and, sometimes, β *multiplier*) is shown in Figure 6-8.

This transistor circuit consists of two transistors with their collectors tied together, and the emitter of the first transistor directly tied to the base of the second (see Figure 6-8 B). I mentioned in the preceding paragraph that a Darlington transistor is sometimes called a "β multiplier." The reason for this name is that the amplification of this amplifier equals the product of the individual transistor's amplification. The total result is a high degree of amplification and a high input impedance.

In the last section, it was said that output transformerless circuits are very popular in audio systems. In this section we have pointed out that Darlington circuits have very good characteristics (good amplification and high input impedance). Now, it follows that two Darlington transistors connected in an OTL configuration should make a pretty good audio amplifier. The pairs of transistors are connected as Darlington pairs. This arrangement is therefore often termed a *Darlington pair complementary symmetry* configuration. Next, because there is

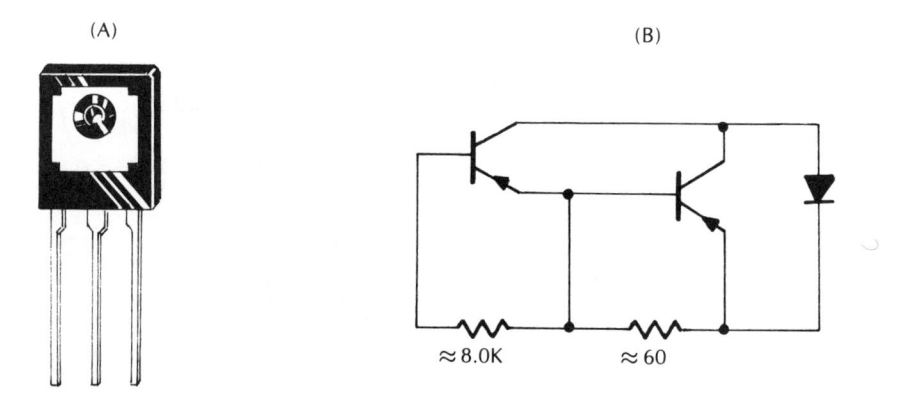

Figure 6-8: A general purpose PNP Darlington complementary 40-watt power transistor (2N6034). (A) shows the case and (B) is the schematic of the device.

HOW TO WORK WITH PRACTICAL AUDIO CIRCUITS

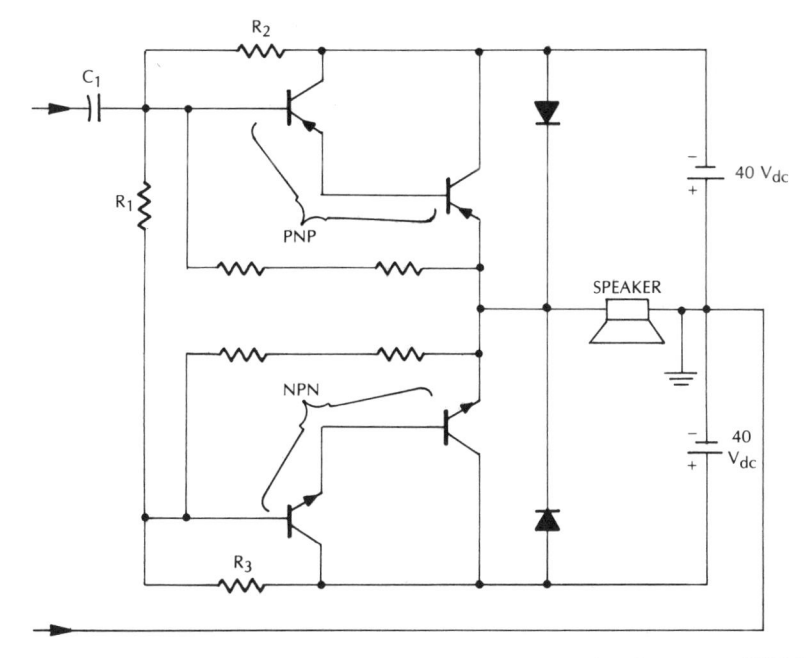

Figure 6-9: Darlington OTL arrangement. Upper Darlington (PNP) could be a 2N6034, and the lower Darlington (NPN), a 2N6037. Maximum ratings: V_{CE} = 40 Vdc, V_{cb} = 40 Vdc, I_c = 4 Adc, power = 40 watts.

a large amount of negative feedback taking place during operation, the percentage of distortion is very low. See Figure 6-9 for a basic schematic of a complementary symmetry Darlington OTL amplifier arrangement.

Audio Amplifier Design Fundamentals

Few things are more rewarding to the electronics experimenter than successfully breadboarding one of his own circuit designs. Fundamentally, there are two steps when designing audio amplifier circuits. First, the analytical (paper, pencil, and mathematics), and then breadboarding. In this section, we will emphasize the analytical; however, you should be able to go straight to the breadboard after finishing your paper designs.

In general, most of the basic design information you will need for a particular transistor can be obtained from a manufacturer's data sheet. Usually you will find that distributors such as Radio Shack include the required data and pin diagrams when you purchase a transistor. Before you actually purchase the solid state components, you must decide how much output power you want and what speaker impedance you will use. Let's assume that you want to drive a 4-ohm speaker (or two 8-ohm in parallel) and have an output power of 8 watts. Now, we can proceed to calculate the required power supply voltage and circuit components.

First, let's use an OTL design, i.e., complementary power transistor—NPN and PNP—for our circuit. This is because, as you may recall, it eliminates an expensive output transformer. Next, find the maximum current needed from each transistor to develop 8 watts in a 4-ohm speaker. The required current is

$$I_c = \sqrt{P/R} = \sqrt{8/4} = \sqrt{2} = 1.414 \text{ A (rms)}.$$

Therefore, the peak collector current that the transistors must handle is

$$I_{cmax} = 1.414 \times I_c = 1.414 \times 1.414 = \text{approx 2A}.$$

To determine the peak voltage you will need from your power supply if peak current is 2 A through a 4-ohm load,

$$V_{max} = I_{cpeak} \times R_L = 2 \times 4 = 8 \text{ volts}.$$

But your workbench power supply should produce about twice this value (8 volts maximum), to allow for the voltage drop across each transistor—meaning it would be best if you chose about 20 volts for your final working voltage for V_{CC}.

The base drive (I_b) for each of the power transistors will be equal to I_{cmax}/β. Let's assume we want a minimum current gain (β) of 40. The maximum I_b should be

$$I_{cmax}/\beta = 2/40 = 50 \text{ mA}.$$

When you are designing an amplifier such as this, you should increase the base drive a bit to prevent clipping. Let's use 50% more. By doing this, you end up with 75 mA. This value of base drive current is what the preamplifier for the power amplifier must handle (75 mA is the required collector current of the medium power driver transistor).

The minimum value of the capacitor you want to use for coupling the output of the power transistors to your speaker (4 ohms, in this example), is found by using the formula

HOW TO WORK WITH PRACTICAL AUDIO CIRCUITS

$$C = \frac{1}{2\pi(R_L)(f_{min})} \text{ (for } -3dB \text{ at } f_{min})$$

where:

$2\pi = 6.28318$
R_L = speaker impedance
f_{min} = the lowest frequency that will be produced at one-half power (−3dB). This means that all audio frequencies below this one will be at less than one-half power.

In our example, using a minimum audio frequency of 20 Hz, we arrive at a value of

$$1/(6.28318 \times 4 \times 20) = 0.01984$$

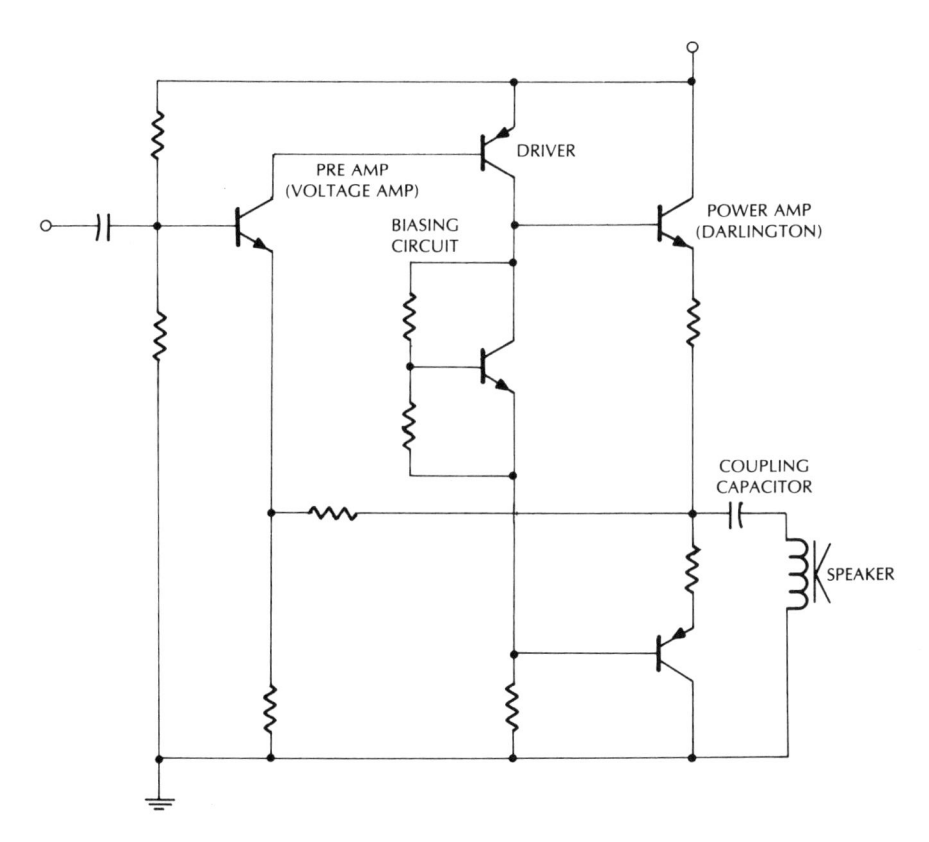

Figure 6-10: Typical complete OTL audio amplifier using transistors in a complementary system. See text for design example of the power amplifier and how to calculate the value of the coupling capacitor.

or very near 2,000 μF, which is what would be best to use. However, you could use a value as low as 1,000 μF without too much loss at the low end of the amplifier output frequency range. Figure 2-4 (A) shows a block diagram of an 8-watt Class B audio amplifier in integrated circuit form (see Chapter Two). A Darlington OTL arrangement with preamp, driver, and power amplifier transistors is shown in Figure 6-10.

You might like to try designing your own output OTL power amplifier. Use the two transistors 2N4918 (PNP) and 2N4921 (NPN). If you use two 8-ohm speakers in series and an output power of 5 watts, your requirements should be

I_{cpeak} = 790 mA, supply voltage (peak) = 12.6 V, and, again assume a minimum beta of 40, power amplifier drive current (base current) = 19.75 mA.

Power Requirements

According to Class B amplifier theory, the power dissipation of each power transistor should be about one-quarter of the output power. In the example in the preceding section (OTL design problem), we used 8 watts. Therefore, in this case, the power dissipation would be 0.25 × 8 = 2 watts. Actually, the amplifier should have a collector current flowing (even with no signal on its input), for efficient operation and negligible distortion. That is, the amplifier should be biased Class AB. Therefore, the power dissipation will be slightly higher than 2 watts. When choosing complementary pairs of transistors, you should choose a pair that is capable of dissipating at least one-quarter of the output power (2 watts, in our example), and preferably more. By the way, any amplifiers that are biased Class A—drivers, for example—are usually not large power dissipators. To calculate the power dissipation, simply use the formula

transistor dc collector voltage × transistor dc collector current.

Three other formulas that may help you are

$P_{in} = P_{dis} + P_{out}$, $P_{dis} = P_{in} - P_{out}$ $P_{out} = P_{in} - P_{dis}$

where

P_{in} = power in watts delivered to the power amplifier

P_{dis} = power dissipated by the power amplifier
P_{out} = power in watts delivered to load (speaker, etc.).

Should you use these formulas, it must be understood that an audio amplifier circuit is very sensitive to changes you might make in the load, particularly power amplifiers. Output power will decrease very rapidly if you vary the load impedance either way (up or down) from your calculated load impedance. Two examples will give you some idea of how much: (1) if the load is twice the output circuit impedance, your output power will be reduced by approximately 50%, and (2) if you place a load of 40% the output impedance on the amplifier, the output power will be reduced by approximately 25%.

Distortion in Audio Systems

It is difficult to enumerate all the types of distortion. However, one specification you will find given for almost every audio system is "Total Harmonic Distortion." No matter how well you design an electronic amplifier circuit, there is always the probability that harmonics will be present. These harmonics combine with the desired frequency and produce undesired distortion (called *harmonic distortion*), as is the case when any two signals are combined—radio frequency, audio frequency, TV frequency, or whatever.

The THD measurement determines the amount of harmonic energy added to the output signal of an audio system by the audio system itself. To measure the THD of an audio system, you must apply a signal (with much better quality than the amplifier being tested can produce) to the input of the device under test, and then measure the harmonic content at the output.

When you test an audio circuit over a wide range of frequencies (usually 20 Hz to 20 kHz) for harmonic distortion, and plot a graph similar to the one shown in Figure 6-11, the result of your work (in percentage) is the THD.

This example shows that there is less than 0.2% THD at any power level between 100 mW and full-rated output power, at any test frequency between 20 Hz and 20 kHz. A modern stereo receiver will generally have a THD specification much better than this: for example, a THD of no more than 0.05%, and

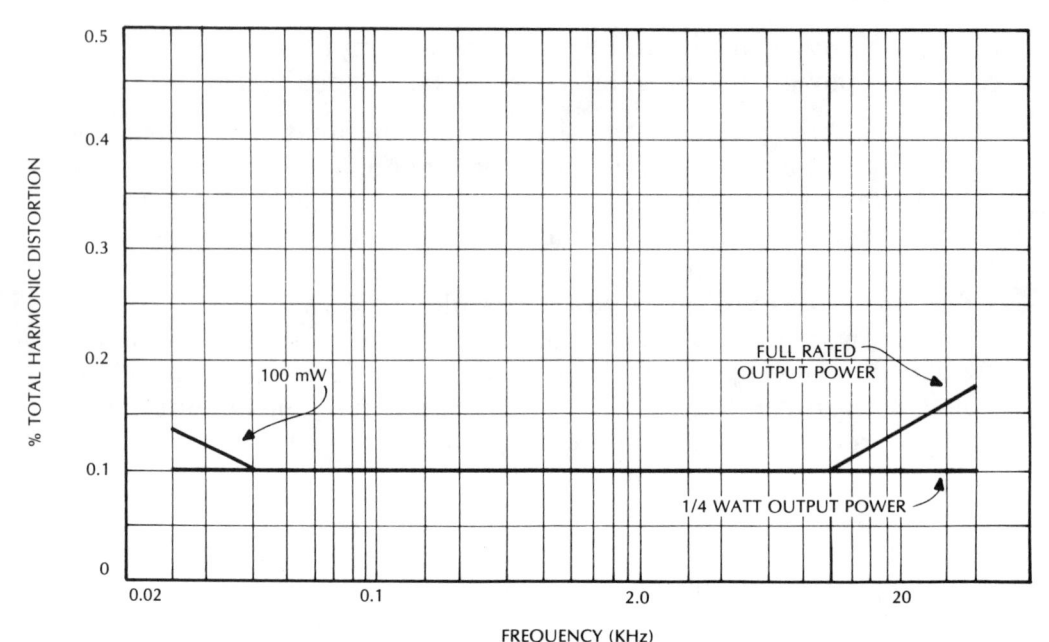

Figure 6-11: THD versus frequency plot of a wideband (20 to 20,000 Hz) test of an audio amplifier.

even better in many cases (for instance, 0.02% over the specified frequency range).

You will find that most THD analyzers attenuate the fundamental test frequency with a notch filter and measure all harmonic energy that may be present throughout the audio band. In this type measurement, THD is

$$\text{THD (\%)} = \frac{\text{harmonics}}{\sqrt{\text{fundamental}^2 + \text{harmonic}^2}} \times 100.$$

The method just described is fine, if all you want is the total harmonic output as distinct from each individual frequency of the output. If your requirement is more demanding, you'll have to use a wave analyzer. In this case, the THD is calculated using the formula

$$\text{THD (\%)} = \sqrt{\frac{\text{2nd harmonic}^2 + \text{3rd harmonic}^2}{\text{fundamental frequency}}} \times 100.$$

In this type of measurement, the wave analyzer is first tuned to the fundamental frequency and the amplitude is recorded. Then the wave analyzer is tuned to the second, third, fourth, and so on, harmonics, and amplitude measurements are recorded for each of these frequencies. These amplitude readings are then used with the formula just given, to arrive at a THD for the equipment under test. Although it is difficult (more work) to measure THD with a wave analyzer, the results are precise. See the section titled "Troubleshooting Audio Amplifiers" for the most common causes of poor frequency response.

Practical Decibel Measurements

Since our ears respond to audio signals in a logarithmic fashion, it is to our advantage to measure audio signal levels in decibel (dB) units. All of us who use an ac voltmeter (from the least to the most expensive), have seen dB scales printed on the face of the meter. However, it is important that you understand that the decibel is a power ratio, and not a voltage ratio. But, voltage ratios (dB_V), do correspond to power ratios (dB_W), if you measure the two voltages across equal load resistances. On the other hand, if the load resistances are not equal, don't you believe those readings! Because, in this case, the voltage values do not correspond directly to the power values.

There are two techniques that lend themselves well to the experimenter who wants to measure attenuation (or gain) in decibels. One, measure each voltage, current, or power level separately and use one of these formulas:

1) $dB = 10 \log P_2/P_1 = 20 \log E_2/E_1 = 20 \log I_2/I_1$

   (when the load resistances are equal);

2) $dB = 10 \log P_2/P_1 = 20 \log (E_2 \sqrt{Z_1} / E_1 \sqrt{Z_2}) = 20 \log (I_2 \sqrt{Z_2} / I_1 \sqrt{Z_1})$

   (when the load resistances are not equal).

The other practical method uses an oscilloscope to make a wideband measurement between various points on a waveform shown on the viewing screen of the scope. Once you make the scale described, simply place it on your scope and read dB's straight off the screen.

Let's assume that you have a Heathkit oscilloscope such as the kit model 10-4105. The graticule on this scope (and all other standard models) has eight vertical and ten horizontal centimeter gradations. See Figure 6-12 for a replica of a standard scope graticule.

To calibrate your scope scale in dB's, use the formula

$$dB = 20 \log \text{max vertical gradations} / \text{selected vertical gradations}.$$

For example, for horizontal line number 4 (vertical axis), your mathematics should be

$$dB = 20 \log \text{max V grad (8)} / \text{selected V grad (4)}$$
$$= 20 \log 8/4 = 6 \text{ dB}.$$

If you do this for each of the Y-axis lines on your scope graticule, you should end up with a scale similar to the one shown on the right in Figure 6-12. To measure the attenuation of any selected point (or points) on a wideband scope display, first set the scope controls until the waveform exactly fills the screen, as shown in Figure 6-13.

To determine the attenuation of any particular frequency within the total display, simple locate the frequency in question, then calculate the attenuation in dB's, as has been explained. Or, for a rough approximation, place the dB points shown on the face of your scope with a grease pencil and read the attentuation straight off the scope screen. Your answer will be in dB's below the peak signal (frequency) voltage.

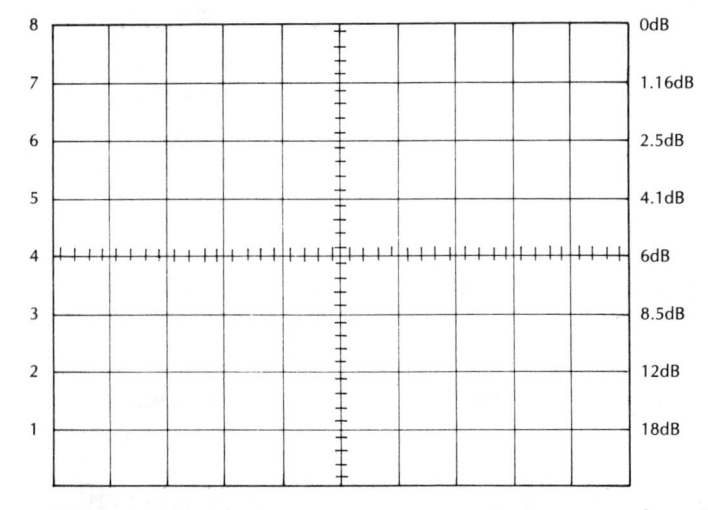

Figure 6-12: Standard oscilloscope graticule showing home-brew dB scale on the right.

HOW TO WORK WITH PRACTICAL AUDIO CIRCUITS

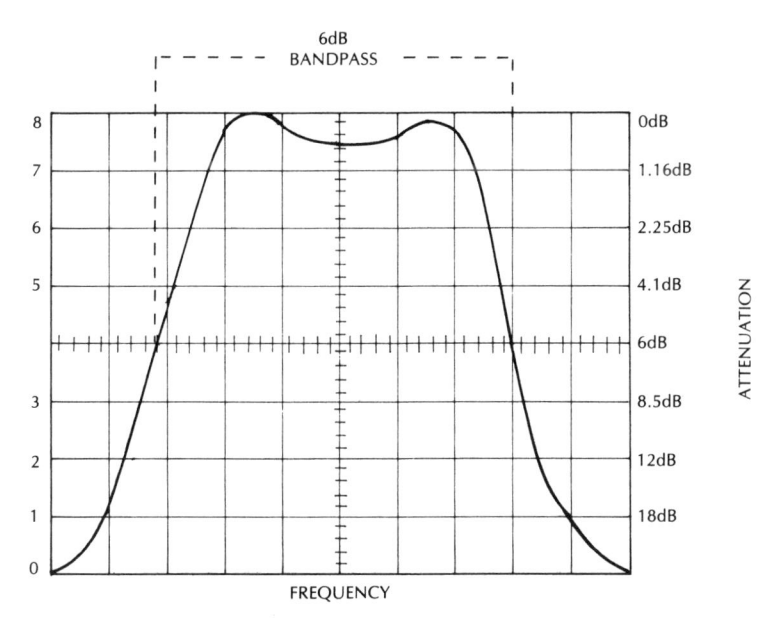

Figure 6-13: Example waveform presented on an oscilloscope, that might be used during a frequency-by-frequency measurement.

Power Measurements

Many audio amplifier specifications include a power bandwidth factor; for example, 250 watts, minimum rms per channel into 8 ohms, with less than 0.025% total harmonic distortion from 20 Hz to 20,000 Hz. The power output of an audio amplifier is found by measuring the output voltage E_o (rms) across the load resistance R_L (Figure 6-14) at any frequency. Or across the entire frequency range, if you are making a wideband power measurement and using the formula E^2_o/R_L to calculate power out.

In some cases, you may encounter conversions of the dBm units. Normally, this is an abbreviation for decibels above (or below) a power level of 1 milliwatt. When you encounter a dBm value, it corresponds to a level of power measure and is then calculated as a ratio to 1 mW. Although it is standard practice to reference dBm values to 1 mW in 600 ohms, it is possible to use other parameters. For instance, let's use 600 ohms with a given ac volts (rms) measured across it, for an example of how to calculate gains using voltage measurements.

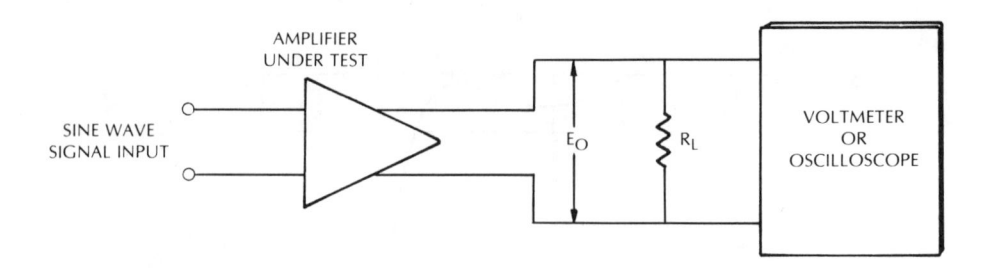

Figure 6-14: Power output measurement. Power output equals E^2_o/R_1.

Assuming an ac volts (rms) measurement is to be made, we can start with this fact: zero dBm corresponds to 0.775 Vrms across 600 ohms. Why? Because $E = \sqrt{PR} = \sqrt{0.001 \times 600}$ = approximately 0.775, where E is voltage, P is power in mW, and R is load resistance.

Next, let's assume you measure 7.75 V (rms) across the load resistance connected to the test amplifier. In this example, our answer is

$$\text{dBm} = 20 \log 7.75/0.775 = 20 \times 1 = 20 \text{ dBm}.$$

In summary, zero dBm corresponds to 0.775 V (rms) across 600 ohms. Similarly, 20 dBm corresponds to 7.75 V (rms) across 600 ohms, and 42 dBm corresponds to 100 V (rms) across 600 ohms.

Impedance Measurements

Dynamic Output Impedance:

The power output measurement wiring diagram shown in Figure 6-14 can be used to measure the dynamic output impedance of an audio amplifier circuit after it is breadboarded and operational. The connections are the same, except that the load resistance (R_L) value will have to be experimented with until you find the maximum power the amplifier is designed to deliver to its load.

After you have found the maximum power point, turn off power to the circuit and remove the resistor you are using for

HOW TO WORK WITH PRACTICAL AUDIO CIRCUITS

R_L. Don't change its value during the process of removal, if it is some form of variable resistance.

Now all you need to do is measure the resistance with your ohmmeter. The value you read is the dynamic output impedance of the amplifier under test. However, this reading is good only at the particular test frequency you used as an input signal to the amplifier. The audio signal source you use for an input must have a *constant amplitude* from about 20 Hz to 20 kHz. Also, don't overdrive the amplifier under test. If you see *no increase* in the output monitor reading when you attempt to increase the input signal level, you are overdriving the amplifier.

Dynamic Input Impedance:

While you have your amplifier on the breadboard, you may want to measure the *dynamic input impedance*. Figure 6-15 shows the wiring diagram for this measurement.

Adjust your signal source to the audio frequency band (or frequency) you want to test the amplifier for. Then adjust the variable resistance until you read the same voltage on voltmeters 1 and 2. Next, turn off power to the amplifier and remove the variable resistance from the circuit (be careful not to change its setting). Now measure this resistance with your ohmmeter. This reading is the amplifier dynamic input impedance. As before, this reading is good only for the frequency used during the measurement.

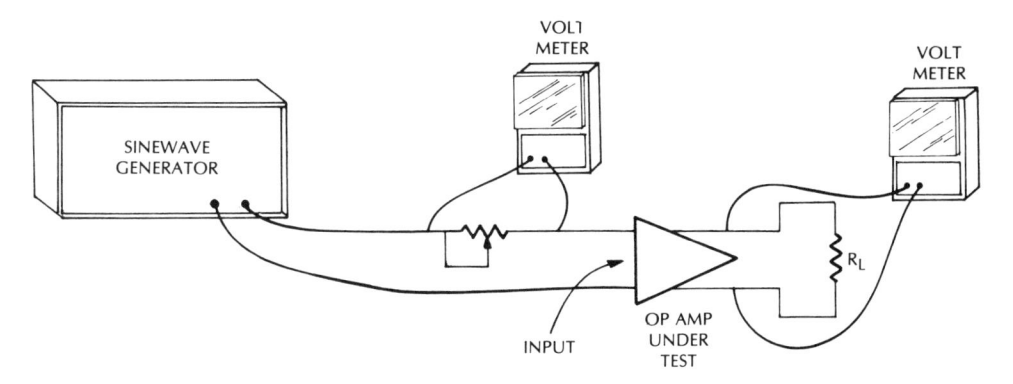

Figure 6-15: Wiring diagram for a dynamic input impedance measurement.

Audio Preamplifiers

The last section discussed input and output dynamic impedance measurements; however sometimes you will want to connect a home-built amplifier circuit to a stage already manufactured. Audio preamplifiers are usually low level signal devices, and therefore every milliwatt of power can be important. If you do not know the input and/or output impedance of two circuits, try to match the impedance of the two stages you are connecting together. There will be maximum power transfer when the impedances are matched. Look for a maximum reading on your output-measuring instrument (ac voltmeter or scope), or listen for maximum output on a speaker, if the preamplifier is driving a power amplifier. Figure 6-16 shows a schematic of a transistor OTL audio amplifier. As you can see, this type of amplifier could possibly use a Darlington complementary transistor (see Figure 6-8 B).

The troubleshooting procedures for these audio amplifiers are explained in the last section of this chapter, but it should be pointed out that if you encounter a Darlington tran-

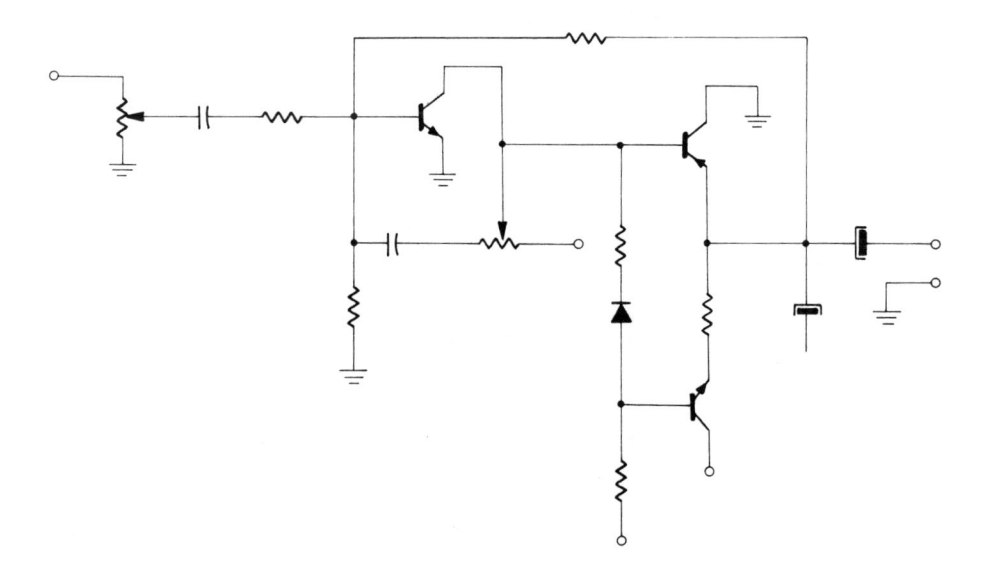

Figure 6-16: Transistor (or Darlington) preamplifier.

sistor, it must be replaced with a Darlington. Also, a Darlington has only three leads (just as has a bipolar transistor), so take care when replacing them (see Figure 6-8 A). Incidentally, a Darlington power transistor schematic can be the same as a Darlington preamplifier. It is simply the power-handling capability of the device that makes the difference. For example: a few hundred milliwatts for the preamplifier and something like 40 watts or more for the power Darlington is normal.

Audio Power Amplifiers

Quite a bit has been said about Darlington amplifiers in the past pages, and this is because many of them are in use in all types of audio systems. So far, we have described a single Darlington transistor audio amplifier and the designs known as *dual-Darlington* circuits. As another example, Figure 6-17 shows a dual-Darlington output circuit that is frequently used to achieve a higher power output than one Darlington could

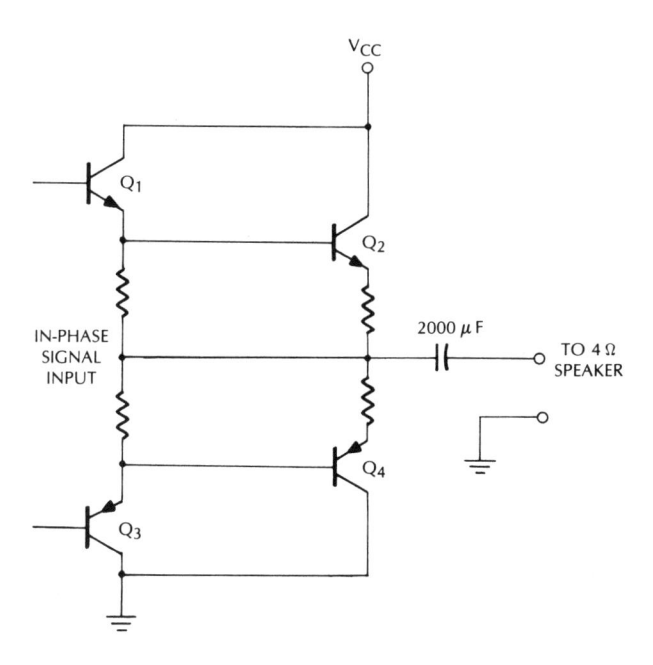

Figure 6-17: Dual-Darlington output schematic. This circuit is used to produce more power out than one Darlington transistor can provide.

provide. Examples of 25 W Darlington transistors are the 2N6043 and 2N6040, which are designed for a load of 4 ohms.

Sometimes a voltage gain is desirable. In this case, the circuit shown in Figure 6-18 may be used. In either circuit (Figure 6-17 or 18), phase inversion is not required preceding the stage, since a positive-going signal on the input will turn transistors Q_1 and Q_2 (in both schematics) on, and Q_3 and Q_4 off. Of course, the reverse will happen when a negative-going signal is placed on the inputs.

If you should desire to use discrete components and construct the circuit shown in Figure 6-18, you could use a transistor such as the MJE340 (you would, of course, require four of them). This transistor's specs are

0.5 ampere, 300 volts, rated at 20 watts, in a T0126 package.

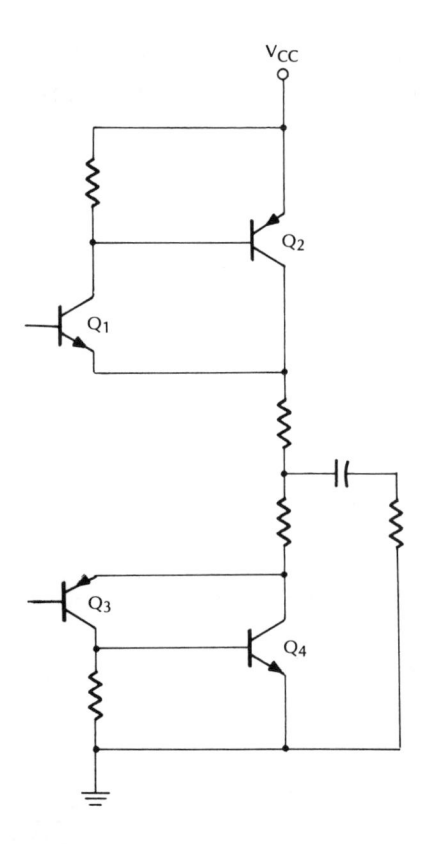

Figure 6-18: Complementary transistors wired to produce voltage gain.

HOW TO WORK WITH PRACTICAL AUDIO CIRCUITS   177

The mounting hardware for this type package is shown in Figure 6-19.

It is very important that you note that several of the complementary output circuits we have described could undergo a heavy current increase through one or more transistors, if there is a short circuit in the attached load (speaker, speakers, resistor, or any external wiring to the load). This can,

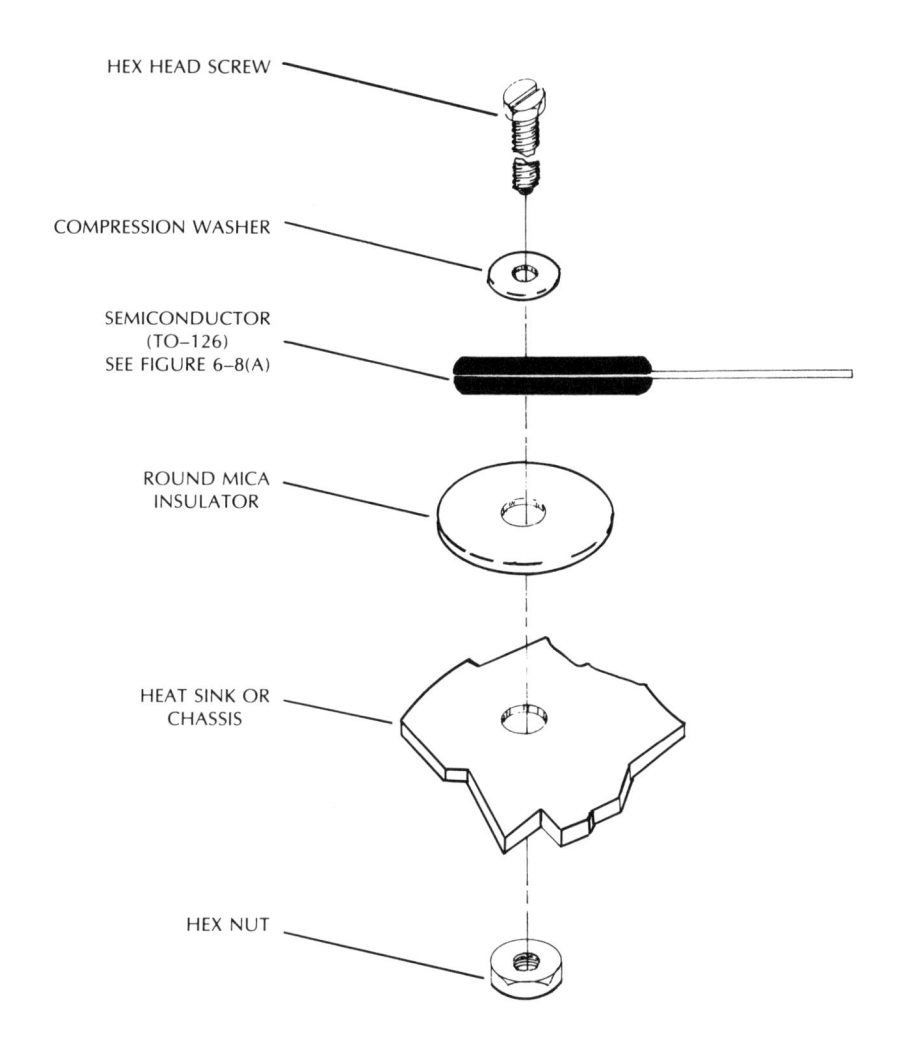

Figure 6-19: Mounting hardware required for a T0126 package.

and usually will, destroy a transistor and/or other components in the circuit. Therefore, when breadboarding a circuit without transistor collector current-limiting resistors, always include some sort of short circuit protection (see Chapter 7). Use a fuse if no other protection is provided in the circuit. The maximum collector current allowed for the transistor will set the current rating of the fuse you should use.

Troubleshooting Breadboarded Audio Amplifiers

When a breadboarded circuit fails to perform as expected, grab your multimeter (DMM). DC, ac, ohms, and current measurements are the backbone of audio amplifier troubleshooting. To pinpoint the trouble, follow these three basic steps:

1. *Analyze the circuit.*
    a. Check your wiring against the schematic.
    b. Check to see that all wiring connections are making good electrical contact.
2. *Test the circuit.*
    a. When troubleshooting transistor circuits (assuming dc voltage is applied to the circuit), start by measuring bias circuit voltages. *Example:* No audio output with zero emitter bias voltage on the output transistor. If you measure zero bias with your voltmeter, immediately suspect the transistor. It's very possible, and probable, that the base-emitter junction is an open circuit.
    b. If you are troubleshooting an **IC** that is plugged into a socket, the easiest way to check the **IC** is to make a simple substitution. *(Note: Never substitute **IC**'s with power applied to the circuit under test!)* This is also true when checking modularized circuits. Should you wish to make operational checks on an in-circuit **IC**, a good place to start is to measure the source-current drain. If a substantially higher or lower value of current than your specs call for is measured, in all probability the **IC** or external components are defective.

3. *Test the components.*
   a. Measure the resistor values (in circuit) using a low-power ohmmeter.
   b. Make a "bridging" test of capacitors. To do this, loosen one end of the capacitor under test (for example, cut a PC conductor at one terminal of the capacitor) and bridge it with a known good capacitor of the same value.

A few notes: Look for any transistor with a very low collector voltage. (This transistor probably has a partial short because it is conducting very heavy.) Or, on the other hand, check for a transistor with a very high collector voltage. In this case, perhaps there is an open circuit somewhere in your breadboarded amplifier. This could be caused by a bad circuit design, a component failure, improper solder job, etc.

Many troubles in a breadboarded amplifier are caused by improper bias. If the faulty operations appear only when you apply an input signal, suspect the bias. Try different bias resistors (emitter, base, and collector). Increase and/or decrease their values by about 10 or 20 percent from your calculated design values. An oscilloscope really helps during these trials. Keep looking at the output signal while adjusting the bias resistor values. If it's impossible to find a proper bias relationship, either your calculated design values are extremely bad (off more than 50%), you have a bad solder job, improper transistor and/or components, or, if you experience oscillations, *excess positive feedback*. One possible cause of this problem is long hookup leads. Try moving the hookup wire to different places on the breadboard (rattle them around a bit) and see if this will stop the oscillations.

CHAPTER SEVEN

Modern Regulated Solid State Power Supplies

In their never-ending quest to produce the *perfect* power supply, designers are continuously developing electronic circuits with a considerably higher level of complexity than is encountered in any system we have ever seen before. The result? Literally dozens of dc power supply regulators and systems in a great variety of shapes and sizes, employing dramatically different operating principles, all vying for the technician/experimenter acceptance.

In this chapter, we will examine the basic types of power supply circuits, placing special emphasis on the latest advances and how they can be used to your benefit.

Regulated Power Supplies

As you learned in Chapter Five, the regulated power supply is perhaps one of the most critical of all experimental breadboard components. Today's *bipolar IC's* often used in experimenters' projects operate from a dc supply with a nominal level of 5 volts and a maximum level that must not be allowed to go above 5.25 V. This is one reason why a regulated supply is important to you.

Operational amplifier circuits, for example, require power sources that provide clean, stable dc. You can use batteries as a source of good dc, but their expense will usually restrict them to portable applications, rather than to workbench setups where commercial ac power is readily available. The most economical way to go is to use a *well-regulated power supply* to convert the commercial ac at your workbench into a dc voltage that is held constant at all times.

There are two basic types of regulated power supplies that can be used with solid state circuits—*bipolar* and *single ended*. The bipolar supply produces a positive and negative output voltage, both removed from ground and having opposite, but equal, voltages. The basic idea for a simple bipolar supply is shown in Figure 7-1.

Notice, the ground shown in the schematic (see Figure 7-1) is a true ground; that is, it can be connected directly to earth ground, and frequently is. *Warning:* Be careful that you do not connect an *artificial ground* and *true ground* together. If you do, sparks are sure to fly! Figure 7-2 shows the output circuit of a dc power supply using an artificial and a true ground.

If you measure the voltage polarity across resistor R_3 (see Figure 7-2), you will find that the top of the resistor is positive in respect to the bottom (true ground). However, measuring across R_2, the voltage will be plus at the top and negative at the bottom. Of course this means that the top of R_3 is both plus and minus, depending on which measurement you are making.

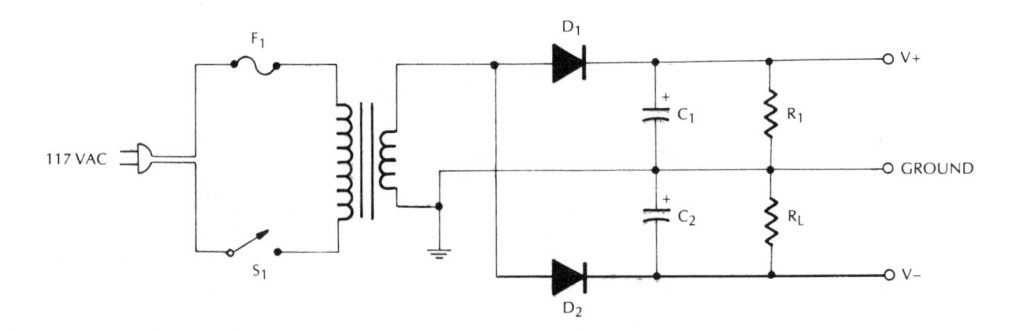

Figure 7-1: Simple bipolar dc power supply circuit.

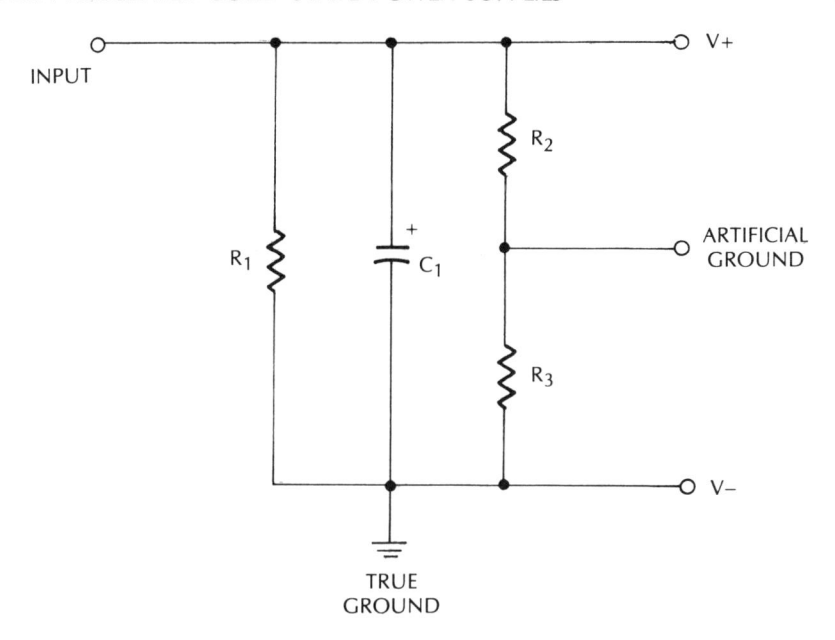

Figure 7-2: DC power supply output filtering circuit showing both artificial and true ground.

Now you should be able to see why one should never connect a true ground to an artificial ground. If you do, you are placing a direct short across R_3 and, as was stated, "sparks will fly." Typically, a circuit such as that shown in Figure 7-2 is designed using a single-ended dc rectifier. A single-ended bridge rectifier circuit connected to the filtering circuit shown in Figure 7-2 is illustrated in Figure 7-3.

Although it is fairly easy to convert a single-ended supply to a bipolar one, there are some cautions that you should be aware of. To begin, your load should be fairly light; that is, the constraints of the power supply components must not be exceeded. As a rule, breadboarded electronic circuits using solid state components consume small amounts of current, but we have discussed several transistors that consume considerable amounts of current. For example, Figure 6-4, in Chapter Six, shows a transistor that consumes 7 Adc when operated in a continuous mode at its recommended collector voltage (30 Vdc, in this case).

If the current drain from either of the power supplies (Figures 7-1 and -3) is very low, and uses a sufficiently large filter capacitor, the pard (Periodic and Random Deviation) on

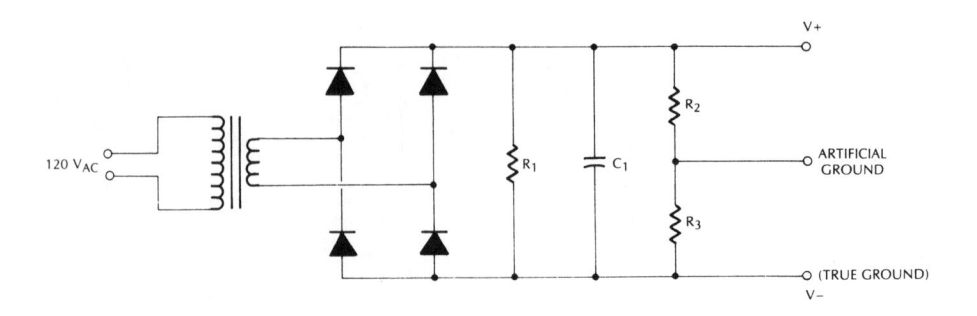

Figure 7-3: Schematic diagram of a bridge rectifier including a basic filter circuit with a single-ended output that has been converted to a bipolar output by use of R_2 and R_3.

the output will be acceptable. However, in some applications even a low-level pard can cause you trouble. Therefore, as a safeguard against pard problems, it's best to use an electronic filter on the output of a bridge rectifier, since an electronic filter will guard against ripple, hum, and noise, as well as spikes on the output.

If you use a zener diode in your dc power supply filter design, it will provide not only the equivalent of a very large filter capacitor (several tens of thousands μF), but also very good voltage regulation. Figure 7-4 contains a basic schematic for a single-ended filter system that includes a zener diode, and

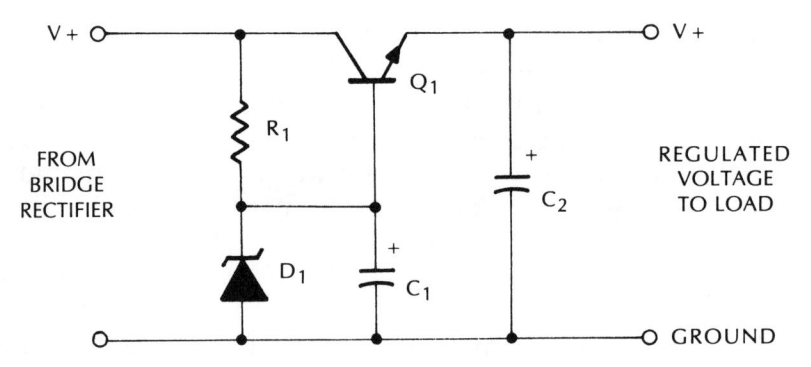

Figure 7-4: Power supply filter using a zener diode to multiply filter capacitance and absorb transients.

MODERN REGULATED SOLID STATE POWER SUPPLIES

will provide a positive dc output. Two of these filters, placed back to back, can be used to provide filtering for bipolar dc supplies. Figure 7-5 shows how this can be accomplished.

IC voltage regulators operate basically in the same manner as the filter circuits we have shown. As illustrated in the preceding paragraphs, the input voltage from the unregulated power supply is applied to the reference circuit and load. The reference (the zener diode in the previous examples) determines the accuracy of the regulated voltage output to the load. In general, IC voltage regulators do not have the power-handling ability of the discrete circuit regulators. However, you can overcome this problem by using external (external to the IC regulator) power transistors.

IC regulators often include special features such as overload protection, overvoltage protection, automatic excessive

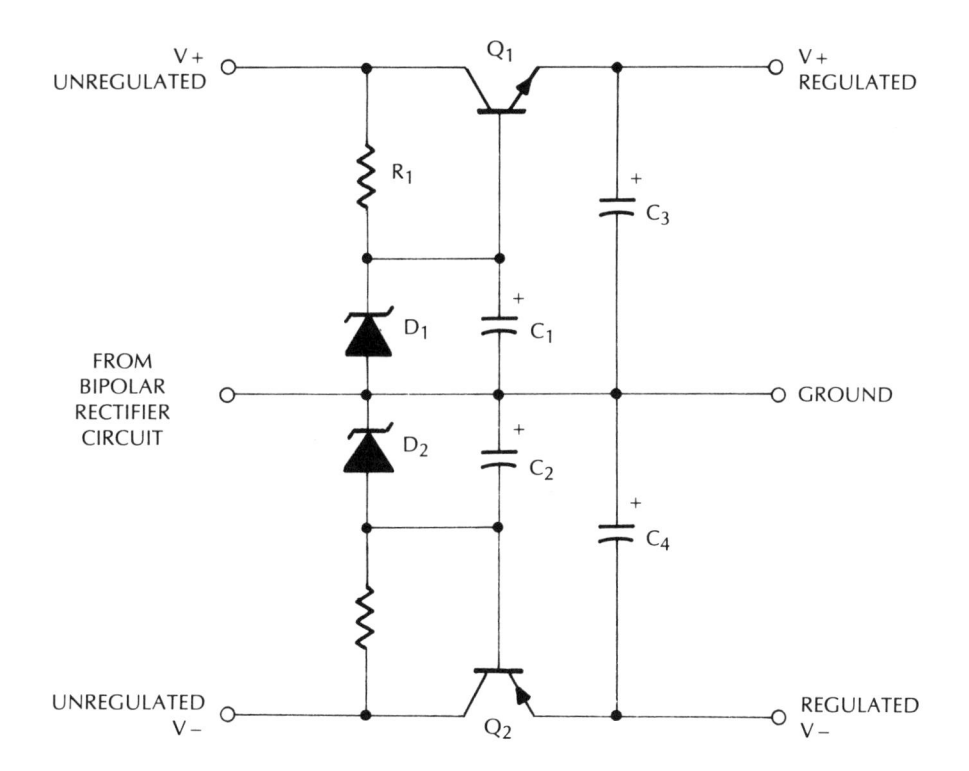

Figure 7-5: Wiring diagram showing how two single-ended dc power supply filters can be connected to create a bipolar dc power supply.

heat shutdown, short-circuit protection and, in some cases, remote control shutdown. As can be easily seen, most of these special functions can be, and usually are, very important to the experimenter.

Basic Concept of Regulation

When you breadboard an electronic circuit and energize it, there will be, in respect to this voltage with no load connected, a variation of the dc power supply's output voltage. One of the characteristics of all voltage regulators is known as *load voltage regulation*. This characteristic refers to the regulator's ability to maintain a nearly constant voltage supplied to the load under all conditions (variable load conditions).

Assuming that your shop temperature is not too hot or cold—and stays constant—and that your workbench ac line voltage remains at the same voltage level during the measurement, the load voltage regulation is expressed as a percentage and calculated (after you have made both measurements required) by using the formula

$$\text{load voltage regulation} = \frac{\text{no load voltage measurement} - \text{full load voltage measurement}}{\text{no load voltage measurement}} \times 100.$$

If you try several different voltage values for no-load and full-load, you will quickly find out that the *lower* the load voltage regulation values are, the better the dc power supply stabilization is. For example, suppose that you have a voltage-regulated supply that is providing a full-load current of 7 amperes to a breadboarded power amplifier circuit. With the amplifier circuit disconnected (no load), the dc power supply output voltage is 38 volts (no load voltage). With the 7-ampere amplifier load connected to the power supply output terminals, the voltage drops to 37.9 volts (full load voltage). The load voltage regulation of your power supply is

$$\frac{38.0 - 37.9}{38.0} \times 100 = \text{approx } 0.26\%.$$

Although this is just an example problem, you should be

aware of the heat created at current levels such as 7 amperes. In general, you are fairly safe using the heat sink methods described in the previous pages (up to current levels of about 5 amperes). But, at current levels such as 7 amperes and above, as was mentioned in Chapter Five, it becomes increasingly necessary to use a blower or fan in addition to the heat sinks (see Figure 6-6, Chapter Six, for an example).

If you measure the load-voltage regulation under different load conditions (more like real world conditions), you will find that the load-voltage regulation value for a basic dc supply will not remain constant. For instance, if you measure the voltage output at 25, 50, 75, and 100 percent of full load, you will find that the relative output (full load to no load) will not change in a linear manner. In fact, as you increase the load, the change in voltage readings will increase. The full-load voltage will decrease as the load is increased. Or, to put it another way, your calculated load voltage regulation will show the power supply to have a higher (worse stabilization) load-voltage regulation.

As has been pointed out, the purpose for the circuits shown in Figures 7-2, 7-3, and 7-4 is to keep the output voltage at a fixed level under all load conditions. You should also note the fact that all of these voltage regulator circuits require a dc input. Incidentally, almost all voltage regulator circuits call for a dc input that is at least 2.5 volts higher than the rated output voltage.

Because of the wide use of computers, the need for high current regulator circuits has increased considerably in the past few years. Figure 7-6 shows a 5-volt dc, 5-ampere regulator, using a MC1469 voltage regulator suggested by Motorola.

Referring to Figure 5-11 (Chapter Five), you will see that this schematic shows only one dc input to the circuit that is based on the MC1469 we are again discussing. However, the circuit shown in Figure 7-6 includes two input voltages (V_{in1} and V_{in2}). The V_{in2} supply is an auxiliary power source used to power the IC regulator, and the heavy load current is obtained from this second low voltage power supply. Although you can connect the collector lead of the pass transistor to pin 3 of the IC, as shown in Figure 5-11 (use a single 9.5 V supply, in this case), using the two supplies will save you about 17½ watts. This circuit can supply current up to 5.0 ampere to loads such as

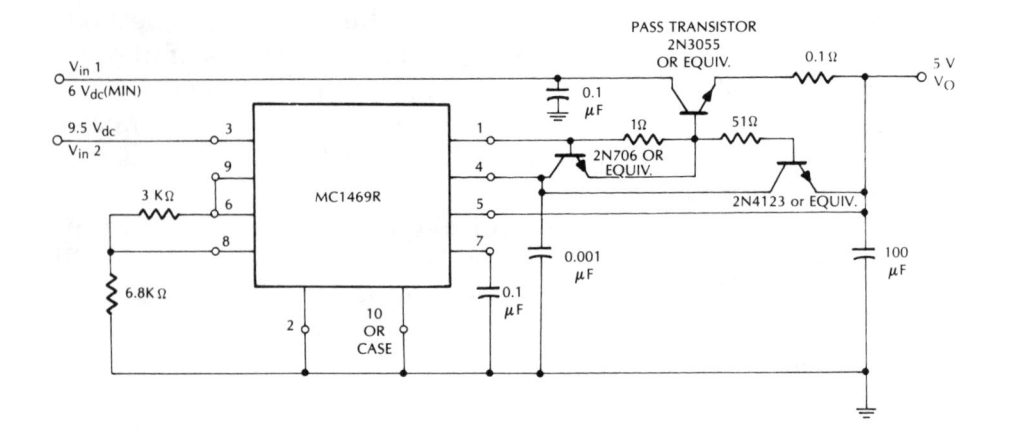

Figure 7-6: Five-volt, 5-ampere regulator based on Motorola's positive voltage regulator IC.

computers, etc., using an input voltage at V_{in1} (pass transistor collector voltage) of 6.0 volts minimum. Furthermore, the pass transistor is limited to 5.0 ampere by the added short circuit current network in its emitter.

To go one step farther, Figure 7-7 is a schematic of how two IC regulators can be wired to form a ±15 V, ±400 mA complementary voltage regulator. For systems such as are shown, you must use a positive and a negative regulator. For example, Motorola's MC1469R (positive) and MC1463R (negative) are two that are available.

The wiring diagram shown in Figure 7-7 is actually what is called a *complementary rocking voltage regulator*. This means that both supplies arrive at the same voltage level simultaneously. As an example, when $+V_o$ equals zero output, $-V_o$ must equal zero output. The MC1469R (the positive regulator) is placed in the tracking mode by grounding pin 6 (one side of an internal differential amplifier) and connecting pin 5 (the other side of an internal differential amplifier) at the junction of the two k ohm resistors (see Figure 7-7). The differential amplifier controls the *series pass transistor* (2N706), assuring that the voltage at pin 5 will always be zero. This, in turn, means that you should measure $+V_o = -V_o$ when pin 5 equals zero.

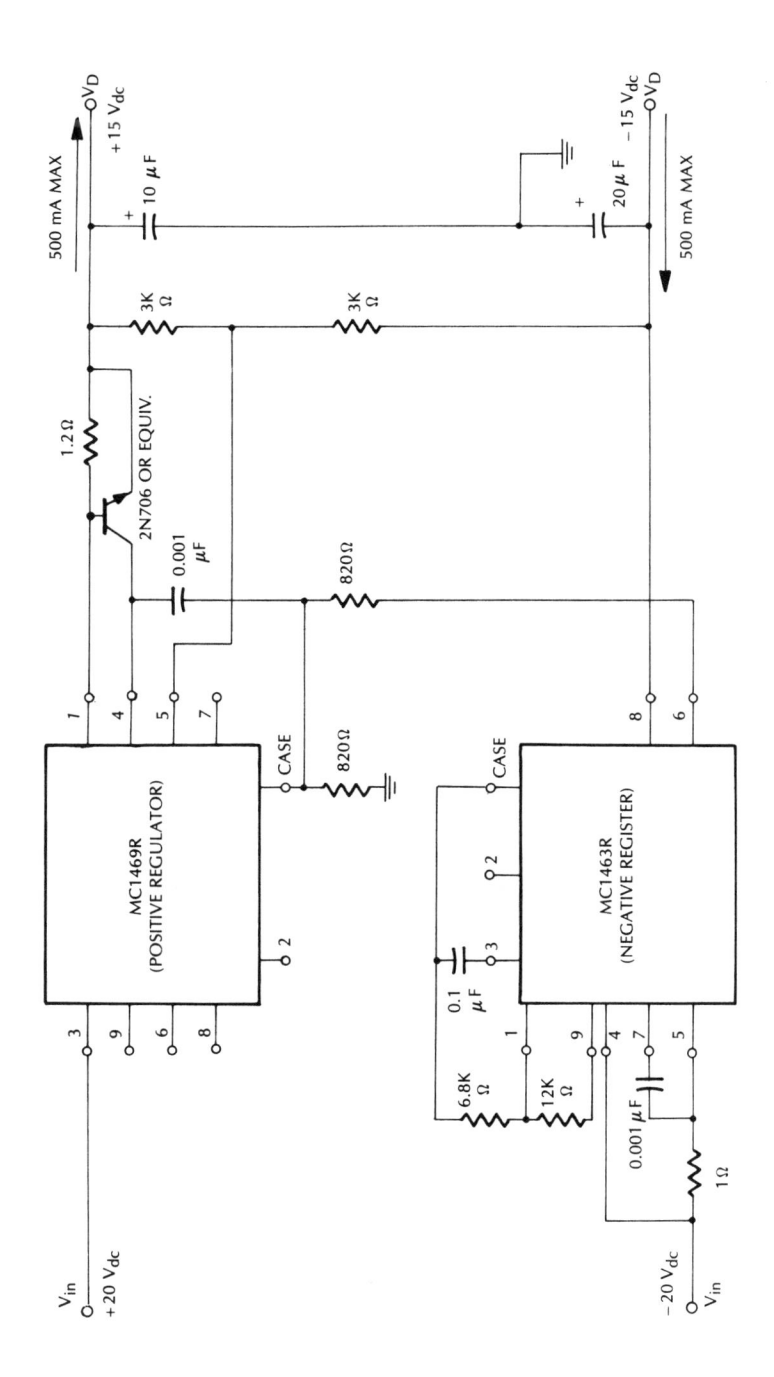

Figure 7-7: Complementary voltage regulator using Motorola's positive and negative regulator IC's, MC1469R and MC1463R.

Line Voltage Regulation

Speaking of a voltage-regulated power supply, line voltage regulation is a measure of the change in output voltage brought about by a specified variation in power line voltage. Let's assume that you want to know what effect power line voltage changes will have on your dc power supply output when the power line voltage drops from 115 (nominal) to 104 Vac, and increases to 125 Vac (your specified range of line voltage variation). The formula is

$$\text{line V reg.} = \frac{\text{output V max} - \text{output V min}}{\text{output V at nom line V}} \times 100.$$

To perform this measurement, you will need some sort of line voltage regulator such as a Variac or a Powerstat. You'll need this item in order to set the line voltage to the desired levels. You will also need some form of load to place on the dc power supply. Usually the load is variable and such that it will drain one-half the full rated current from the dc supply under all conditions (changing line voltage).

Now, using a nominal line voltage of 115 Vac, and saying that you measure 0.25 output voltage change from the lowest setting (104 Vac) to the highest setting of the line voltage regulator (125 Vac) with respect to 25.0 volts output, and your current drain is 50% of the rated maximum, your answer is

$$\text{line V reg (\%)} = 0.25/25.0 \times 100 = 1\%$$

or, ±½% from the nominal dc voltage (25 Vdc output). It should be pointed out that in this example we have been using constant 50% maximum rated current drain from the dc power supply. This is not a best-case measurement (no-load would be best case), nor is it worst case. Full load rated current drain is the worst case. In other words, your power supply may have considerably worse line voltage regulation when measured at the full-rated current drain. Or, on the other hand, much better at zero load. Furthermore, if the line voltage is varied to extremes—much less than 90 Vadc or much greater than 130 Vadc—it's very possible that you will damage your dc power supply.

Load Current Control

As you know, many IC voltage regulators include some form of automatic current control, which is very desirable from the viewpoint of the experimenter. We have shown that Motorola's MC1469R regulator IC can be wired in such a manner that it will, by choosing the correct values when designing the circuit (for example, the value of resistor R_{SC} shown in Figure 5-11), limit the current flow to a certain value.

The current-limiting circuits in regulators such as this one (see Figure 5-11) control load current by sensing any increase in voltage drop across R_{SC}. The wire connecting the right-hand side of resistor R_{ss} to the emitter of transistor Q_2 is the sense line. What happens is that the transistor is cut off any time the load current exceeds the predetermined value, which in turn develops a voltage drop across the resistor and causes the bias of the transistor to reach cutoff value (in this case, the circuit is called an *error amplifier*).

If you use a supply that delivers 5 amperes or more to the load, it is considered good building practice to use conductors, just as short and large as practical, between the dc supply and the load. The reason for this statement is that you want to avoid the I_R heat loss that would result from long and small wires at such high currents.

Using Protection Techniques

We have just dealt with one important protection technique—the current-limiting circuit—that has the prime responsibility of protecting the series pass transistor (transistor Q_1 in Figure 5-11). All series pass transistors have to pass the total load current of that dc supply and, of course, can get very hot internally. This means that if you choose a marginal transistor for the job, it is very likely to short out prematurely and place the full output voltage of the bridge rectifier (or whatever type of rectifier you are using) on the main power bus of your breadboard, computer, etc. The scheme shown in Figure 7-8 can protect your equipment from this type of disaster. The circuit

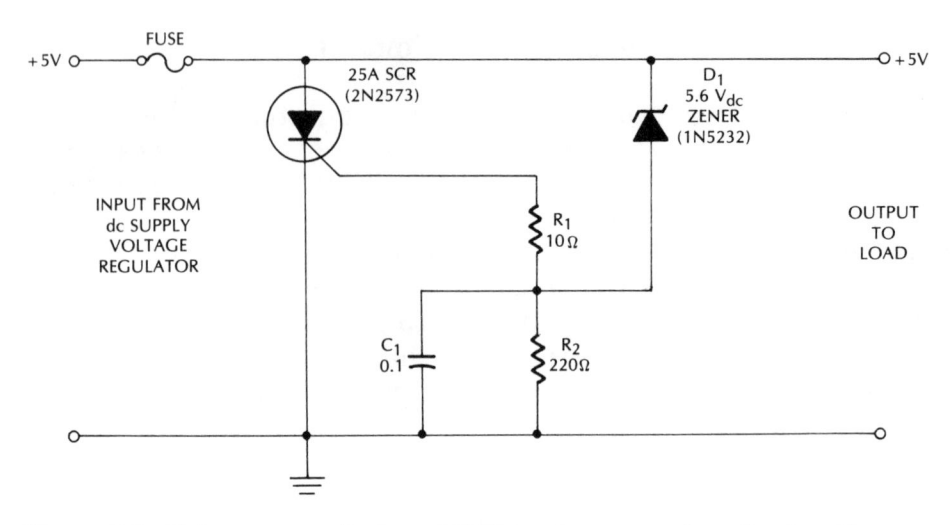

Figure 7-8: Using a zener diode and SCR to place a crowbar short on the circuit, to blow a fuse in case of overvoltage.

uses a high current SCR, and a fuse, in the event of excessive load current in your expensive equipment, is deliberately caused to open. This brute force approach is greatly speeded up by the *crowbar* action of the SCR.

Diode D_1 is a Motorola 5.6 volt zener (rated at 1 watt), and will not pass current until the output voltage reaches this level. Therefore, as long as the output voltage remains where you want it (at 5 Vdc), the fuse will not blow. But if the series pass transistor in your dc regulated supply shorts and the supply voltage shoots up (that is, up to 6.5 Vdc), then D_1 will conduct. This, in turn, permits current to pass to the gate of the 2N2573, causing the SCR to appear as a short circuit between the +5 Vdc line and ground and, of course, blows the fuse. This may seem like overprotection but it could save you a lot of components (and much money), especially when working with experimental regulators tied to many dollars worth of IC's.

Switchmode Regulators

The variable resistor action of a series pass transistor discussed earlier is widely used today to regulate the output voltage of a dc supply. But another popular way to control or stabilize output current or voltage is to use a *switching device*, in place of the series pass element, to vary dc output level.

Because the power supplies and regulators we have been discussing (linear types using discrete components) are simple devices to build, many experimenters have used them in their voltage regulator designs. However, because of certain advantages over the linear circuits, switchmode power supplies have become very popular. Some of the advantages are as follows:

1. The series pass type regulator must have the input voltage higher than the output but, by the use of simple jumper wires, the output voltage in a switching regulator can be stepped up, stepped down, or inverted to the opposite polarity.
2. Perhaps the most important advantage: all switching regulators have high efficiency for all input and output conditions (up to 90%), and are excellent to use where size is a factor.

Nevertheless, there are disadvantages to go along with the advantages. For example, the following list includes some of the main characteristics of switching type regulators and contrasts their performance with that of the linear units:

1. In general, switchmode regulators' response time to rapid changes in load current is poor compared with the series pass units we have discussed in the previous pages.
2. Unless you use input filters, switchmode regulators can generate electrical noise and electromagnetic radiation.
3. The advantages of the switchmode regulators are
   a. They can be built small and lightweight.
   b. Often, thermal considerations are of little importance.
   c. They can serve the unique function of a dc-to-dc transformer.
   d. They can be driven with poorly filtered dc.

This all translates into power supplies with small transformers, minimum cooling, and inexpensive operating cost (low power consumption and high efficiency).

Basic Switchmode Configurations

Unlike linear regulators, switchmode regulators are based on using transistors in a nonlinear fashion. The unregulated input voltage is "chopped" with a saturated transistor and this energy is stored in an inductor and capacitor. This stored energy is then supplied to the load as needed. Figure 7-9 shows a switchmode step-down regulator (more properly called a *pulse-width modulated step-down converter*).

Regulators of this type (switching) provide switching after rectification, and therefore are sometimes called *dc switch regulators*. In general, the dc load current is interrupted at a fixed frequency rate determined by the value of an external resistor, R_T, (pin 6), and an external capacitor, C_T, (pin 5). See Figure 7-9.

The internal linear oscillator frequency is determined by

$$f_{osc} = 1.1/(R_T C_T).$$

Using the values given for R_T and C_T in Figure 7-9, the frequency of the internal sawtooth oscillator works out to be approximately 23.4 kHz.

In the circuit shown, the control lines, pin 4 dead time control (D.T.), pin 13 output control (O.C.), pins 9 and 10 (emitters of two internal output transistors used in the process of pulse-width modulation), are all tied to ground (pin 7).

Now, let's examine switching regulators such as this one (and other positive regulators) in the most general case. As you have seen, the dc load current is chopped at an audio frequency rate and we can now say that it is controlled by variation of duty cycle. A subsequent inductor is required (see Figure 7-9, 1.0 mH @ 2A inductor) to average the dc level. It also helps remove the switching frequency (about 23.4 kHz, in our example) and its harmonics. The basic diagram in Figure 7-10 illustrates the fundamental concept of dc switching regulators.

Figure 7-11 shows two other possible arrangements of a switching regulator: a step-up, where the output voltage is greater than the input voltage, and an inverting type, which will produce a negative output for a positive input.

If one of the points is more positive than the other, it is easy to see (Figure 7-11) that load current must be flowing from

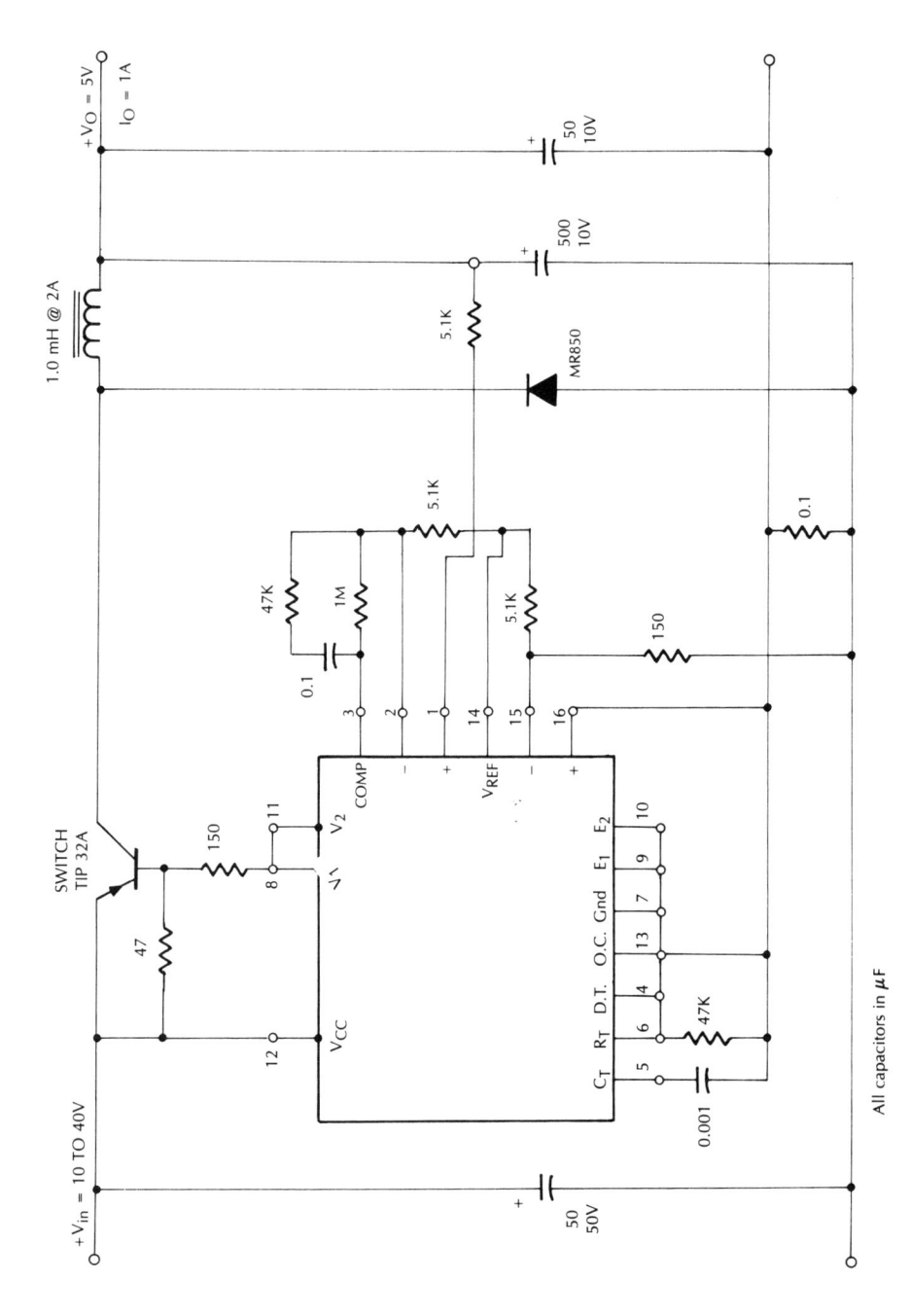

Figure 7-9: Pulse-width modulated step-down converter.

196 MODERN REGULATED SOLID STATE POWER SUPPLIES

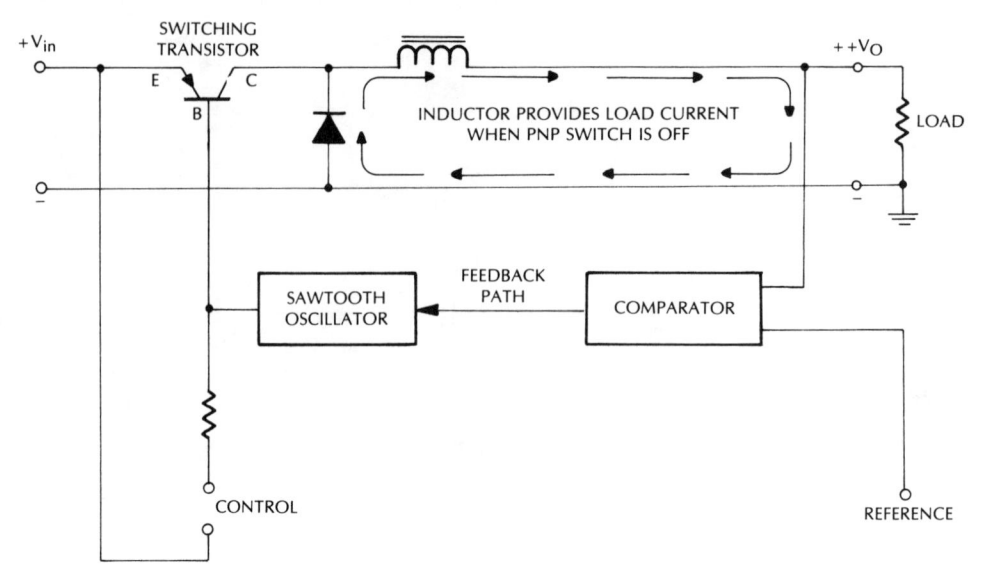

Figure 7-10: Basic functional diagram of a step-down dc switching regulator.

the left-hand side of the diagram ($+V_{in}$) to the right-hand side ($++V_o$). On the other hand, the electron flow in the inverter is in the opposite direction, from $-V_o$ to $+V_{in}$, and down through the inductor because, in this case, ground is more positive than the regulator output ($-V_o$).

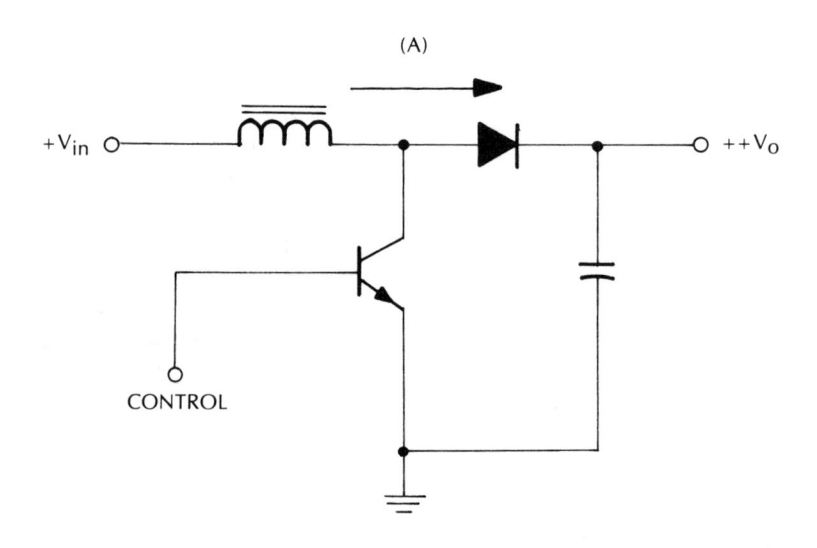

(A)

(B)

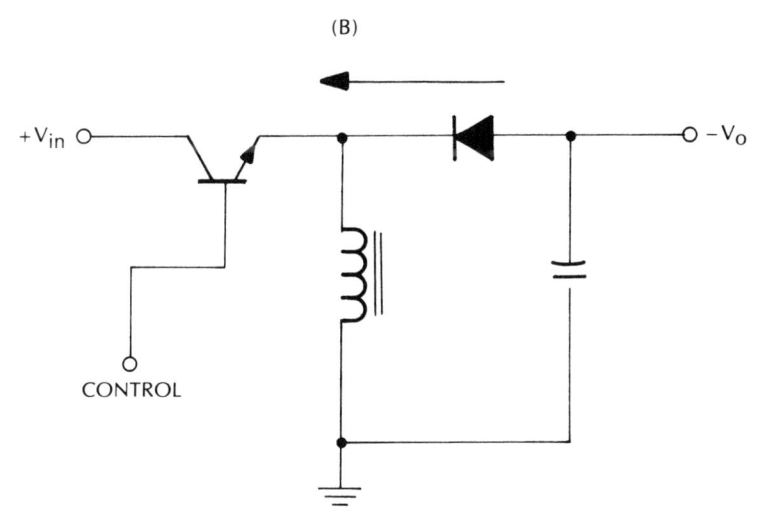

+V_in ○ ——   ○ -V_o

CONTROL

Figure 7-11: Basic diagram of a step-up and inverting type switchmode regulator. (A) step-up, (B) inverting.

Currently, there are various monolithic switchmode regulator IC's on the market. However, these three—step-down, step-up, and inverting types—are the basic regulators that you will most often come into contact with.

Practical Guide to Troubleshooting Solid State Power Supply Circuits

The dc power supply that feeds the regulator is usually a fairly simple circuit and can be tested with only a few pieces of test equipment. A good ac/dc voltmeter is essential. A dc ammeter is also very handy, and, at times, a wattmeter can be helpful. One other piece of equipment, the oscilloscope, is needed to check control pulses and waveshapes found in modern power supplies.

By far the most common rectifier circuit used in today's dc power supplies is the solid state full-wave bridge. We have discussed the need for a good power supply in Chapter Five; however, let us now analyze the various troubles you may encounter while using these devices. Figure 7-12 is the working schematic of the three basic solid state rectifier circuits. Note the simplicity of the stack ac-to-dc rectifiers which convert the ac to pulsating dc.

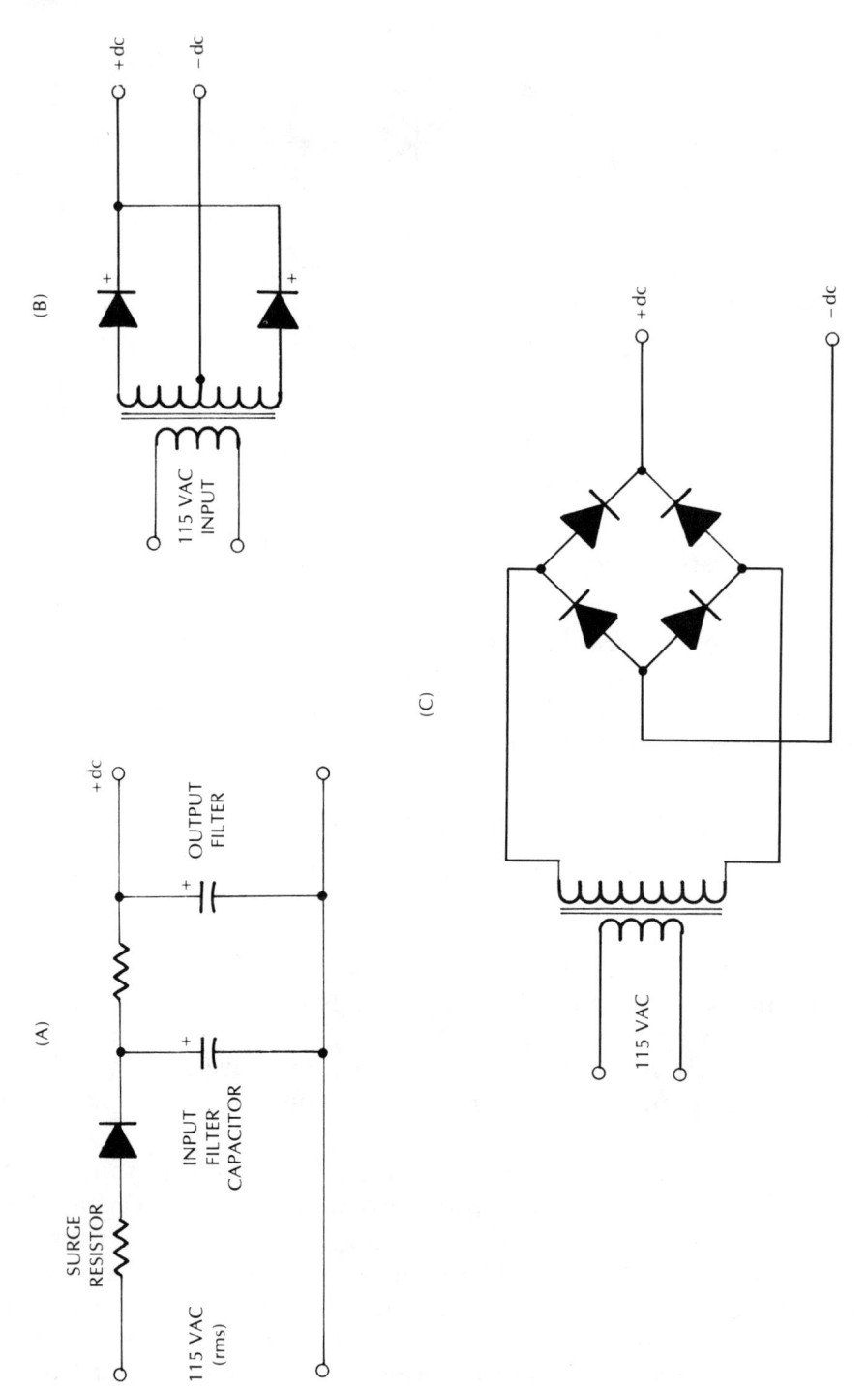

Figure 7-12: (A) shows a half-wave rectifier, including filter; (B) is a full-wave and (C) is a full-wave bridge.

The basic function of a rectifier circuit is to convert alternating current into a pulsating direct current. In the case of a dc-to-dc converter, which was discussed earlier, direct current is converted to direct current at a different voltage (usually higher, see Figure 7-11). In any case, you can check a power supply function very simply by measuring the output voltage. Nevertheless, you can get into trouble if you are not aware of some of the pitfalls. For example, what voltage should you measure on the output of the half-wave rectifier circuit shown in Figure 7-12 (A)?

Our ac input is shown as 115 Vac *rms*. This is what our ac voltmeter will read because it is calibrated in rms. Now, when the ac is rectified it comes out as a series of pulses, but the input filter capacitor will charge up to the *peak value*. A small amount of arithmetic will show this value:

peak value = 1.414 × rms = 1.414 × 115 = 162.6V.

We would see something near 160 volts average dc if there were no load on the circuit. But place a load on the output, and you'll measure something less; perhaps close to 135 Vdc or less. Incidentally, if you need a negative dc output voltage, simply reverse the diode and filter capacitors.

The next circuit, the full-wave rectifier, usually is used with a power transformer, as shown in Figure 7-12 (B). *Note:* The diode cathodes are tied together and provide the +dc output. The transformer center tap is the −dc lead and usually is grounded, as shown. In general, the voltage output of the full-wave rectifier follows the same rules as the half-wave rectifier in that, in most applications, the average dc output voltage is higher than the rms input line voltage. Of course, this is only true provided a step-up transformer is used in the circuit—which was the case in many older pieces of equipment such as TV sets, etc.

It's all but impossible to speak of modern solid state regulated dc power supplies without pointing out that the bridge rectifier circuit shown in Figure 7-12 (C) is, by far, the rectifier most widely used today. Although this circuit requires four diodes, using this hookup will still save you money. First of all, the transformer secondary does not have to be center tapped. The common cathode connection is, as you can see, the +dc terminal and the common anode is the −dc connection. Another convenient feature of this circuit is that, by grounding the appropriate terminal of the bridge, you can develop a dc voltage of whichever polarity you want.

In a full-wave bridge rectifier circuit with capacitance filtering on the output (or a half-wave and standard full-wave, for that matter), if you measure a low output voltage, it's possible that the trouble is an open or leaky filter capacitor. The way most technicians check one of the capacitors is to bridge another capacitor across the one under suspicion. If the dc voltage returns to normal, the problem is over. A word of warning: *always turn power off before bridging one capacitor with another.* If you bridge a large electrolytic capacitor across another in-circuit capacitor, it may cause a large surge of current that will, in turn, damage other solid state components, or perhaps a very expensive computer circuit.

If bridging a capacitor does not work, check the rectifier diodes. A bad diode will also produce a low output voltage. If you have a scope, check the pard (ripple voltage) because an open diode will cause a large increase in ripple voltage. A shorted diode will cause a pronounced drop in output voltage (to about half normal), and will probably blow a fuse or trip a circuit breaker.

Sometimes bridging a capacitor with a known good one will not remove all ripple voltage. In this case, more than likely the capacitor you are checking is not open or shorted, but is leaky. This is not true for all capacitors found in power supplies (for example, input capacitors. See Figure 7-12 A). If you place scope leads across this capacitor, in a modern solid state bridge rectifier such as those found preceding **IC** regulators, you'll probably see about 10 volts peak to peak.

If you are selecting replacement electrolytics, the voltage rating of the capacitor should be *at least* 1.3 times the dc output of the rectifier circuit. As a rule, when in doubt about the value of an electrolytic capacitor, use a larger value than you think you need.

When the output voltage goes down every time you attach a load to the supply, or the equipment does not seem to draw enough current, a wattmeter will give you much information. For example, let's assume you are checking a piece of electronic equipment that has a rating plate on the back of the chassis, and this lists the rated power as 200 watts. If your

MODERN REGULATED SOLID STATE POWER SUPPLIES

SYMPTOMS	PROBABLE CAUSE
No ac current.	Ac input voltage. Defective transformer. Open circuit breaker or blown fuse.
Rectifier circuit draws less current than normal (no load).	Open diode. Open resistor. Partially open transformer winding.
Rectifier circuit draws more current than normal (no load).	Shorted diode. Leaky capacitor. Partially shorted transformer winding.
Low output dc voltage (no load).	Open input filter capacitor.

Table 7-1: Troubleshooting chart for dc power supply rectifier circuits.

wattmeter reads 100 watts, it means that there is an open or a weak circuit in the equipment. On the other hand, a reading of 300 watts means there is a partial short (something leaking current). About 450 watts means you have a short and the circuit breaker should have tripped (or a fuse should have blown). Table 7-1 is a troubleshooting chart for various symptoms and their probable causes.

Today you'll find that working with modern dc power supplies will require more use of an oscilloscope than has been true in the past. For instance, a scope is a *must* when you are checking the control pulses used in the switching supplies discussed in previous pages. These pulses must have the correct waveform and go to the correct place in the circuit (see Figure 7-9).

Another point to remember is that, since the operation of solid state components can be seriously affected by small voltage changes, it is very important that all voltages to your breadboard be well regulated and ripple free. *When in doubt, check with your scope!* In summary, first check the ac line voltage. Next, check the dc output voltage and then measure the current drain. Take it one step at a time, refer to Table 7-1, and your troubleshooting will be much easier. Remember, remove the load from the supply and then check, stage by stage, the rectifier circuit and then the regulator.

Voltage Regulator for Photovoltaic Power Systems

Many experimenters like to work with photovoltaic cells such as the inexpensive silicon solar cells available on the surplus market. But, as you may know, they do not produce enough power to charge lead-acid or nickel cadmium batteries used in lighting systems. And they are difficult to work with, especially when it comes to soldering their connecting leads.

If you are serious about providing a high efficiency lighting system for your home, operating a refrigerator, television or water pumping system, perhaps Motorola's MSP23E20 (a 20-watt photovoltaic module), or MSP43E40 (a 40-watt photovoltaic module) will fill the bill as a basic module for your solar system. Both the 20-watt and the 40-watt units contain 33 high quality silicon solar cells to provide 12 volts battery charging. As with any solar cell, large-area silicon cells work best. In one of our examples, the 40 watt module, the cells are large 4" × 4" squares.

When you connect the output of your photovoltaic to batteries for energy storage, it would seem that that is all there is required. After all, energy from the sun is free, so who cares how much current flow the device produces? It's not that easy. You should have a voltage regulator to prevent overcharging your batteries. Basically, this is the same principle as that used in your automobile. One regulator specifically designed to do this job is the MSR12, a shunting device for use with a 12-volt photovoltaic power system. Figure 7-13 is a simplified solar battery charger designed around the MSR12.

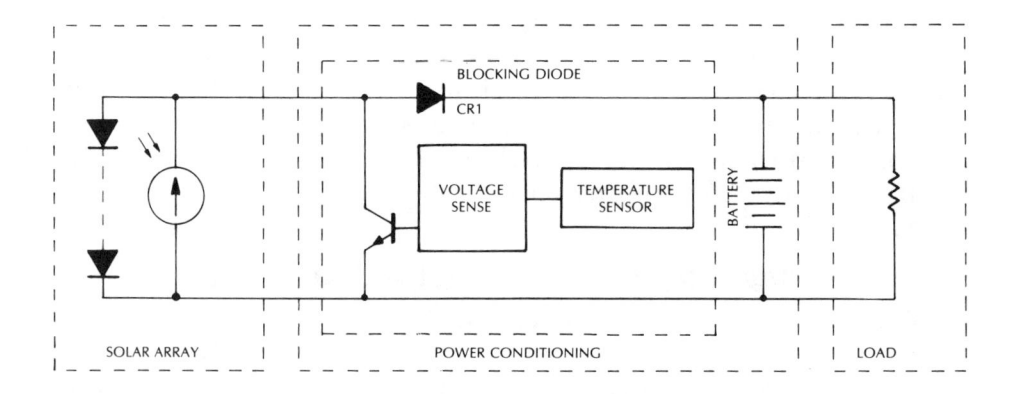

Figure 7-13: Simplified solar battery charger using a voltage regulator.

MODERN REGULATED SOLID STATE POWER SUPPLIES

Referring to Figure 7-13, you will notice a temperature sensor in the right-hand section of the block diagram of the power-conditioning circuit (the regulator). This self-contained temperature sensor regulates battery voltage based on battery temperature. Therefore, the regulator is intended to be mounted in the battery storage area very close to the batteries themselves. Figure 7-14 is a graph of the regulator output versus sensor temperature.

The blocking diode (see Figure 7-13) is incorporated in the regulator to prevent battery discharge through the solar array during the periods of no sunshine. The transistor, Q_1, short-circuits the solar array when the battery reaches full charge. The reason for this is that the solar array will then dissipate its full output power and the user does not have to install a massive shift regulator. It should be pointed out that these photovoltaic arrays can be operated in a "short-circuit" condition for an indefinite period, without damage to the unit. In case you are wondering, the battery is protected from discharge during the short-circuit periods by the blocking diode CR_1. Figure 7-15 shows how the system is interconnected. (A) is a single voltage regulator and (B) is a dual regulator and array system.

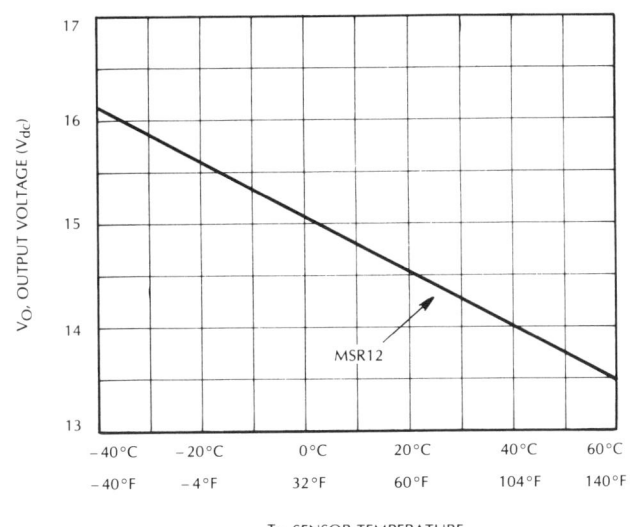

Figure 7-14: Solar system voltage regulator voltage output versus an internal temperature sensor (MSR12 voltage regulator).

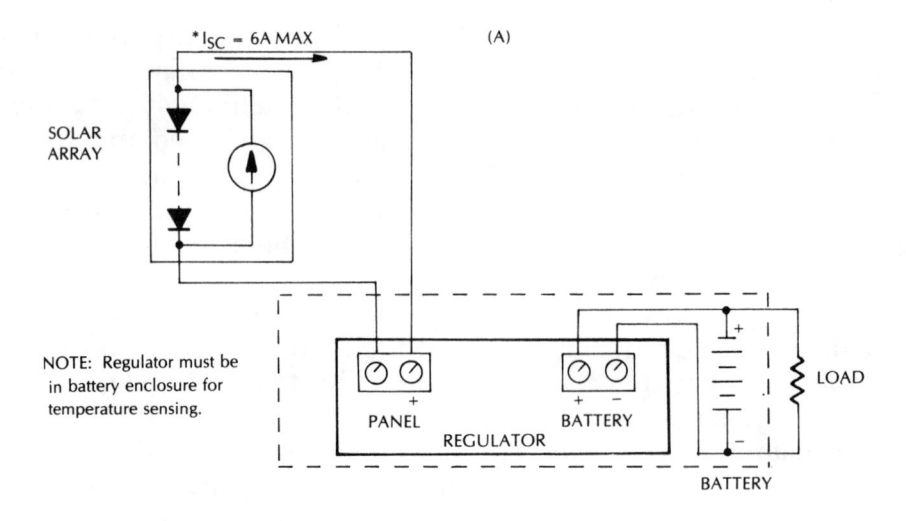

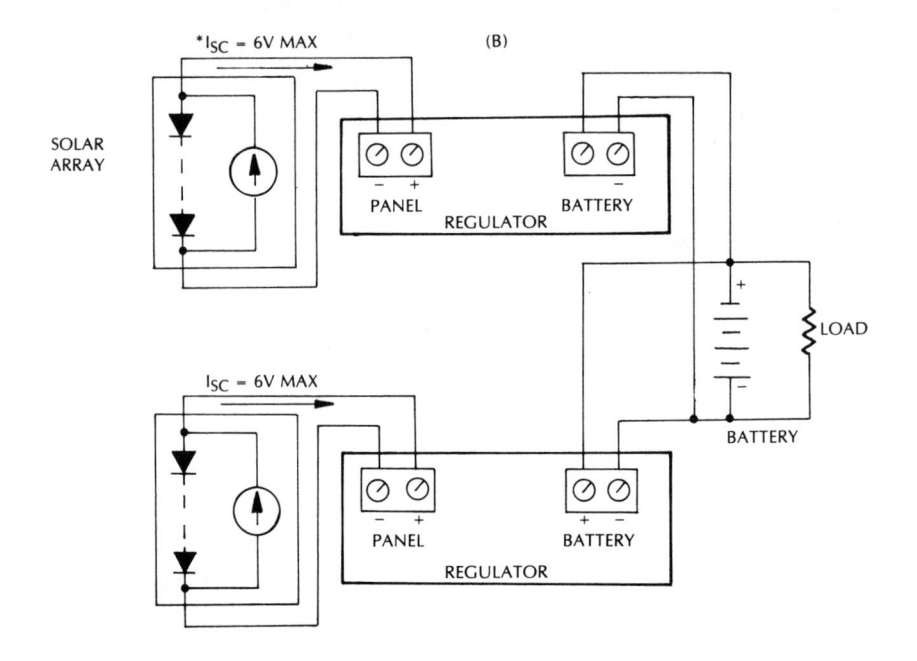

Figure 7-15: Systems interconnected using a solar array, regulators, and battery.

CHAPTER EIGHT

How to Interface to Contemporary IC Applications

You cannot just connect any **IC** to any circuit and expect the system to operate properly. In fact, in a practical sense, microprocessors and microcomputer chips are useless without some provisions for properly interfacing them with the outside world. As was explained in Chapters Three and Four, there must be provisions for getting logic data into and out of them in some meaningful fashion that is compatible with the manufacturer's specifications for the **IC**.

Then, too, there is a growing need nowadays for interfacing an ever-increasing number and variety of low-cost decimal and hexadecimal keypads and other mechanical devices with digital circuits. Also, the experimenter must take special care when attempting to work IC's of different logic families (**TTL**, **CMOS**, etc.) into the same digital system.

In this chapter, you'll find practical methods that will be very helpful to you when working with input/output voltage levels, input/output current levels, plus input/output stages such as debouncing circuits, various decoding systems, and analog-to-digital conversion techniques.

Analog-to-Digital / Digital-to-Analog Conversion Techniques

One necessary stage in any digital system that processes information originating in analog form is the analog-to-digital (A/D) converter. It is used to convert analog electrical current or voltage levels to representative digital words in digital subsystems. As one would imagine, digital-to-analog conversion is the generation of analog current or voltage levels in response to digital words.

Take, for example, a 3½-digit digital voltmeter (DVM). The instrument's input section is analog (inputs for volts, milliamperes, ohms, etc.), but the output (decoder/driver and display section) is digital. The A/D converter fills the gap between these two subsystems.

Several companies manufacture a set of IC's that can be used to construct an analog-to-digital converter subsystem. One such system uses Motorola's MC1505 and MC14435 IC's. The MC1505 uses the proven dual-ramp A/D conversion technique. The MC14435 is the digital logic section of the converter and is used to produce the complete 3½-digit DVM function. A functional diagram with pin connections using these two IC's is shown in Figure 8-1.

Due to its simplicity, the dual-ramp (also called *dual-slope*) conversion technique is very popular. Referring to Figure 8-1, the MC1505 (the dual-ramp converter) is an integrating A/D converter in which the analog signal is converted to a proportional time interval, which is then measured digitally. In this system, an integrator (pins 6 and 7) is used as an input to a comparator (output at pin 9). There are, also, an on-chip voltage reference, a pair of voltage/current converters, a current switch, and associated control and calibration circuitry. Only two capacitors and two calibration potentiometers, zero calibration (pins 4 and 5), and full-scale calibration (pin 11 and ground), plus V_{CC} (+16.5 Vdc *maximum rating* for the MC1505) and ground, are required for breadboarding this pair of IC's. The *maximum* V_{CC} rating for the MC14435 is +18 Vdc. See Figure 8-2 for the pin connections and functional block diagram of this IC.

The A/D converter IC is also used to convert analog currents and voltages to represent digital words for use in computers and microprocessors. In fact, now that MCU's and MPU's

INTERFACING TO CONTEMPORARY IC APPLICATIONS

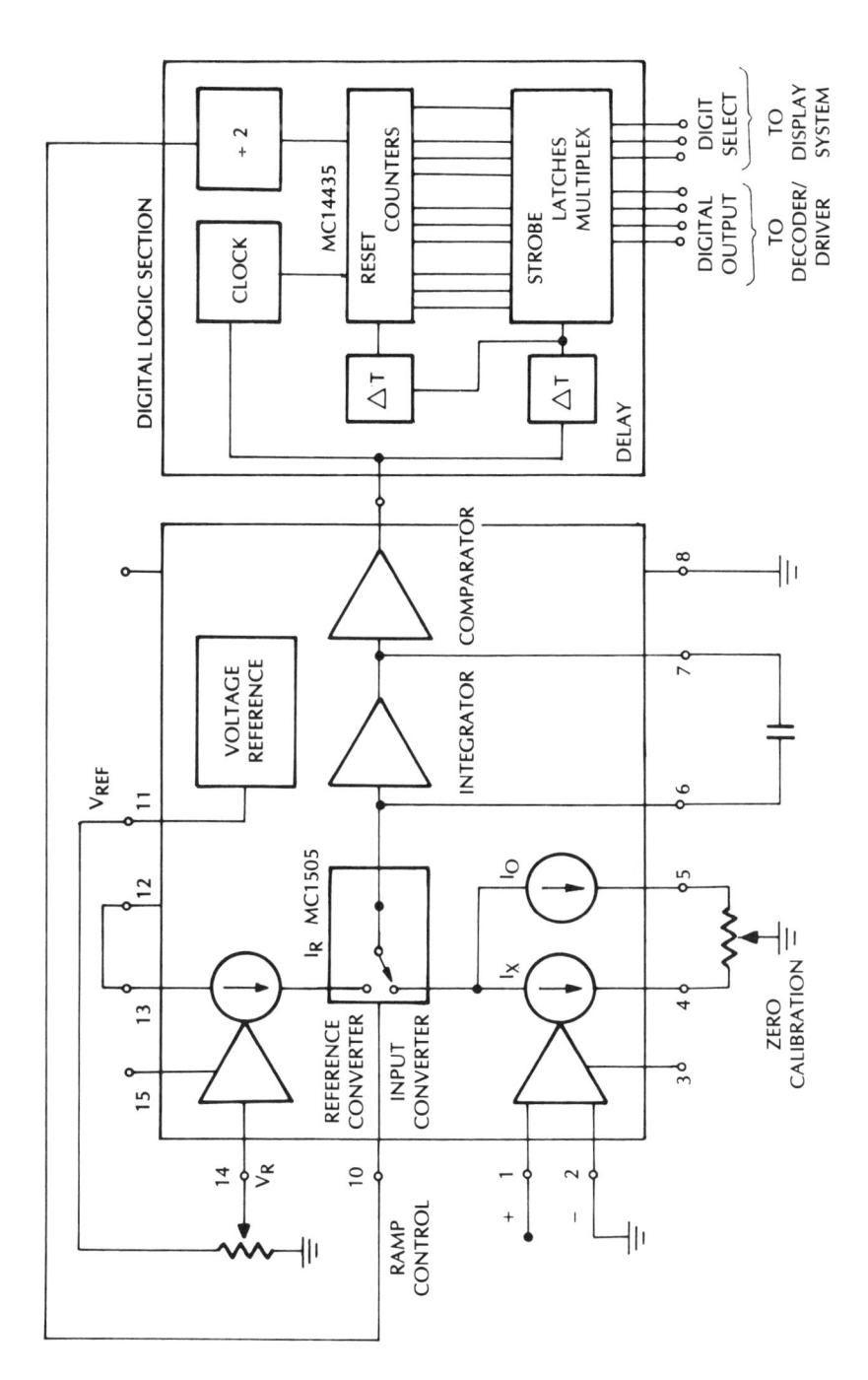

Figure 8-1: Analog-to-digital converter subsystem functional diagram and pin connections, using MC1505 and MC14435 IC's (see Figure 8-2 for MC14435 pin connections).

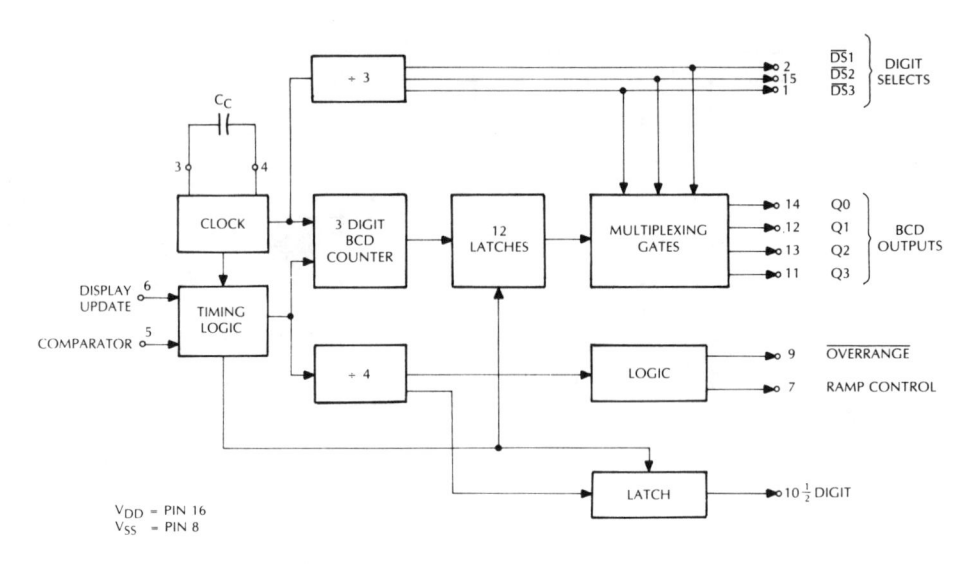

Figure 8-2: Block diagram and pin numbers for the MC14435, A 3½ digit A/D logic subsystem (see Figure 8-1 for the complete converter subsystem).

are so widely used, A/D converters have become even more popular than they were when most were used in control systems, instruments, and the like. But these A/D converters are not of any use until they're interfaced to a MPU or MCU.

There are several approaches to this job. First, in some cases (such as in the design of the MC68705R3, an 8-bit EPROM microcomputer unit), the 8-bit A/D converter is implemented on the chip. Up to four external analog inputs (via port D, in this case) are connected to the A/D through a multiplexer. Second, if you have a MCU, it may require only that you plug in an I/O cable to the input slug on the back of your computer. But, if you are breadboarding, you may have to interface directly to the MPU or MCU, and this requires a knowledge of the MPU control signals (see Chapters Three and Four).

As in MPU interfacing jobs, there are two approaches that you can take: the I/O route or the memory-mapped way. Chapter Four explained the I/O based system. For an example, see Figure 4-7 for typical port I/O circuitry. It is possible that

INTERFACING TO CONTEMPORARY IC APPLICATIONS 209

you may have to interface with the bus lines inside your computer, or directly to the MPU, as has been explained. In any event, you will have to provide some means to identify the control signals and decode the I/O port addresses.

In the memory-mapped systems, the output of the A/D converter is seen as another location in memory. Or, to put it another way, the term "memory-mapping" is telling you that the CPU will see the A/D converter as a memory location.

If you are breadboarding and using a MPU, it will probably be necessary for you to build your own ports. In this case, it will require an address-decoder circuit so that the MPU will know when your program is calling for the A/D converter you are interfacing to the MPU. Figure 8-3 is an example of an inexpensive and simple address-decoder that may be used either as an I/O port address-decoder, or in memory-address decoding applications.

The heart of this address-decoder is a low-cost 8 input

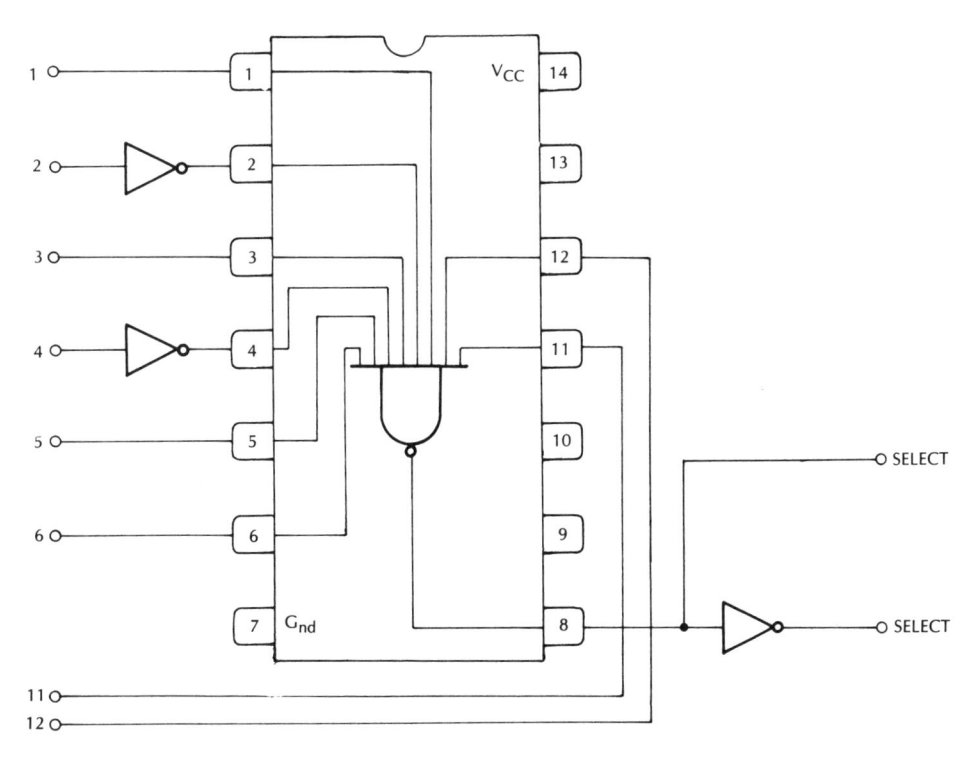

Figure 8-3: Inexpensive and simple address-decoder that can be used in I/O port systems or memory-mapped systems.

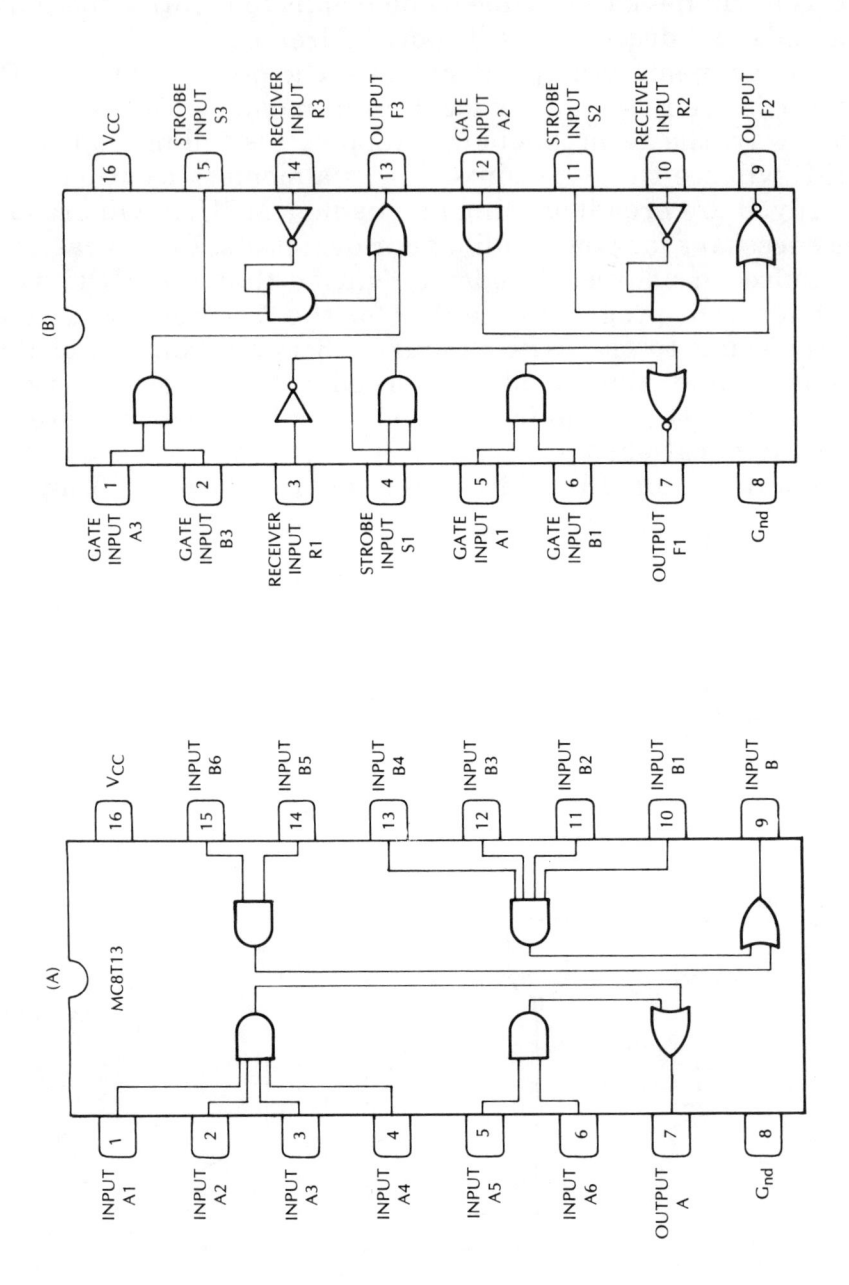

Figure 8-4: General purpose I/O interface IC's. (A) dual-line drivers, (B) triple-line drivers.

positive NAND gate, the 7430. A logic 1 input voltage is required at all eight input terminals (pins 1, 2, 3, 4, 5, 6, 11, 12) to ensure logic 0 at the output (pin 9). A logic 0 input voltage is required at any input terminal to ensure a logic 1 level at the output. What you want, then, is to make those inputs high only when the desired address is present. This will probably require one or more inverters. Any inverter will do (for example, a 7404) as long as it's **TTL** compatible.

What happens is that only with a certain 8-bit binary input signal (selected by you) will the output of this circuit be all highs (11111111) and, in turn, select the A/D converter output as an input to the MPU, etc. The output of the NAND gate is labeled $\overline{\text{Select}}$ (all 0's). When you pass the **IC**'s output signal through the inverter, as shown, it is complemented to form all positive going data bits. One of these signals may be used to tell you that the address is selected correctly.

The circuit just explained is a simple device that you may want to use to select an A/D converter. But there are also *general purpose* I/O interface **IC**'s that are readily available on the market today. For example, Motorola's MC8T13/14 combination is specified for general **TTL** systems applications. Figure 8-4 shows these two **IC**'s: dual-line open emitter drivers (the MC8T13) and triple-line receivers (the MC8T14). Both are general purpose I/O interface devices.

Understanding Phase Shift and Compensation

The basic rules for **IC**'s, in dynamic input impedance measurements, are essentially the same as for the audio circuits discussed in Chapter Four (also see Figure 6-5 for a wiring diagram for a dynamic impedance measurement). However, the impedance presented by any reactance and resistance combination (RC or RL) changes with frequency, thus altering the characteristics of any **IC** (or discrete component, for that matter). To give you some idea of why this is true, Table 8-1 lists the formulas for calculating impedance and phase angle of a reactance and resistance network.

Any inexpensive hand-held calculator that has engineering math capabilities can, in just a few minutes, produce an answer to any problem requiring the use of these formulas. But

IMPEDANCE (Z) AND PHASE ANGLE (∅)		
SERIES CIRCUIT	Z =	∅ =
R–L	$\sqrt{R^2 + X_L^2}$	arc tan X_L/R
R–C	$\sqrt{R^2 + X_C^2}$	arc tan X_C/R
R–L–C (NON-RESONANT) REVERSE X_L AND X_C IF X_C IS LARGER	$\sqrt{R^2 + (X_L - X_C)^2}$	arc tan $\dfrac{X_L - X_C}{R}$
PARALLEL		
R–L	$\dfrac{(R)(X_L)}{\sqrt{R^2 + X_L^2}}$	arc tan R/X_L
R–C	$\dfrac{(R)(X_C)}{\sqrt{R^2 + X_C^2}}$	arc tan R/X_C
CURRENT (SERIES CIRCUIT) I	$\dfrac{E_{applied}}{Z}$	
CURRENT (PARALLEL CIRCUIT) I	$\sqrt{I_R^2 + I_X^2}$	
IMPEDANCE (PARALLEL CIRCUIT) Z	$\dfrac{E_{applied}}{I_{total}}$	

Table 8-1: Formulas for resistance and reactance combinations.

(using these formulas), your answers will be only rough approximations. Proper phase compensation (correcting for proper phase angle, etc.) of such **IC**'s as **OP AMPS** is, at best, a difficult trial-and-error job. Nevertheless, bandwidth, slew rate, output voltage swing, output current, and output power of an **OP AMP** are all interrelated and are dependent on compensation.

Motorola and other manufacturers will usually recom-

INTERFACING TO CONTEMPORARY IC APPLICATIONS

mend one or more methods for compensation when using one of their **OP AMP IC**'s. Many **OP AMPS** have terminals provided for connection of external compensation components to the internal circuits. For example, the LM301 is one type of general purpose **OP AMP** that has external connections that allow the user to choose the compensation capacitor best suited to his needs. Figure 8-5 shows a circuit called a *Standard Compensation and Offset Balancing Circuit*.

In summary, the bandwidth, slew rate, output voltage swing, output current, and output power of an **OP AMP** are all interrelated. These characteristics are frequency-dependent and depend upon *compensation*. Some **OP AMPS** are internally compensated and others are designed so the user can compensate them, depending upon his needs. There are several

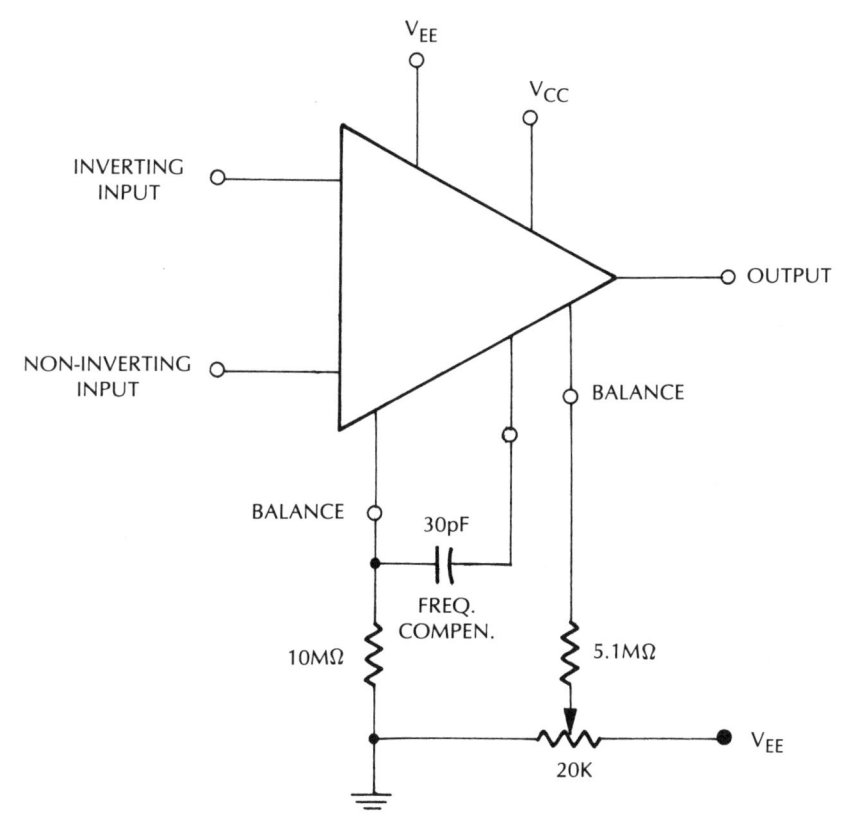

Figure 8-5: Standard compensation offset and balancing circuit recommended by Motorola for their LM3021 and similar OP AMPS.

methods of compensating an **OP AMP,** but whichever method is used, it requires a knowledge of the **OP AMP** circuit characteristics and it is usually best to use one of the manufacturer's recommended methods. This will be found on most data sheets.

Output Interfacing

While all input interface circuits must be designed around the **IC**'s input voltage and current specs, output interface circuits must take into consideration the **IC**'s output voltage and current loading (for example, fan-out rating of a digital **IC** and recommended speaker impedance in the case of an audio power amp). *Note:* In case you don't remember, fan-out refers to the number of parallel loads within a certain logic family (**TTL, CMOS, MOS,** etc.) that can be driven from one output of a digital **IC**.

More and more digital techniques are finding applications in experimenters' circuits, and one of the most popular output devices for digital circuits is a light-emitting diode (LED). Both the single LED and 7- or 8-segment LED numeric display assembly are in wide use (Chapter Two includes a description of these optoelectric devices and their operation). Figure 8-6 shows two discrete LED indicator circuits you can use; however, all 7- or 8-segment displays need some form of decoder to operate properly, as you will remember.

While the experimenter can use the interface-to-LED circuits shown in Figure 8-6, and they are adequate for breadboarding one-of-a-kind circuits, if he is interested in interfacing several LED displays to a system it's better to use LED driver **IC**'s. For example, the MC75491 (a quad LED segment driver) and the MC75492 (a hex LED digit driver) are designed to interface **MOS** logic to common cathode LED readouts in serially addressed multidigit displays. **IC**'s such as these two are called *multiple light-emitting diode (LED) drivers*. The MC75491 and MC75492 are shown in Figure 8-7.

A typical use for these **IC**'s is for interfacing an **MOS** calculator **IC** (chip) to a standard 7-segment multidigit display. Before we look at a demonstration circuit that takes advantage of these multiple LED drivers, let's first review a standard 7-segment layout and associated decimal display.

IC's (such as a calculator chip) used to drive these dis-

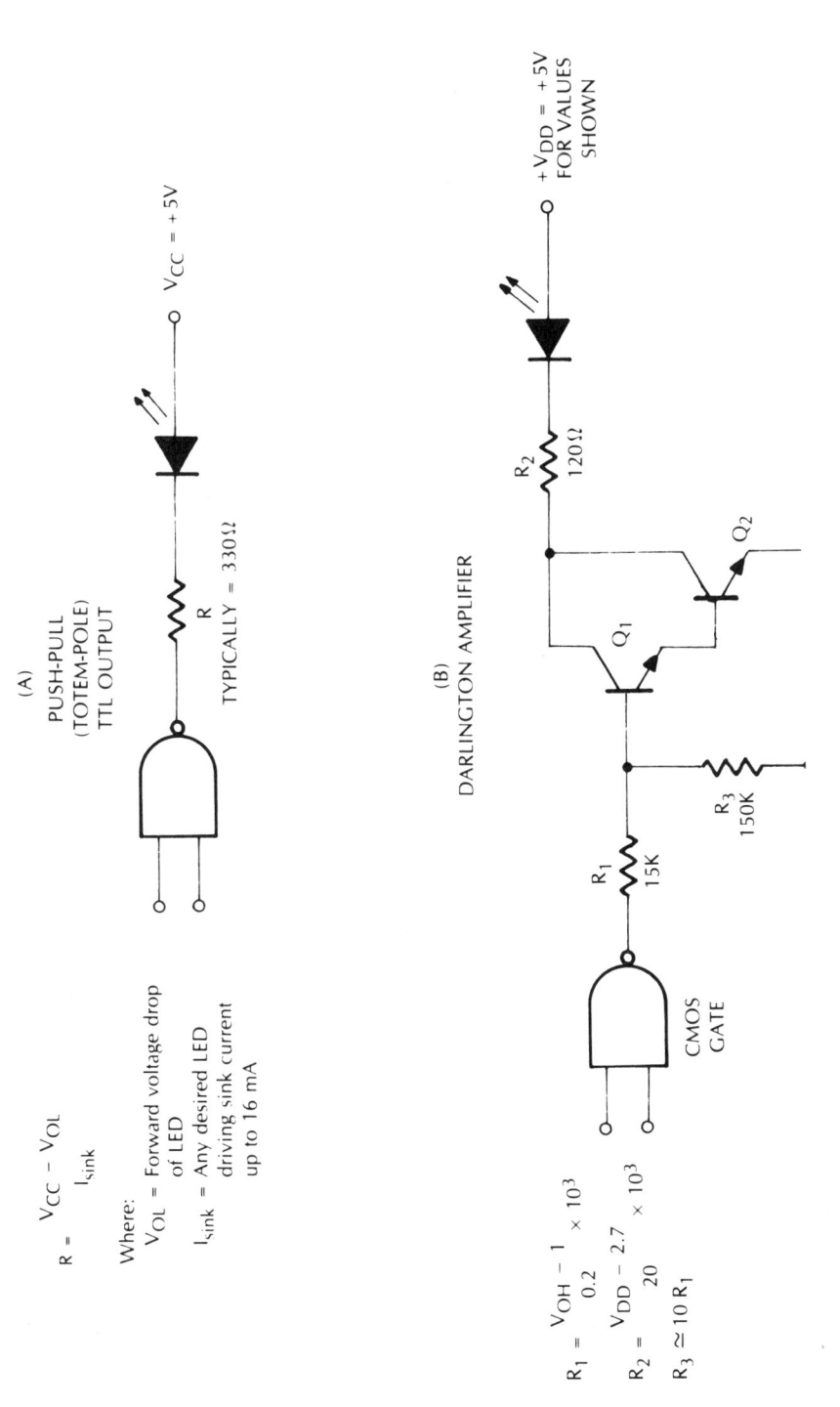

INTERFACING TO CONTEMPORARY IC APPLICATIONS 215

(A)
PUSH-PULL
(TOTEM-POLE)
TTL OUTPUT

$$R = \frac{V_{CC} - V_{OL}}{I_{sink}}$$

Where:
V_{OL} = Forward voltage drop of LED
I_{sink} = Any desired LED driving sink current up to 16 mA

(B)
DARLINGTON AMPLIFIER

$R_1 = \frac{V_{OH} - 1}{0.2} \times 10^3$

$R_2 = \frac{V_{DD} - 2.7}{20} \times 10^3$

$R_3 \simeq 10\, R_1$

Figure 8-6: (A) TTL-to-LED, (B) CMOS-to-CMOS, interface. Note: In general, CMOS gates cannot supply enough current to light an LED directly.

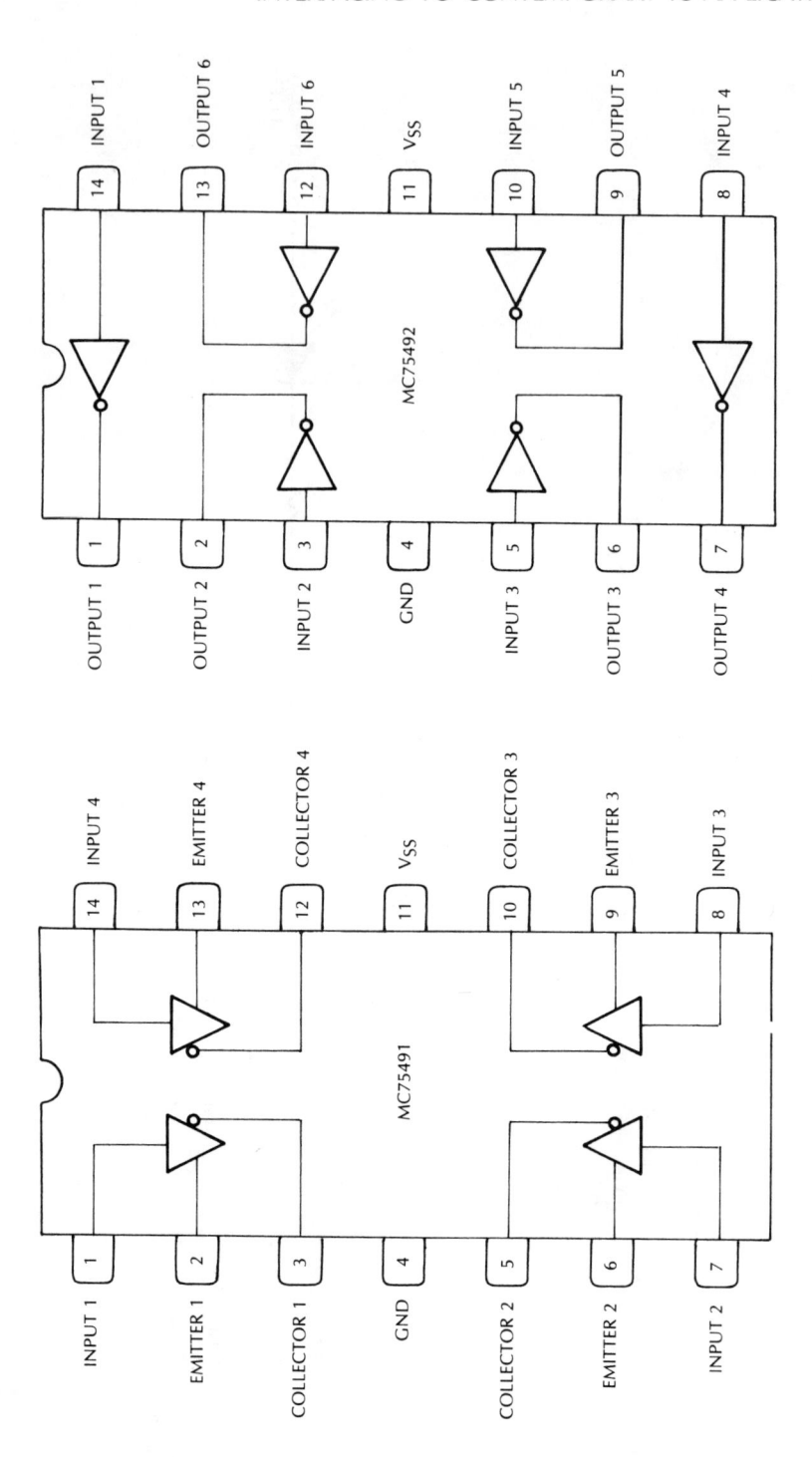

Figure 8-7: Connection diagram for the MC75491 and MC75492 multiple LED drivers.

INTERFACING TO CONTEMPORARY IC APPLICATIONS 217

plays have output pins labeled A, B, C, D, E, F, G, and DP (Decimal Point). In turn, these output pins must turn on certain segments of the 7-segment display to display a certain decimal number. For convenience, a standard 7-segment decimal display is shown in Figure 8-8.

Now, keep in mind that each digit of a certain decimal display must utilize one entire standard 7-segment display unit for each of the digits displayed. For instance, to display the decimal number 520042 would require six individual 7-segment LED displays to be turned on in accordance with the segment letters shown in Figure 8-8.

The **MOS** calculator chip-to-LED interface example circuit suggested by Motorola, and shown in Figure 8-9, is a stripped-down block diagram. This example uses time multiplexing of the individual digits in a visible display to minimize display circuitry. Multiplexing is the best system to use for displays requiring a large number of 7-segment displays. Incidentally, one can also use BCD-to-7-segment decoder **IC**'s (which may be less expensive) where only three or four decimal numbers are to be displayed (see Chapter Two). The system shown can display up to 12 digits with decimal point, using only two MC74591's and two MC74592 drivers.

Referring to Figures 8-7 and 8-9, notice that the MC75491

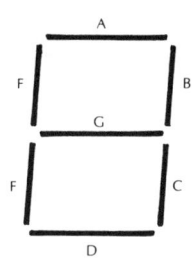

Figure 8-8: Standard 7-segment layout and associated decimal display.

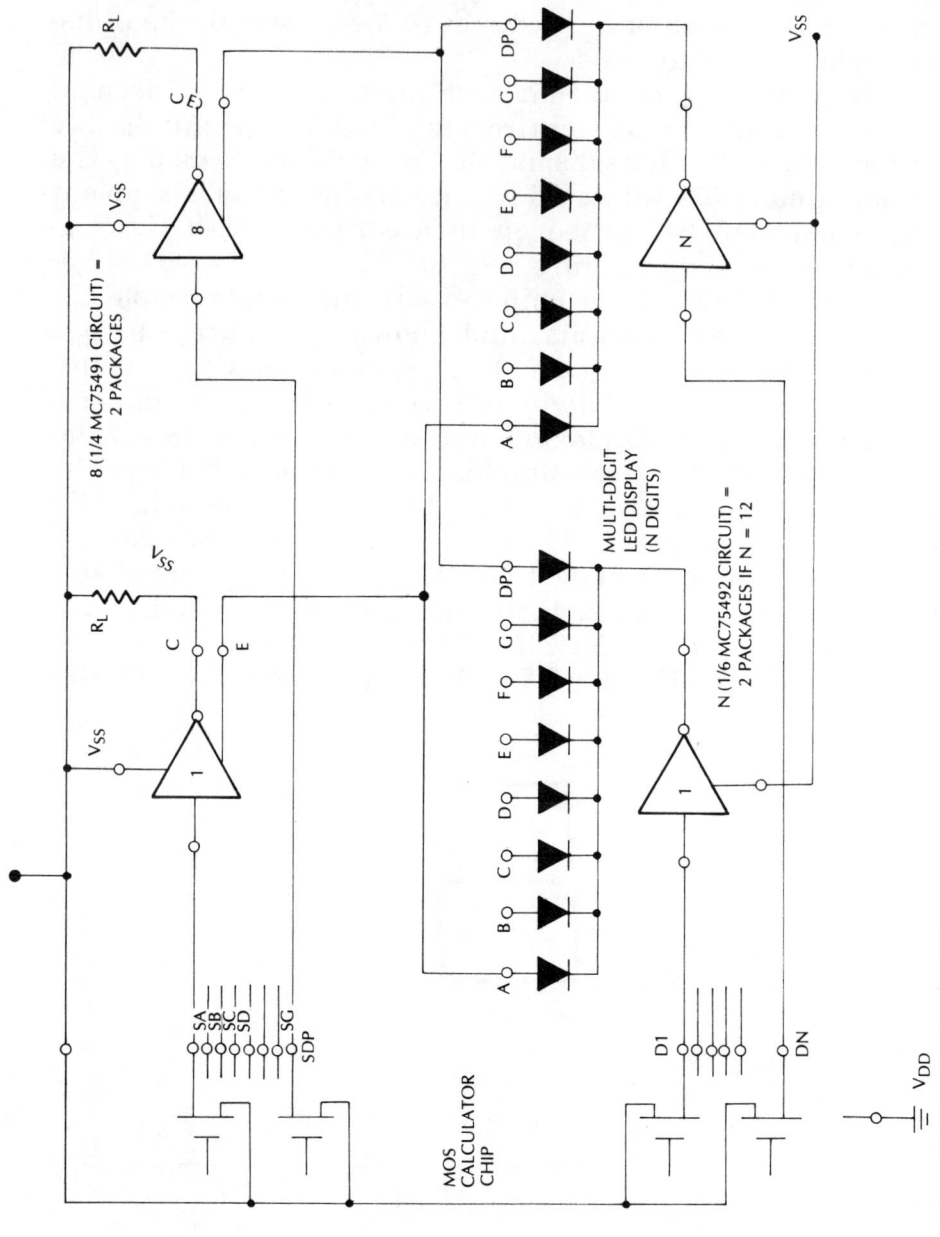

Figure 8-9: MOS calculator chip 7-segment display interface circuit.

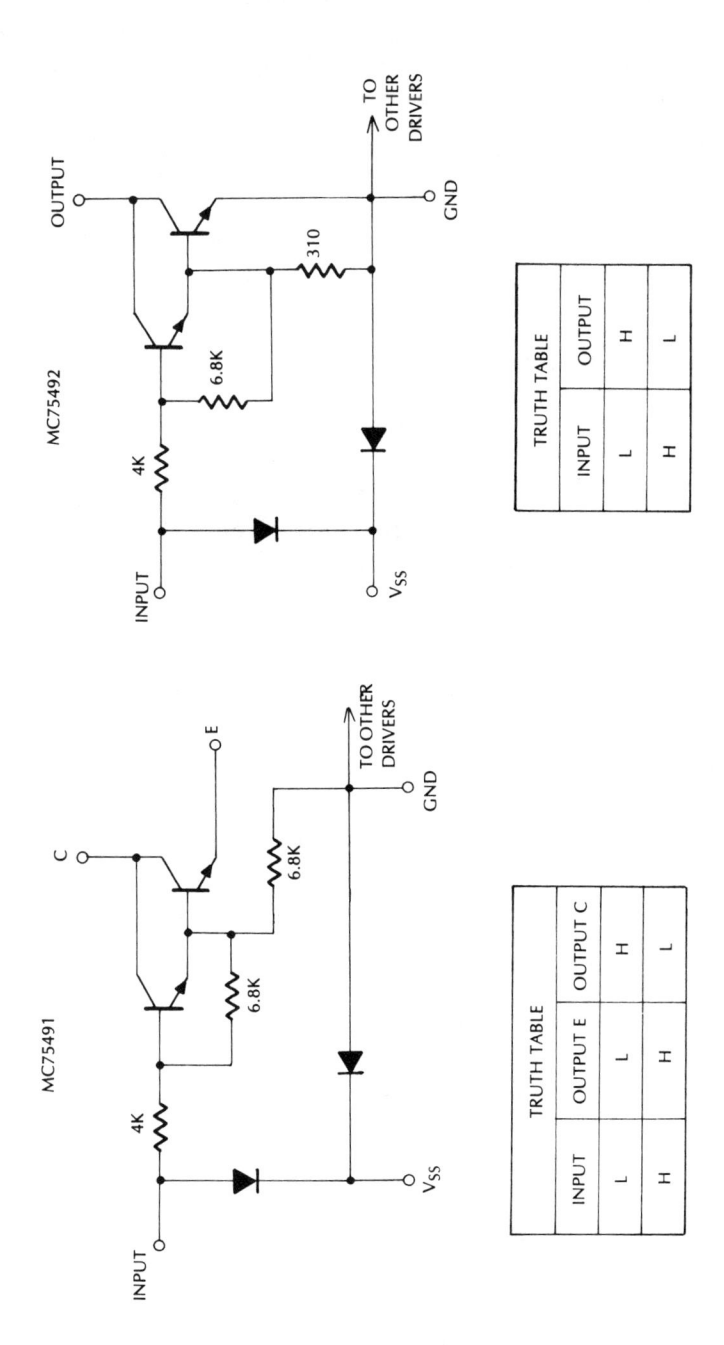

INTERFACING TO CONTEMPORARY IC APPLICATIONS 219

Figure 8-10: Schematic and truth table for a single LED driver within the MC74591/2 IC.

IC's have two outputs (C and E, the collector and emitter of an internal transistor), and the MC75492 has only one (the collectors of two internal transistors wired together). The MC75491 driver unit outputs are connected to the individual LED display's anodes and the MC75492 driver outputs are connected to the LED's cathode. Note, this is a common cathode LED display configuration, therefore it requires logic highs to turn on the segments. An individual driver schematic and truth table for each of these IC's is shown in Figure 8-10.

LED Indicator/CMOS Driver Circuits

In some cases, it is necessary to limit the amount of current through an LED with a limiting resistor (R), as shown in Figure 8-11. Also, you will see that we have included both common cathode and common anode LED circuits in this illustration. The drivers are **CMOS IC's** with only a single internal unit of the **IC** being shown.

The formula for finding the necessary value of the current-limiting resistor is

$$R = (V_{DD} - V_F)/I_F$$

where V_F is the forward voltage drop measured across the conducting LED (typically, about 1.7 volts). I_F is the desired current flow during the LED operating period (typically 10 mA).

Example:

$V_{DD} = 11.7$ volts, $V_F = 1.7$ volts, $I_F = 10$ mA
$R = (11.7 - 1.7)/0.01 = 1000$ ohms.

You will find that 1000 ohms is recommended for many applications that interface an **LED** to a **CMOS IC**, and a current-limiting resistor is frequently required.

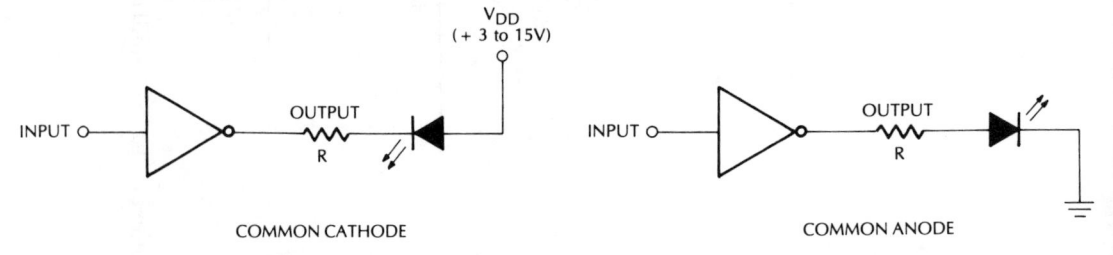

Figure 8-11: LED indicator driver circuits with current-limiting resistors.

INTERFACING TO CONTEMPORARY IC APPLICATIONS   221

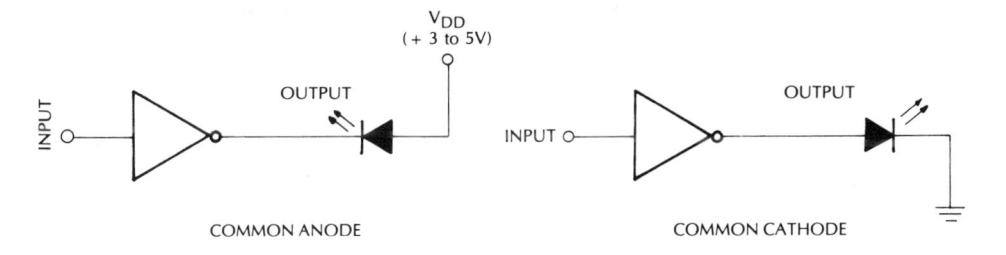

Figure 8-12: How to connect a common anode and common cathode LED circuit to a CMOS IC output terminal with V_{DD} in the 3-to-5-volt range (in this case, no current-limiting resistor is required).

There are IC's that do not require a current-limiting resistor when their output is driving an LED circuit, as was shown in Figure 8-9. Two methods to connect an LED directly to a **CMOS** output circuit are shown in Figure 8-12.

Driving High-Voltage/Current Peripherals

To drive a peripheral often requires more current and/or voltage than an IC or discrete component can deliver. Figure 8-13 shows a circuit that can be used to drive a discrete tran-

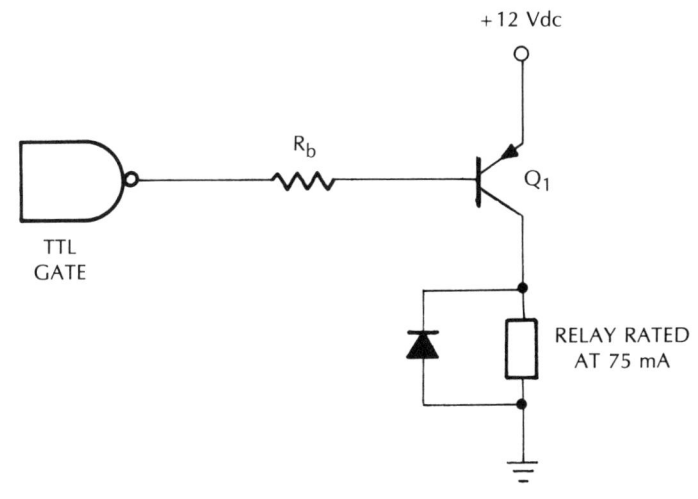

Figure 8-13: Interfacing a low current (16 mA sink current) TTL gate to a relatively high current load. The diode must be included to prevent current surges caused by the sudden opening and closing of the relay.

sistor amplifier circuit that will provide a considerable *current* gain.

As you may know, the sink current level of most **TTL** gates and many other **TTL** devices (Schmitt triggers, etc.) is rated at 16 mA. Incidentally, **TTL** buffers and the like usually drain off much more energy from the system. Therefore, we can take advantage of these fairly large sink currents that these **TTL** devices drain off at logic 0. In general, you will find that 16 mA is more than enough to drive the transistor you want to use into saturation.

Now, remembering that the exact relationship between collector current and base current (called *beta*) is expressed as

$$h_{FE} = \beta = I_c / I_b,$$

let's assume that the transistor we are using has a beta of at least 12 (a conservative assumption, in almost any case). It follows that our transistor circuit can drive a load up to 192 mA, with a gate sink current of 16 mA (I_b, in the formula).

Since our amplifier has a current gain of 12, the load current (I_L) equals 12 I_{sink}. Knowing this, the value of R_b can be determined if you assume a certain value of load current. The formula, in this case, becomes

$$R_b = V_{CC} - (V_{OL} + V_{eb}) / {}^1\!/_{12} \text{ of the load current.}$$

As an example, let's say that we want to drive a relay coil having a current rating of 75 mA and a resistance of 160 ohms. Using our formula and assuming the transistor emitter base

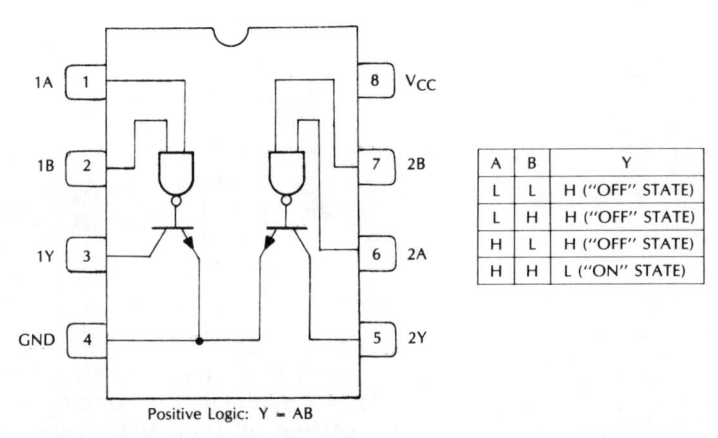

Figure 8-14: MC75461 dual high-voltage peripheral drivers.

INTERFACING TO CONTEMPORARY IC APPLICATIONS

voltage drop (forwarding bias voltage drop) and the other circuit voltage drops add up to about 1 volt, and the voltage on the transistor emitter is +12 Vdc as shown, in this case

$$R_b = 12 - 1/(0.083)(75 \text{ mA}) = 1.8 \text{ kohm}.$$

An **IC** that may be used to drive relays and lamps, and as a line driver, **MOS** driver, or as a buffer, is the MC75461, a positive **AND** gate array. The other **IC**'s are **NAND, OR,** and **NOR** gates; i.e., the 62, 63, and 64, respectively, all have exactly the same pin connections as shown for the MC75461. But obviously, the truth tables are not the same. Each of these peripheral drivers contains a pair of **TTL** gates, with the output of each gate internally connected to the case of a transistor. See Figure 8-14. They are also **TTL** compatible and have 300 mA (max) output current capability.

Testing a Peripheral Driver

You can use the wiring diagram shown in Figure 8-15 to test one of these devices (the MC75461 or similar **IC**). A few points to remember during this test, or any other, for that matter, are listed here:

1. You must observe all operating restrictions. Reversed polarity, excessive supply voltage, and drawing too much supply current can destroy an **IC**. For example, Figure 8-15 lists a maximum load current (I_{OL}) of 300

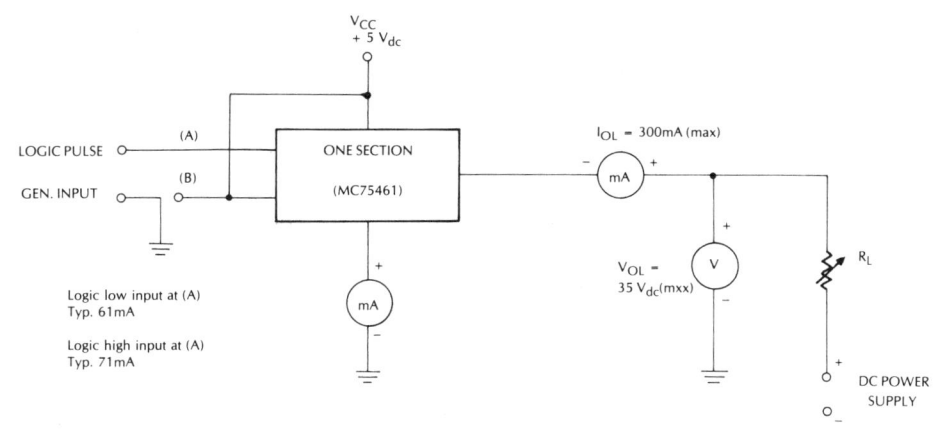

Figure 8-15: Test circuit for a MC75461 IC or similar peripheral driver. Each input is tested separately.

mA, and a maximum load voltage (V_{OL}) of 35 Vdc. *Under no circumstance should you exceed these values.*

2. *Never* remove or insert an **IC** into a socket with power applied to the circuit.

3. In general, do not connect an input pulse generator to an **IC** with power to the circuit off, *particularly if the **IC** is a **CMOS** device.*

4. When working with **CMOS IC**'s, all unused **IC** inputs should be connected to V_{DD} or V_{SS}. If you are having troubles with a circuit (such as instability or the drawing of too much current), check for unterminated terminals on the **IC**.

Bus Interface

You will remember, from Chapters Three and Four, that a *bus* in a computer is usually 8, 16, or 32 lines used as a path

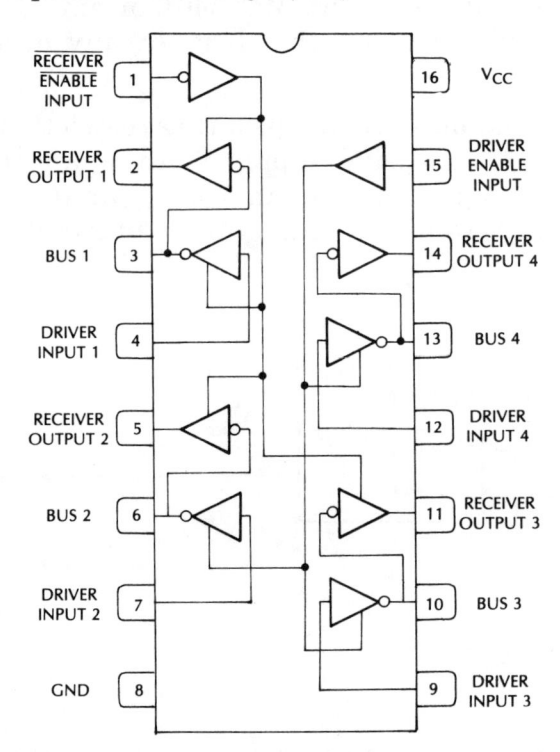

Figure 8-16: Data bus extender. Quad, bidirectional, with 3-state outputs.

INTERFACING TO CONTEMPORARY IC APPLICATIONS   225

over which information is transmitted from any of several sources to distinct destinations. You may see the term *highway* used when a bus interconnects many system components. Also, you will encounter the term *handshake bus*. This is used to indicate a bus that is used for connecting the MPU-based system to some peripheral.

If you are building a system using a standard 8080 or M6800 MPU, the maximum number of IC's you can connect to it (without using data bus extenders) is 3 RAM's, 3 PIA's, and 1 ACIA. Figure 8-16 shows the logic diagram and pin numbers for Motorola's MC6880 data bus extender.

This family of IC's (the MC6880 is an inverting type and the MC6889 is a non-inverting type) is designed to extend the limited drive capabilities of the NMOS type 6880 and 8080 MPU's. The maximum input current of 200 μA at any of the data bus extender input pins assures proper operation despite the limited drive capability of the MPU chip. Figure 8-17 shows a MPU bus extender application using Motorola's components.

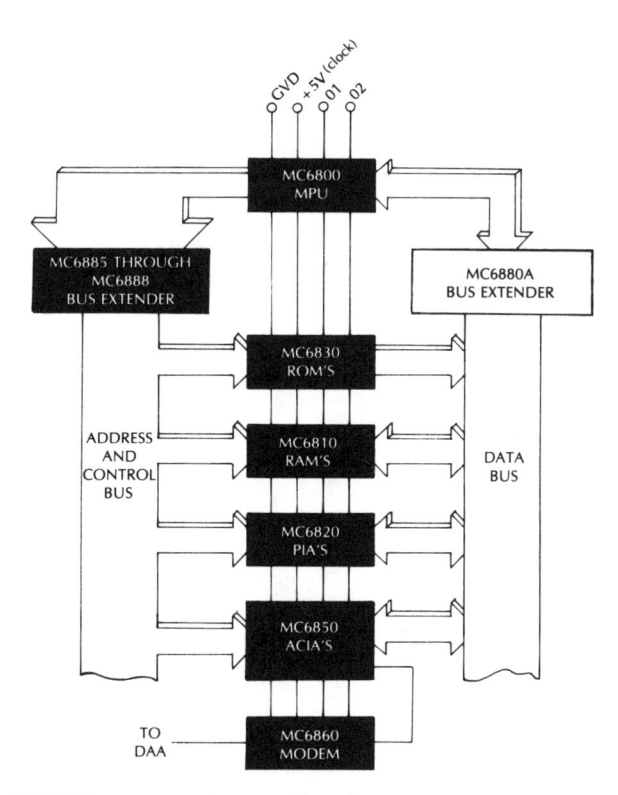

Figure 8-17: MPU bus extender application.

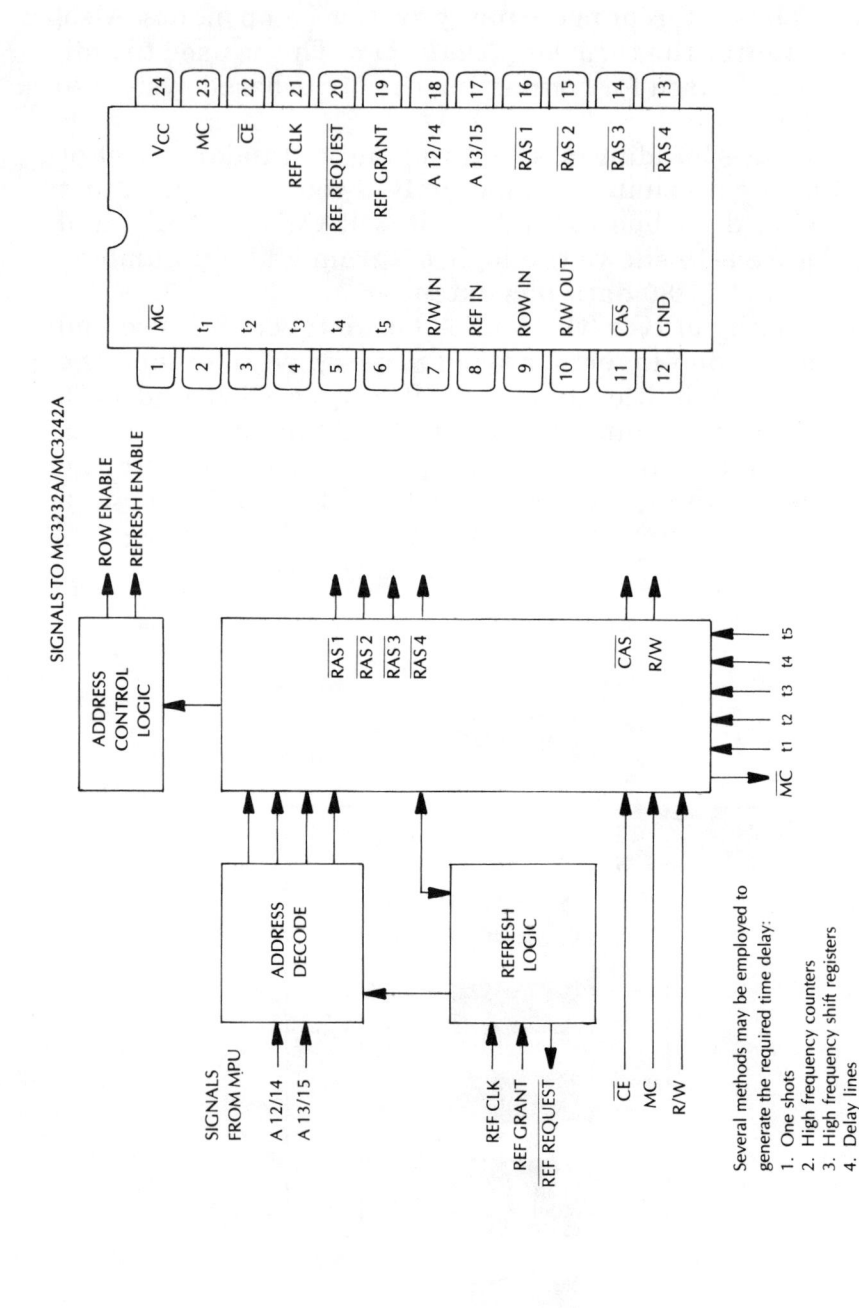

Figure 8-18: Block diagram and pin connections for the memory controller MC3480.

MPU Memory Interface

In Chapter Three, we described a microcomputer that required only two **IC**'s: the MC6846, a **ROM**, I/O timer, and the MC6802 **MPU** (see Figure 3-4). However, that system has limited memory. By using an **IC** such as the MC3480, a memory controller, it's possible to control popular 16-pin 4k, 16k, or 64k dynamic **NMOS RAM**'s in an **MPU** system designed around **MPU**'s such as the M6800. The controller, with an oscillator, will also generate the necessary signals required to insure that the dynamic **RAM**'s are refreshed for the retention of data (see Figure 8-19). Figure 8-18 contains a block diagram and the pin connections for the MC3480.

To understand how an **IC** such as this one does its job, you must have a description of the functions at each pin and, as you can see, this is a 24-pin **IC**. Table 8-2 is the pin description information for this memory controller.

PIN DESCRIPTION TABLE

Name	No.	Function
RAS1*	16	Row Address Strobe pins which connect to each of the dynamic RAMs to latch in row address on memory chips.
RAS2	15	Decoded to 1 of 4 during cycle. All 4 go low during refresh cycle.
RAS3	14	
RAS4	13	
CAS*	11	Column Address Strobe pin which connects to each dynamic RAM to latch in column address.
R/W Out*	10	This pin signals the dynamic RAM whether the RAM is to be read from or written into.
Row En	9	Row Enable output which goes to the MC3232A (MC3242A). It signals the Address Multiplexer that the lower half (Row Addresses) or the upper half (Column Addresses) of the address lines are to be multiplexed into the dynamic RAM address inputs. A Logic 1 on this output indicates the Row Addresses, and a Logic 0 indicates Column Addresses.
Ref En	8	Refresh Enable Output. A Logic 1 signals the Address Multiplexer that a refresh cycle is to be done, and a Logic 0 indicates that address multiplexing should be done.
CE	22	Chip Enable Input. A Logic 1 on this pin disables all chip functions, except that of Refresh and the MC output. CE must be low during t1 low-to-high transition to initiate R/W cycle. Once t1 is initiated, the cycle is independent of CE.
R/W In	7	The Read/Write input pin receives information from the M6800 as to the direction of data exchange in the dynamic RAM. It transmits a Logic 0 to the R/W output for a Write Cycle and a Logic 1 for a Read Cycle.

Table 8-2: Pin description table for the MC3480 dynamic memory controller.

Table 8-2 (continued)

A13(A15) A12(A14)	17 18	Upper Order Address lines from the M6800. These two inputs decode to four signals controlling the four $\overline{RAS}$ outputs. A14 and A15 apply to 16k RAMs.
MC	23	Memory Clock input from MC6875 clock or other signal source. The rising edge of MC must occur after the rising edge of t1 to avoid aborting the refresh cycle. When MC rises, it resets an internal flag that will terminate refresh at the end of the current cycle. Failure to reset the flag forces the 3480 to refresh every cycle thereafter. MC can be connected to t2 or t3 in noncritical applications.
$\overline{MC}$	1	The buffered complement output of MC. It is a buffered output which may be used to drive the circuitry creating the time delays used on inputs t1 through t5.
t1 t2 t3 t4 t6	2 3 4 5 6	These pins are external timing inputs to sequentially select the outputs to be enabled. They are positive-edge triggered inputs. Assuming a Read/Write cycle is to be executed, a positive edge on t1 forces a Logic 0 on one of the four $\overline{RAS}$ outputs as determined by the A12/14, A13/15 inputs. After a delay, a positive edge on t2 causes Row En to go to a Logic 0, providing address-multiplexing information to the MC3232A or MC3242A. t3 enables the $\overline{CAS}$ output and it goes low. t4 enables the R/W output and it goes low, assuming the R/W input was low. t5 resets all the outputs to a Logic 1 (with the exception of $\overline{MC}$, $\overline{Ref\ En}$, and $\overline{Ref\ Req}$). The inputs t1, t2, t3, and t5 are daisy-chained, so they must be sequentially driven to obtain the desired output signals. t4 can be driven at any time after t1.
Ref Clk	21	The 32 kHz (64 kHz) Refresh Clock signals this pin that another refresh cycle is required. It is a positive-edge triggered input, and upon triggering the $\overline{Ref\ Req}$ pin goes to a Logic 0.
$\overline{Ref\ Req}$	20	The Refresh Request output acts as an input to the MPU system, requesting a refresh cycle. This output has a 5-k ohm pullup resistor to the V_{CC} supply to allow wire-ORing if desired.
Ref Grant	19	Through the Refresh Grant input, the MC6875 initiates a refresh cycle. This input is positive-edge triggered and is enabled only after the $\overline{Ref\ Req}$ pin has gone low. This allows the MC3480 to discern between a Refresh Grant or a DMA Grant even though they appear on the same line. When employing both dynamic memory (refresh) and DMA in a microprocessor-based system with a combined Refresh/DMA Request control on the clock, provision must be made for holding off a DMA request during a refresh period (and vice versa). If this provision is not made, clock stretching (cycle stealing) will continue indefinitely and dynamic microprocessor data will be lost. The positive edge on Ref Grant causes Row En output to go low and Ref En output to go high. This signals the MC3232A (MC3242A) that a refresh address is required. The refresh cycle occurs with the succeeding pulses on t1–t5. A positive edge on t1 causes $\overline{Ref\ Req}$ to go high and all the $\overline{RAS}$ outputs to go low. A positive going edge on t2 causes no change in the outputs, since it controls the address multiplexing (Row En) during the Read/Write cycles. There is no output change when t3 and t4 go high because no $\overline{CAS}$ or R/W signal is needed during refresh. A positive edge on t5 resets the $\overline{RAS}$ and Row En to a Logic 1 state, and REF En to a Logic 0 state, ready for the next Read/Write cycle.
V_{CC}	24	+5.0 V supply. A 0.1 F capacitor is recommended to bypass pin 24 to ground.
Gnd	12	System Ground.

* These outputs are designed to drive the highly capacitive inputs of multiple dynamic RAMs/(150 pF for $\overline{RAS}$ outputs, and 450 pF for $\overline{CAS}$ and R/W outputs). Consequently, these outputs have no short-circuit limit and must be handled accordingly. Good high capacitance load-driving techniques usually include a 10–ohm or greater series damping resistor. It is highly recommended that this be done on $\overline{RAS}$, $\overline{CAS}$, and R/W outputs of the MC3480. The effect of these series damping resistors on rise and fall times must be included in timing considerations.

NOTE: All other outputs are LS/TTL totem-pole configuration unless otherwise noted.

INTERFACING TO CONTEMPORARY IC APPLICATIONS

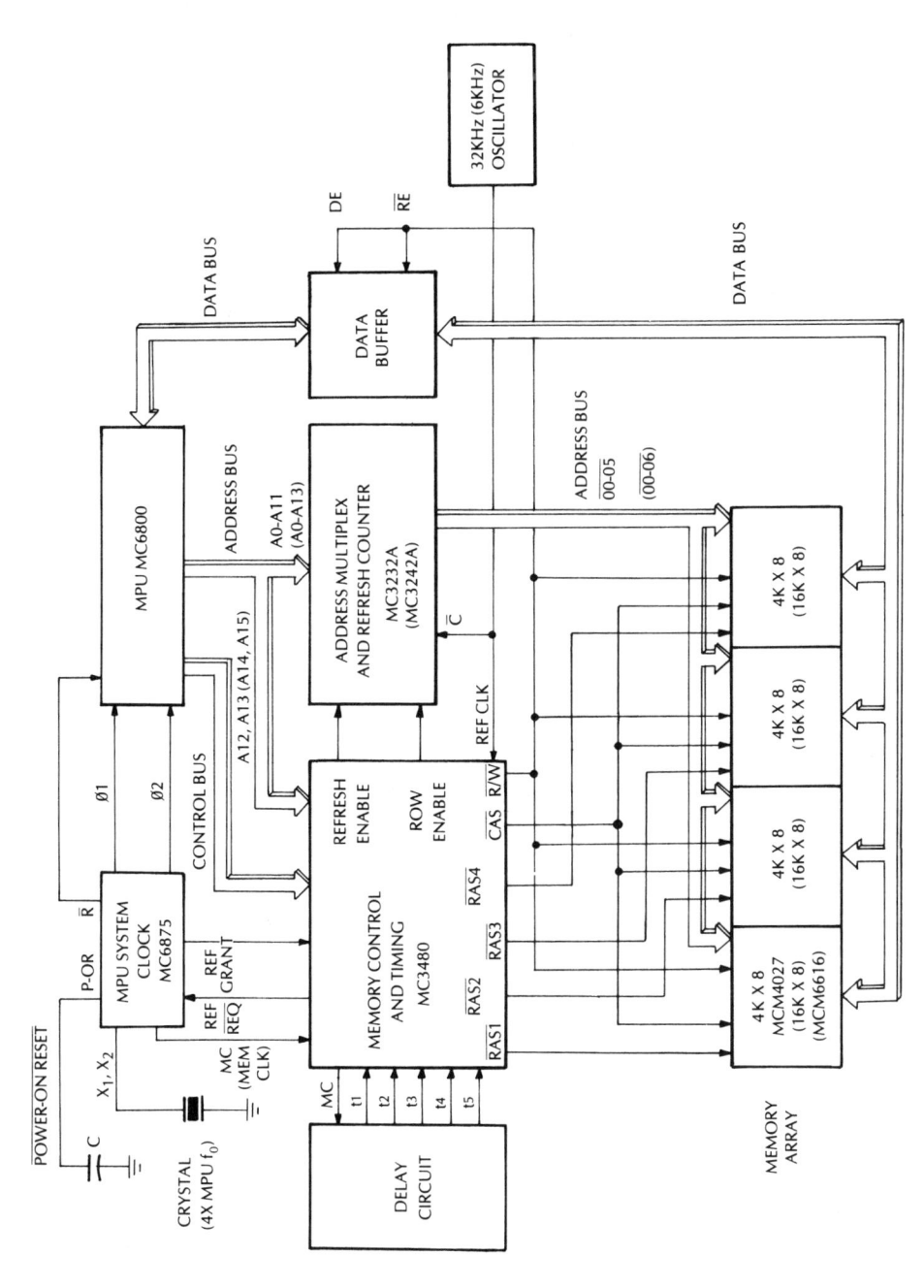

Figure 8-19: Example of a 16k × 8-bit memory system for the M6800 MPU.

230   INTERFACING TO CONTEMPORARY IC APPLICATIONS

You will note, as you read Table 8-2, that there are other **IC**'s required to complete an entire system. In fact, for a 16k × 8-bit memory system for the M6800 or similar **MPU**, it requires a clock such as the MC6875, address multiplex and refresh counter such as the MC3232, and a buffer delay circuit, etc. Figure 8-19 is a typical example of a complete 16k × 8-bit memory for the M6800 or similar **MPU**.

There are two oscillators used in this example: (1) the systems clock (MC6875), which requires a crystal operating at a frequency four times the **MPU** frequency rate, and (2) a 32- or 64-kHz oscillator used to provide the necessary signal so that the dynamic memories will be refreshed in order to prevent loss of stored data. A simple 32-kHz oscillator that can be used in the application illustrated in Figure 8-19 is shown in Figure 8-20. This oscillator is built using a LM311 **IC** voltage comparator and a simple RC network.

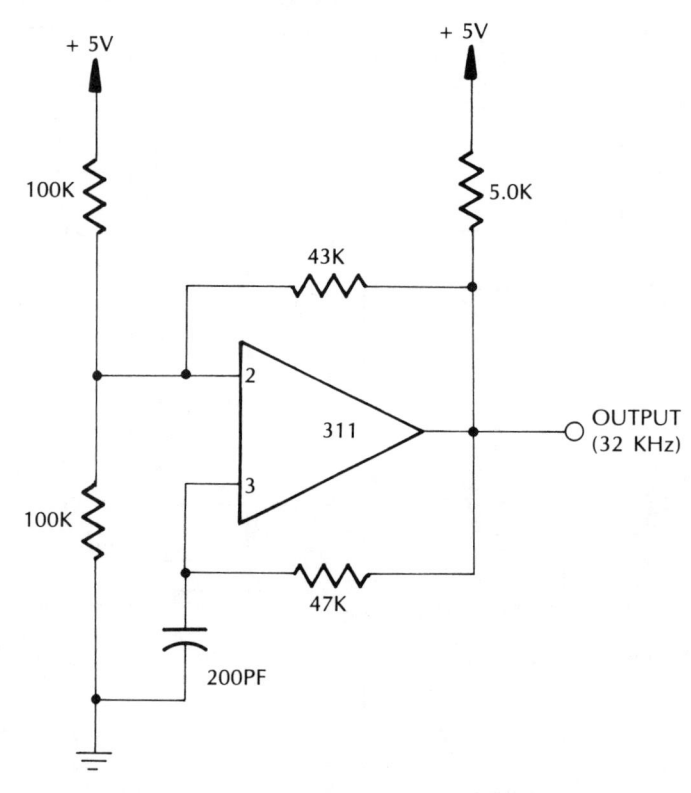

Figure 8-20: Oscillator (32 kHz) suggested by Motorola, for use with the example memory system shown in Figure 8-19.

INTERFACING TO CONTEMPORARY IC APPLICATIONS 231

Peripheral Interface Using Drivers and Receivers in Computer Applications

There are many line drivers and receivers for computer/terminal applications. You will usually find them listed in pairs in sales catalogs (a driver and a receiver). For example, the MC1488 IC is a *quad driver* with output current limiting and the MC1489 is a *quad receiver* with an input voltage range of ±30 volts. Figure 8-21 illustrates a typical application using this pair of IC's. Incidentally, note that the input shows MDTL (DTL is an abbreviation for diode transistor logic and the M standards for Motorola). Also, this driver is compatible with all Motorola MTTL logic families.

It is also possible to adjust the input threshold voltage for the MC1489 receiver IC. This is particularly important because it makes possible excellent interfacing between MOS circuits and MDTL/MTTL logic systems. Figure 8-22 shows how this may be done by placing an external resistor (R) in the IC circuit.

In this application, the input threshold voltage is adjusted by using a certain voltage (to the external response control pin) and a predetermined resistor value that must be

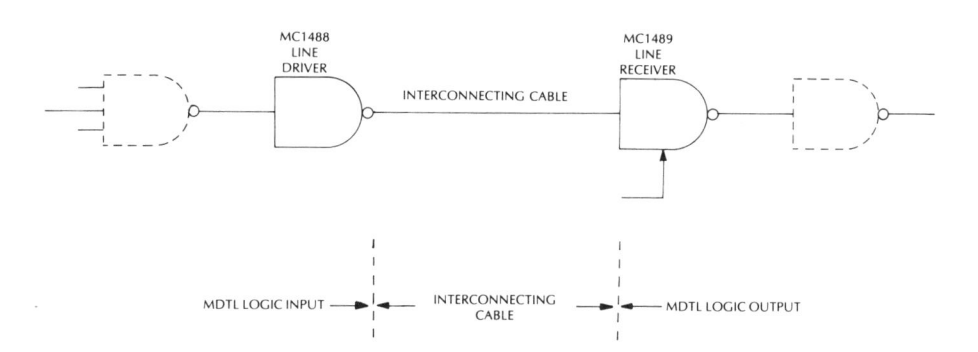

Figure 8-21: Typical application using Motorola quad driver and receiver interface IC's.

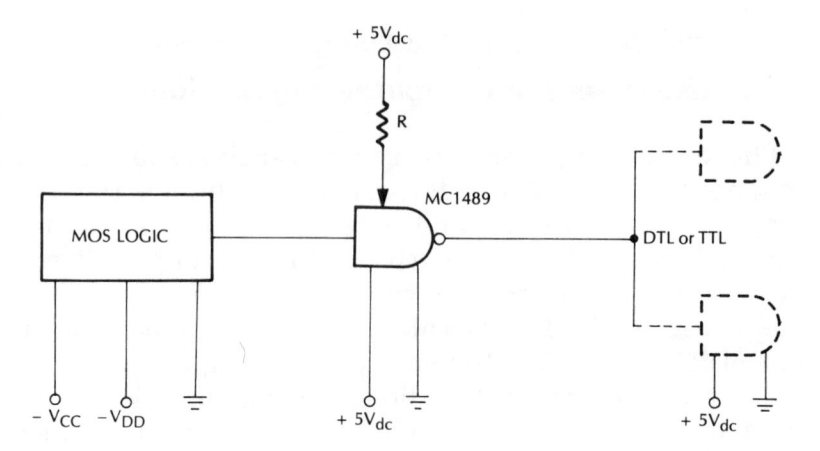

Figure 8-22: Using an MC1489 to interface MOS logic to DTL or TTL logic.

adjusted (the voltage and resistor values), until the threshold voltage falls to about the center of the **MOS** device's voltage logic level. Figure 8-23 illustrates the input threshold voltage adjustment for one of the receivers in the quad receiver. Each receiver has one output terminal, one response control terminal, and one input terminal (see Figure 8-24).

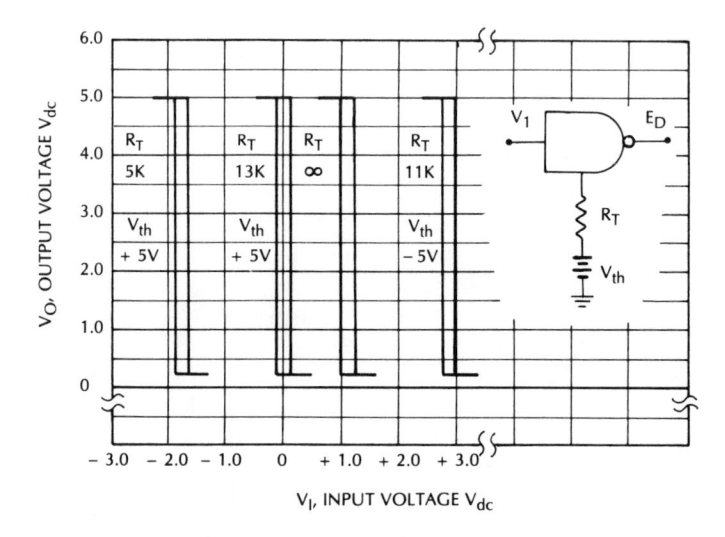

Figure 8-23: Input threshold voltage adjustment chart for a MC1489 line receiver.

INTERFACING TO CONTEMPORARY IC APPLICATIONS

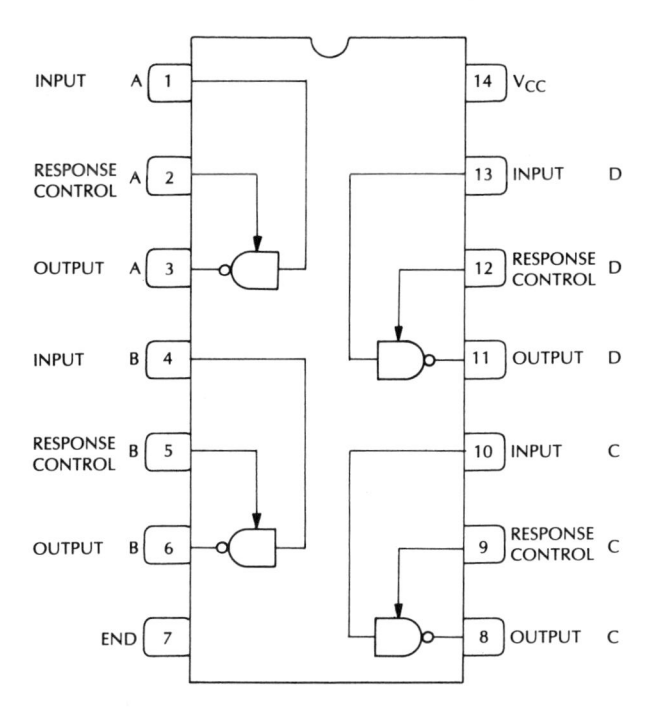

Figure 8-24: Logic diagram and pin configuration for the MC1489L quad line receivers.

CHAPTER NINE

Practical Guide to Modern Power Transistors, Thyristors, and Optoelectronic Devices

As we all know, staying on the leading edge of high technology isn't easy. This chapter has just about everything you'll need to know about today's power transistors—bipolar and **MOSFET**'s. It contains some of the latest solid state power devices. Also included are state-of-the-art components such as optocouplers and programmable unijunction transistors (**PUT**'s).

Power transistors and thyristors (such as SCR's and triacs) are useful in an extremely wide range of applications. In the following pages, you will find step-by-step approaches to understanding these solid state devices in almost any circuit design you may encounter.

Darlington Transistors: the IC's of the Power Field

In the last chapter, under the subheading "Driving High-Voltage/Current Peripherals," it was pointed out that driving peripherals often requires more current and/or voltage than some IC's can deliver. There are Darlington amplifier transistor IC's that are specifically designed for amplifier and driver applications where high gain is essential. As an example, there are the NPN D40C1, 2, 4, and 5 Darlingtons. The D40C2 and 5 produce a minimum current gain of 40k at an I_c of 200 mA, with V_{CC} = 5 Vdc. *Note:* DC current gain = β = h_{fe} = I_c/I_b at a certain ambient temperature (usually 25°C). Incidentally, the maximum h_{fe} for all four of these IC's is 60k, with an I_c of 200 mA @ 5 Vdc. Another example of a two-transistor Darlington IC, the 2N656, 7, and 8 series, is shown in Figure 9-1.

All Darlingtons are basic common collector (emitter followers) amplifiers. When the Darlington is used as a common collector, the output impedance is approximately equal to the load resistance (R_L). In this configuration, the load (whatever it is, speaker, resistor, or feeding a line such as coax, etc.) is taken from the emitter circuit.

Next, the input impedance. In the common collector mode of operation, the input impedance is approximately equal to $h_{fe}^2 \times R_L$. The reason for the high current gain of Darlingtons is the fact that the h_{fe} is approximately equal to the average h_{fe} of the two transistors, squared.

Darlington IC's are not limited to a single Darlington amplifier. Several Darlingtons can be used in one package. A state-of-the-art example of this is Motorola's peripheral interface driver array ULN2803, an "octal Darlington array." See Figure 9-2 for pin connections and a single element Darlington

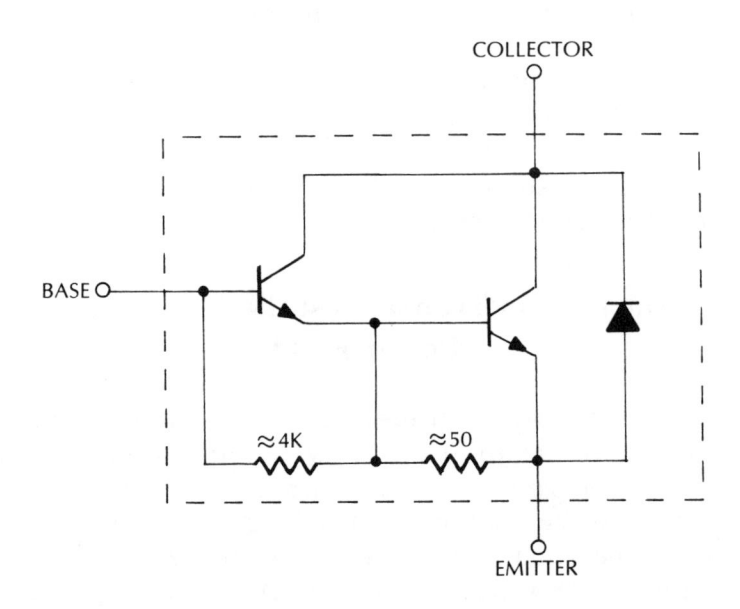

Figure 9-1: NPN silicon power Darlington transistor (120 watts).

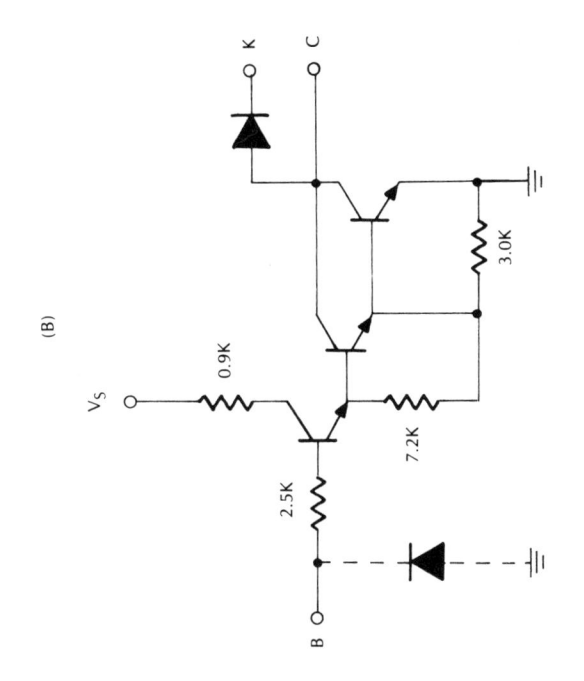

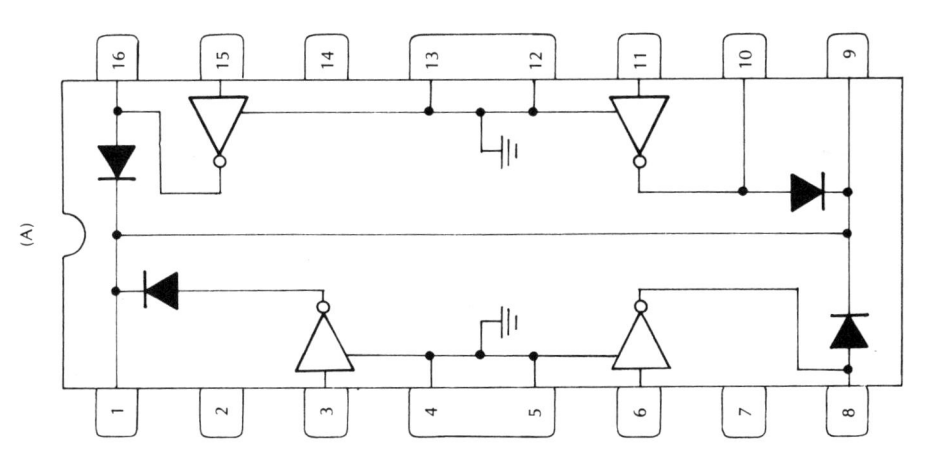

TRANSISTORS, THYRISTORS, OPTOELECTRONIC DEVICES 237

Figure 9-2: Octal Darlington array. ULN2803, with output clamp diodes.

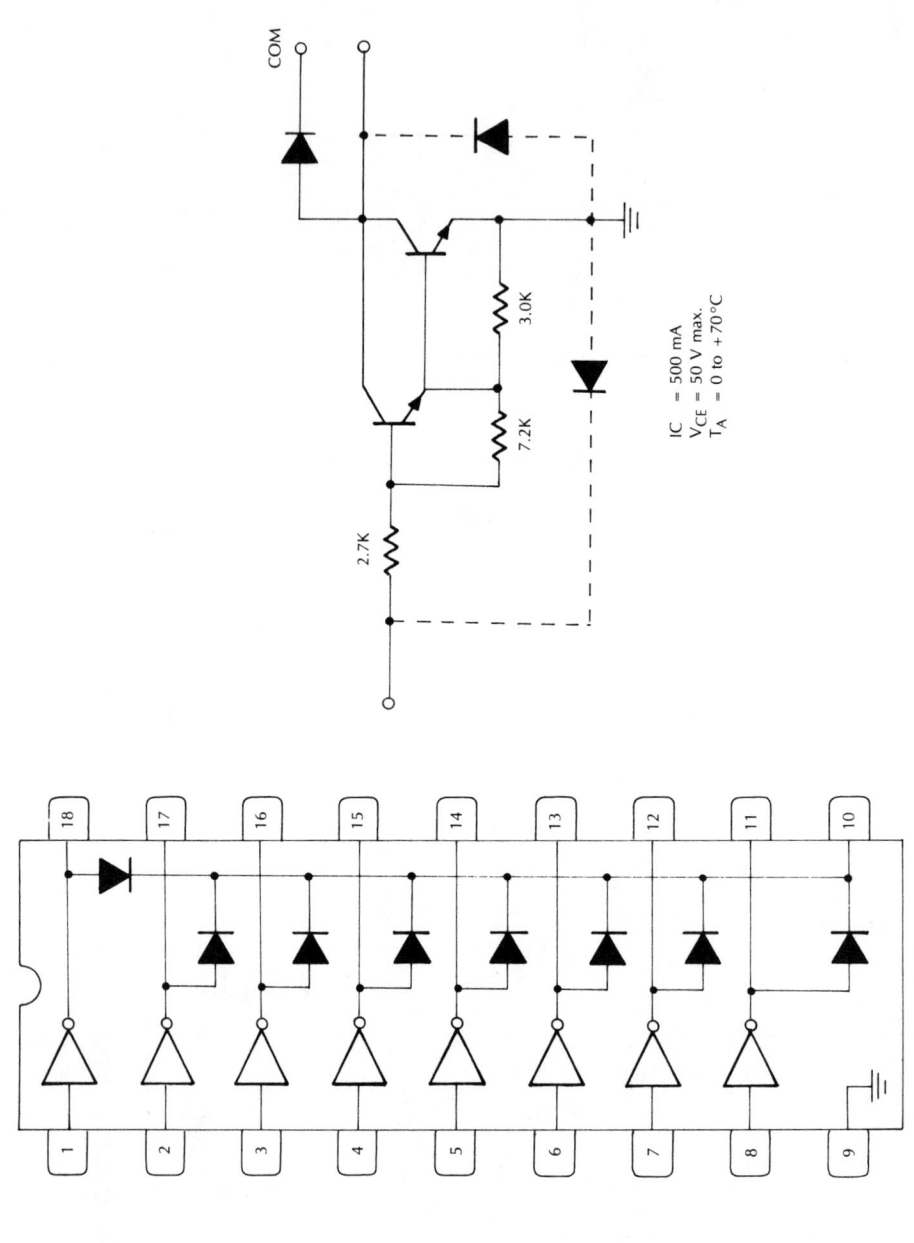

238   TRANSISTORS, THYRISTORS, OPTOELECTRONIC DEVICES

I_C = 500 mA
V_{CE} = 50 V max.
T_A = 0 to +70°C

Figure 9-3: Quad driver array using Darlington circuit designed to utilize three transistors.

schematic for this **IC**. Furthermore, Darlington circuits need not be limited to two transistors. Three (and even four) transistors can be used in the Darlington circuit. An example of this is Motorola's circuit in Figure 9-3 (B). This **IC** is a quad driver array (ULN2968), 1.5 V, V_{CE} = 50 V max (see Figure 9-3 A for logic diagram). The circuit is essentially a common emitter Darlington followed by a common collector Darlington.

Power MOS FET's

It has been pointed out that **CMOS NMOS IC**'s are now being used in many applications where **TTL** once ruled the roost. Recently, Motorola and other semiconductor manufacturers brought out a new generation of power field-effect transistors that are invading the field of bipolar power transistors in much the same way.

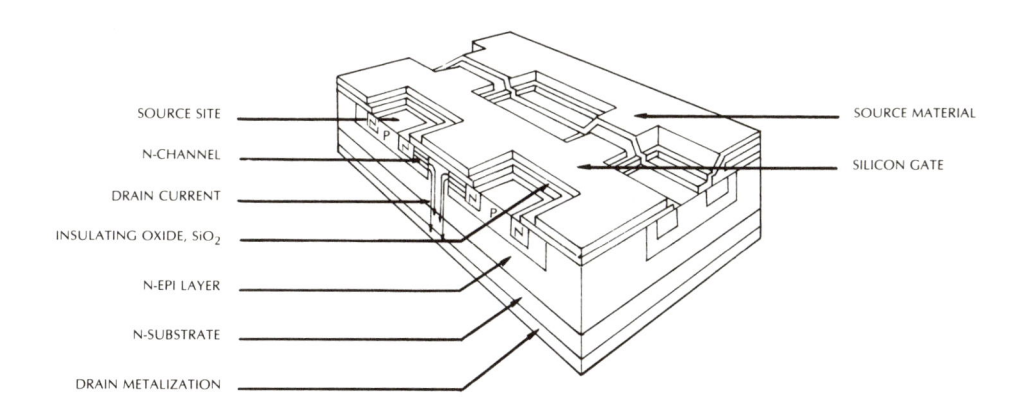

Figure 9-4: Pictorial view of Motorola's TMOS power field-effect transistor. Note vertical current flow in FET structure. This offers a low resistance path and permits smaller chip size, resulting in a major breakthrough in the MOS power transistor. See Figure 9-5 for actual case and schematic for these devices.

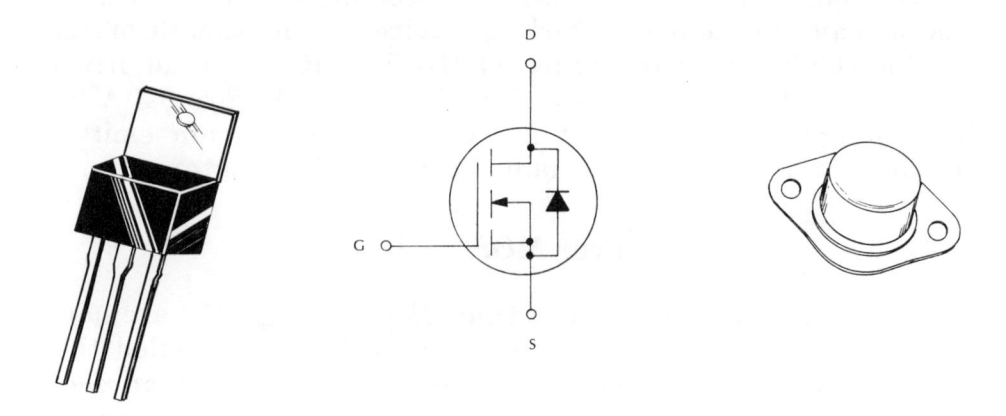

Figure 9-5: Schematic and package types for Motorola's MTM, MTP, N-channel TMOS power FET's.

These power transistors are known by several names (depending on the manufacturer and manufacturing technique): Motorola's TMOS power **MOSFET**, Siemen's SIPMOS, and International Rectifier's **HEXFET**, plus others. But, whatever they are labeled, in general they all rely upon vertical current conduction through the transistor. Figure 9-4 is a pictorial view of Motorola's TMOS power **FET**. Note the label *drain current* and arrows pointing out the direction of *drain current* flow. These four arrows (two coming from the right side, two coming from the left side, and then down) seem to form the letter T; whence the name used by Motorola...*TMOS*. See Figure 9-5 for schematic and package types.

TMOS and Bipolar Power: Advantages and Disadvantages

It's important to realize that, in some cases, you are better off using bipolar power transistors rather than **MOS** power transistors. Here are some of the advantages and disadvantages of these power TMOS **FET**'s:

1. The **FET**'s have high-impedance input that could provide a better impedance match for the driving signal source you are interfacing to.
2. Bipolar power transistors are a current-driven device. **FET**'s are voltage driven. It's possible to use simpler drive circuits in some applications using a TMOS power transistor. To put it another way, TMOS **FET**'s are easy to drive and interface because of their low voltage drive requirements.
3. **MOS** power transistors have significantly faster switching speeds than bipolar. Nanosecond speeds are possible with the TMOS transistors.
4. The forward conduction ("on") resistance may be as small as a few tenths of an ohm. Typically, it's from about 0.25 to 2 ohms for the IRF, MTM, and MTP series.
5. TMOS power transistors are more stable than bipolar transistors. This is because as chip temperature rises, current drain decreases.
6. The principal disadvantage of **FET**'s, in respect to bipolars, is that they are susceptible to electrostatic discharge, as has been pointed out in previous chapters. The TMOS **FET** is no different. For this reason it also requires special handling. You will find an application for these IC's in Chapter Twelve.

Introduction to Thyristors

Thyristors are basically a bistable device comprising three or more junctions. In general, they consist of two generic component categories—silicon controlled rectifiers (SCR's), and triacs (a bidirectional rectifier, essentially two SCR's in parallel). Some of these switches (triacs) are designed to be triggered directly by microprocessors and microcontrollers. Motorola's M4C228-2—10 family is well suited for interfacing personal computers with line-operated appliances and similar devices. Figure 9-6 shows seven schematic symbols representing typical thyristors used in today's electronic industry.

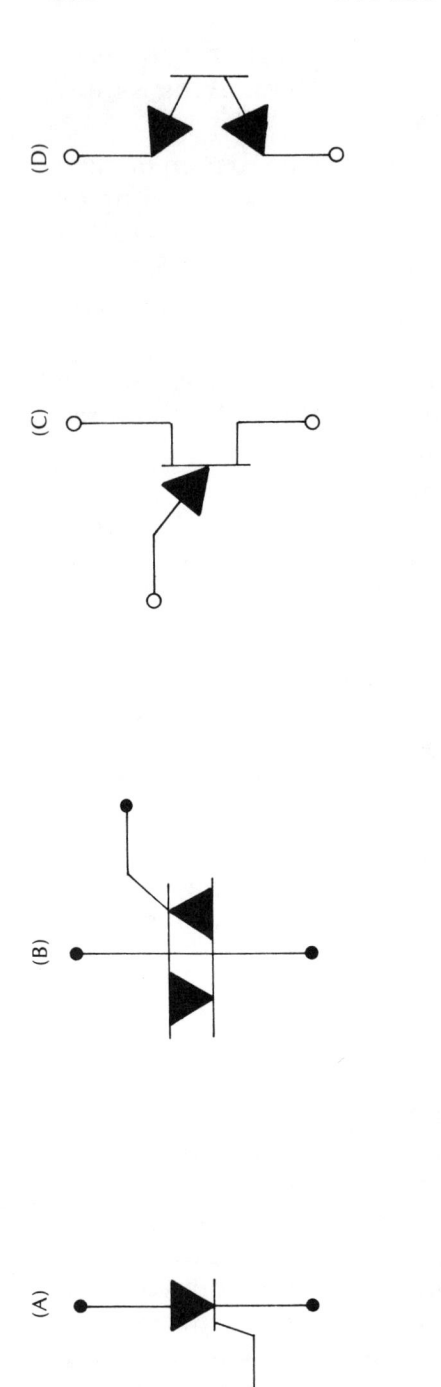

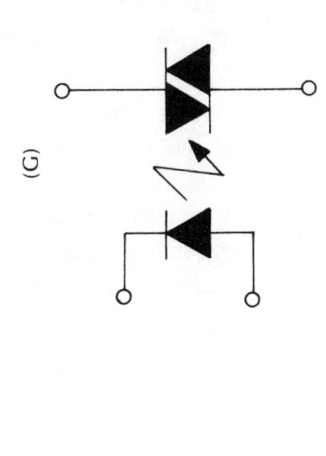

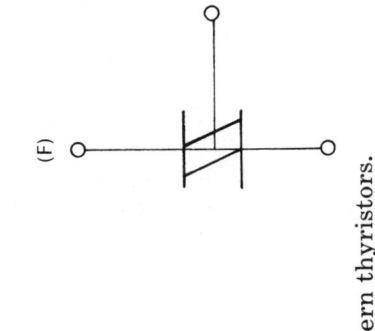

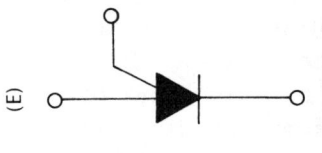

Figure 9-6: Schematic symbols of modern thyristors.

Unijunction Transistors (UJT)

Having briefly described SCR's and triacs, let's look at the others. The UJT (Figure 9-6 C) is a unijunction transistor. These are usually a highly stable IC used for general purpose *trigger applications* and, as a general rule, not expensive. You can also use these transistors in pulse-generating (oscillators) and timing circuits.

Bilateral Triggers (DIAC)

The next symbol shown (Figure 9-6 D) is known as a DIAC. Although the symbol looks like an incomplete transistor (because there is no base lead, and there appear to be two emitters), it is not. In reality, the device is a bidirectional breakdown *diode* that conducts only when a certain breakdown voltage is exceeded. For example, a 1N5758A DIAC (also called a *bilateral trigger*) has a switching voltage range of 16 to 24 volts (both directions). The switching current (both directions) is 100 μA, and the switching voltage change (both directions) is 5 volts. One application for this DIAC could be the control circuit for a triac.

Programmable UJT's

The PUT symbol (see Figure 9-6 E), is depicting a *programmable unijunction transistor*. These transistors physically look like a UJT (which is the same as almost any bipolar transistor; i.e., three leads out of a metal or plastic case) and are similar to UJT's, except that they can be programmed with external "program" resistors—usually labeled R_1 and R_2. Figure 9-7 shows a PUT with "program" resistors R_1 and R_2.

When calculating the value of the gate voltage (V_S) and determining the value of V_B shown in the formula for V_S in Figure 9-7, it is important to refer to the maximum ratings of the PUT. This example is for the 2N6027 and 2N6028 programmable unijunction transistors (40 volts, 375 mW). The low "on" state for these devices is 1.5 volts maximum at a forward conducting current (I_F) of 50 mA. Some useful maximum ratings for these transistors are listed in Table 9-1.

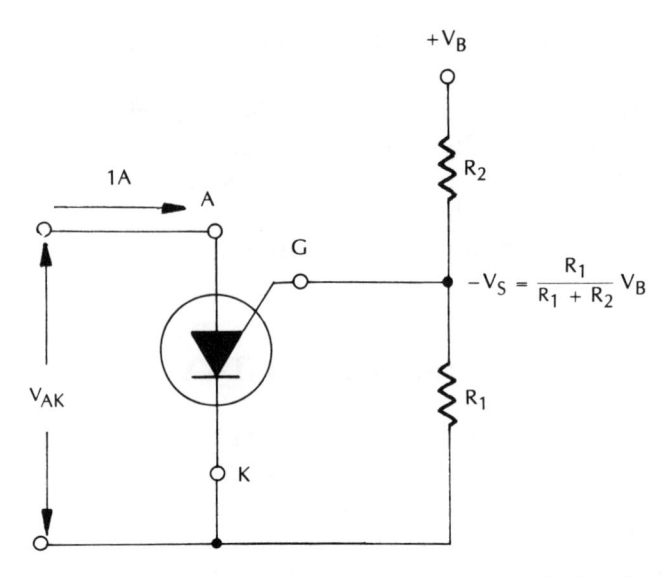

Figure 9-7: Programmable unijunction transistor (PUT) with "program" resistors R_1 and R_2.

MAXIMUM RATINGS				
RATING	SYMBOL	VALUE	UNIT
Power Dissipation Derate above 25°C	P_f $1/\theta_{JA}$	300 4.0	mW mW/°C
DC Forward Anode Current Derate Above 25⌅ 25°C	I_T	150 2.67	mA mA/°C
D DC Gate Current	I_G	± 50	mA
Repetitive Peak Forward Current 100µs Pulse Width, 1.0% Duty Cycle 20µs Pulse Width, 1.0% Duty Cycle	I_{TRM}	1.0 2.0	Amp Amp
Non-Repetitive Peak Forward Current 10µs Pulse Width	I_{TSM}	5.0	Amp
Gate Gate to Cathode Forward Voltage	V_{CKF}	40	Volt
Gate to Cathode Reverse Voltage	V_{GKR}	−5.0	Volt
Gate to Anode Reverse Voltage	V_{GAR}	40	Volt
Anode to Gate Voltage	V_{AK}	± 40	Volt

Table 9-1: Maximum ratings for the PUT 2N6027/8.

Silicon Bidirectional Switches (SBS)

Next, let's look at the SBS symbol (Figure 9-7 F). The letters SBS are an abbreviation for Silicon Bidirectional Switch. The application for this solid state device is similar to the DIAC we have explained, but it has gate electrodes that permit synchronization. Two types are MBS4991 and MBS4992. The main differences between the two are switching voltage and current ratings, plus other electrical characteristics such as forward on-state voltage, holding current, etc.

These bidirectional switches are also called *Bidirectional Diode Thyristors* and they are designed for full-wave triggering in triac phase control circuits, half-wave SCR triggering applications, and voltage level detectors.

Optically Coupled Triac Drivers

Figure 9-6 G, the symbol for an optically coupled triac driver, contains an *infrared* LED and a bidirectional photo-detector in one isolated plastic DIP. These optically isolated devices are used as a triac driver and, typically, have a high isolation voltage (I_{SO}) rating: V_{ISO} = 7500 volt minimum.

Motorola's MOC3030 and MOC3031 are six-pin **IC**'s that are designed for use with a triac in the interface of logic systems, to equipment powered from 120 Vac power lines. There actually are many applications for these devices and all related **IC**'s (there are quite a few on the market today). For instance, computer viewing screens (CRT's), printers (hard copy machines), and almost any experiment or similar application where protection of expensive equipment and/or personnel is desirable, could use an optical isolater.

Optoisolators

In the typical LED/phototransistor optoisolator, a transparent glass window called a *dielectric channel* isolates and insulates the LED chip from the phototransistor. By this means, an isolation of 25,000–75,000, etc., volts is readily achieved. There are quite a few different source/detector combinations, as has been pointed out. Most sources are LED's, but not all. Incidentally, it is very easy to make your own optoisolator (see Chapter Twelve for the details). **IC**'s such as the MOC3030 can be shown schematically. See Figure 9-8.

It bears repeating that one of the most important applications for optoisolators is high-voltage isolation for ex-

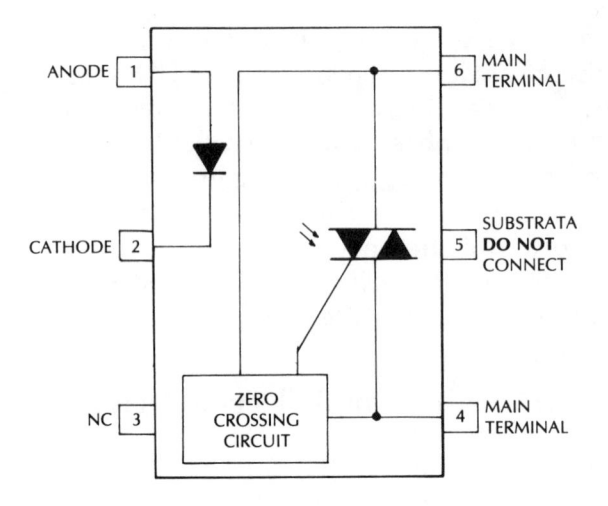

Figure 9-8: Schematic and pin diagram for a MOC3030 and MOC3031 opto coupler zero crossing triac driver.

pensive equipment such as computers and for personnel safety. Figure 9-9 is an illustration of how you might use a MOC3030 in a hot line switching application circuit.

In this circuit, Motorola has shown the load connected to the neutral side of the 120 Vac line. However, you can place the load in either side if you remember that the load will be "hot" in reference to neutral (usually ground) if you place it in the hot side of the 120 Vac line. *Note:* Although this can be done, it is *not* recommended!

The value of R_{in} shown connected to pin 1 (the anode of the internal diode—see Figure 9-9) can be calculated if you

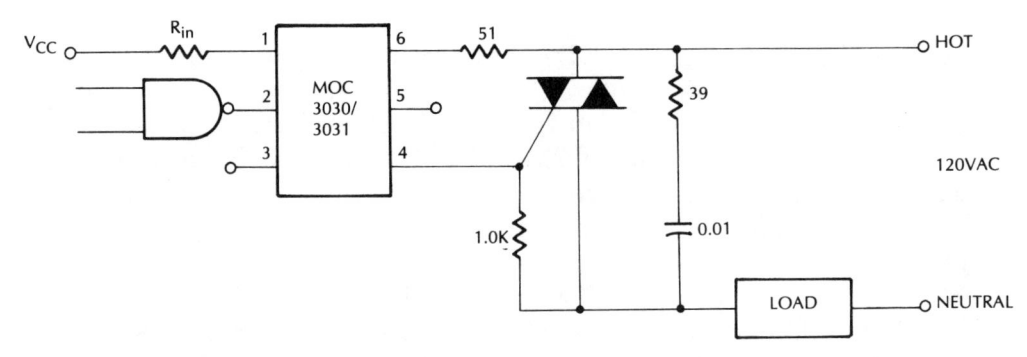

Figure 9-9: Typical circuit using a MOC3030/31 when hot line switching is required.

TRANSISTORS, THYRISTORS, OPTOELECTRONIC DEVICES

know the continuous forward current rating of the IC port. The forward current (I_F) is equal to the rated trigger current (LED trigger current, I_{FT}), which is 30 mA for the MOC3030, and 15 mA for the MOC3031. Also, the resistor and capacitor connected in series (the 39 ohm resistor and 0.01 μF capacitor) may not be needed with some triacs and loads.

Using Opto Coupler/Isolators for TTL to MOS Interface

The 4N35, 6, and 7 opto coupler/isolator with transistor output that is constructed with a NPN phototransistor and PN infrared emitting diode may be used as a **TTL** to **MOS** (P-channel) interface. Figure 9-10 shows how this could be done.

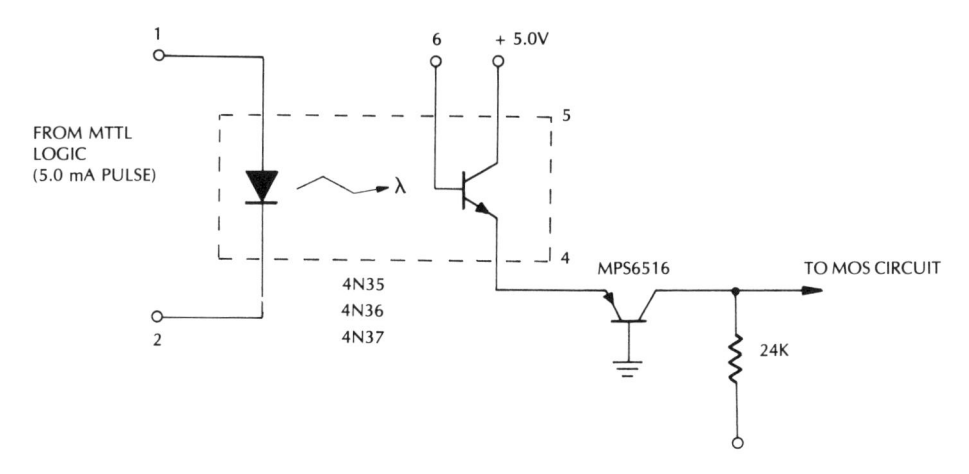

Figure 9-10: Using an opto coupler/isolator to interface TTL to MOS (P-channel).

How Semiconductor Phototransistors Operate

As an experiment, remove the metal top on a fairly large power transistor. This will require a hacksaw. Cut completely around the transistor (very carefully) until you can lift the metal top off the transistor. Don't destroy what's inside the transistor's metal package! This is the chip and it is what you will need for this experiment.

Next, connect your current reading meter to two of the transistor terminals. Set the meter to read a low current value (a few milliamps). Then shine a strong flashlight directly into the exposed chip. Or, better yet, place the chip and meter hookup in direct hot sunshine. At this point, you should see a current reading on your meter. If not, change your meter leads to two other terminals on the transistor. In some cases, you may have to use a small flame as a light source (a kitchen match, candle, etc.).

The reason for suggesting strong sunlight or a small flame is that when light (infrared or visible) of the proper wavelength impinges on a semiconductor material such as a PN or NP junction, the concentration of charge carriers will be found to increase. Bright sunlight has both visible and infrared frequencies over a broad spectrum, and a burning object (such as a match) also radiates visible and infrared frequencies.

Briefly, the actual operation of a phototransistor depends on the biasing arrangement and light frequency. For instance, if a PN junction is forward biased, the increase of current flow through the junctions will be relatively insignificant when the proper light frequency is impinged on it. On the other hand, if the same junction is *reverse biased*, the increase in current flow will be considerable and is a function of the light intensity. Therefore, *reverse bias is the normal mode of operation*.

Now, if the PN junction we have been discussing is made the collector base diode of a bipolar transistor, the light-induced current is the transistor base current. The base lead of the transistor can be left as an open terminal, or it can be used to bias up to a steady state level. As you know, a change in base current can cause a significant change (increase) in collector current. Thus, light stimulation causes a change in base current, which in turn causes an increase in collector current and, considering the current gain (h_{fe}), a relatively large increase.

All *silicon* photosensors (phototransistors, etc.) respond to the entire visible radiation range as well as to infrared. In fact, all diodes, transistors, Darlingtons, triacs, etc., have the same basic radiation frequency response (which actually

TRANSISTORS, THYRISTORS, OPTOELECTRONIC DEVICES

peaks in the infrared range). That is why manufacturers such as Motorola manufacture infrared-emitting diodes. See Figure 10-2 in Chapter Ten for an infrared-light-emitting diode spectral response.

Photo Darlington Amplifiers

Figure 9-11 shows the internal schematic of a photo Darlington amplifier with an output circuit (R_L) included. These plastic NPN silicon photo Darlington amplifiers have a clear plastic case that permits visible and near-infrared light to impinge on the silicon photo detector. They have extremely high radiation sensitivity and are very stable. A less expensive device that will do basically the same job is the phototransistor or photodiode. However, phototransistors are only moderately sensitive compared to the photo Darlingtons. The claim to fame that the photodiodes have is their high speed (for example, 1.0 ns).

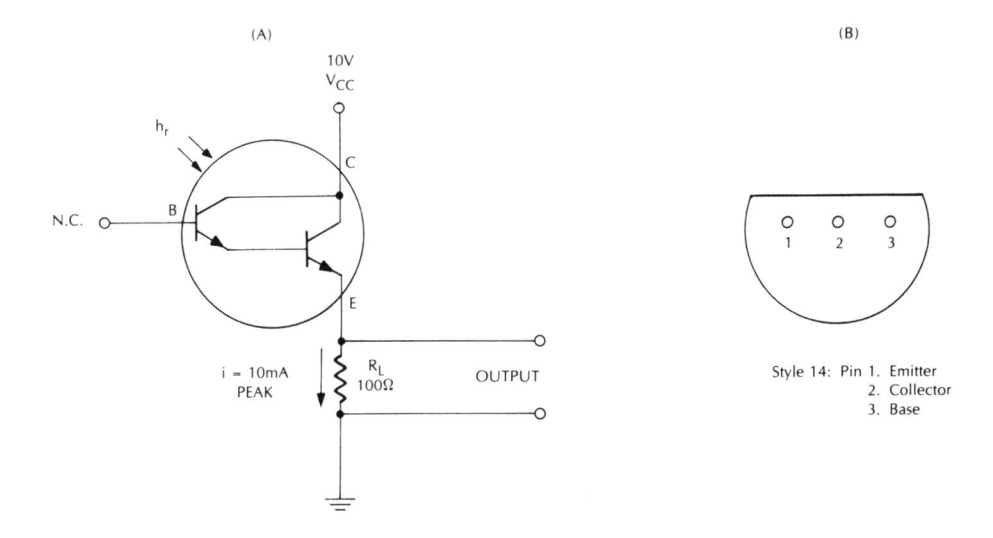

Figure 9-11: Photo Darlington amplifier 2N5777 through 1N5780 (12, 25, and 40 volt). (A) Schematic diagram. (B) Package with pin identification.

Practical Mounting Guide for Power Semiconductors

It is important that you realize that for current and power ratings of a power transistor, thyristor, or any other semiconductor devices such as low current, lead mounted transistors, **MOS IC**'s, small signal and other devices not requiring heat sink mounting, some form of heat dissipator is required to prevent the component's internal temperature from rising above the manufacturer's rated value. If a heat sink and/or fan is not used, you are running the risk of destroying a power semiconductor device.

Poor mounting practices by technicians, experimenters, and hobbyists can result in failures of power semiconductors. This is particularly true when any of the various plastic packaged semiconductors are used. For instance, bending the leads to fit a socket has caused the untimely death of many an electronic component. Figure 6-19, Chapter Six, shows the mounting hardware for a TO-126 packaged semiconductor. These packages are called *Thermopad*® plastic power packages, by Motorola. An important feature of mounting this type of package is the compression washer shown. For your convenience, the compression washer is again shown in Figure 9-12.

The life span of semiconductors packaged in this manner depends largely upon whether you use a compression washer, and how well you burr the hole you drill into the mounting surface. *The heat sink cannot be perforated board*. It *must* be a *metal base*. Also, do not tighten the machine screw any more than is needed to apply the proper force to the bell compression washer (*don't* flatten the washer excessively).

You will find that machine screws and nuts are very good fasteners for all types of semiconductor packages that have a hole in them. But, again, use a compression washer! To be truthful, you can use sheet metal screws in some cases, but be careful that they do not lift the semiconductor package off your heat sink. To improve contact between the package and heat sink, it is frequently better to use one of the thermal greases on the market. *Note:* In some cases, manufacturers recommend that you do not use the thermal compounds when mounting a

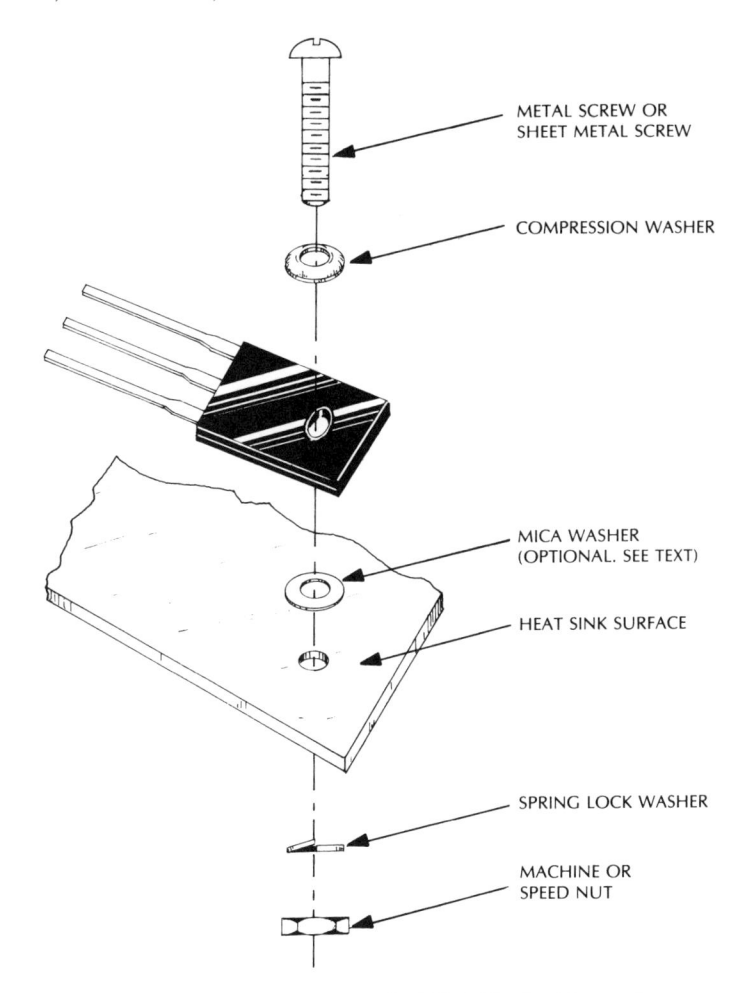

TRANSISTORS, THYRISTORS, OPTOELECTRONIC DEVICES

Figure 9-12: Machine screw mounting of a TO-126 plastic package power semiconductor.

certain device. This information is included in the semiconductor spec sheets.

Specific data for each class of package is given in separate data sheets for each package, and manufacturers usually supply mounting hardware for all power packages. Nevertheless, as a general rule you can do your own package mounting of medium power semiconductor devices if you are careful and follow the instructions on the manufacturer's spec sheet.

TO-3 PACKAGE

INSULATOR

HEAT SINK

PLUG-IN TERMINAL

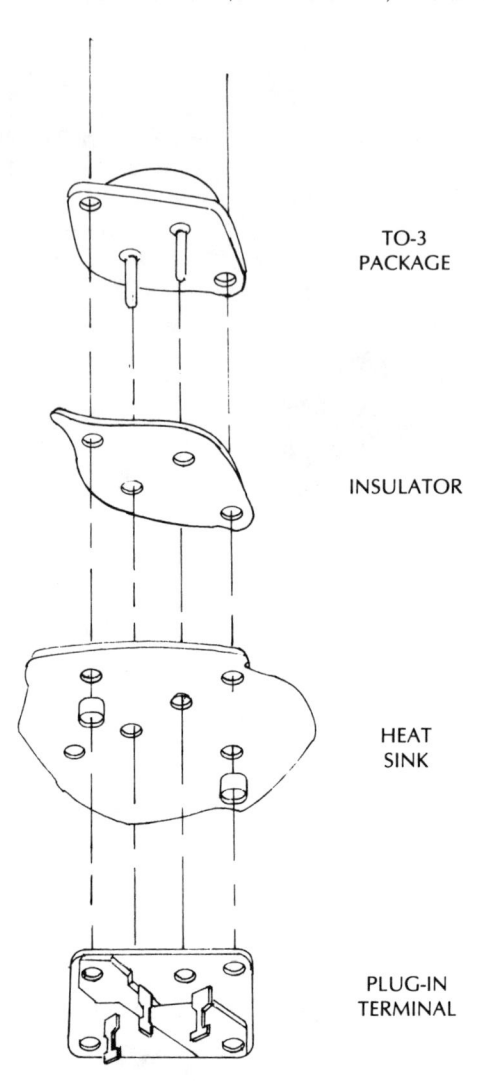

Figure 9-13: Isolating a flat-mount power semiconductor from a chassis. Notice, the heat sink is also not connected to the chassis. The socket center plug-in terminal is the case (anode) connection.

Guidelines for Electrical Insulation of Power Semiconductors

The problem of insulating the case (usually, the anode) of power semiconductors is a common one for many experimenters. For the very best results, it's best to isolate the entire assembly—semiconductor device and heat sink—from the chassis (ground). An example of this type of isolation is shown in Figure 9-13.

There are times when you will find that it is not practical to isolate the entire assembly, as shown. In these cases, use an insulator between the chassis and semiconductor. But, in this type of mounting, in almost every case a thermal compound (such as Radio Shack's "heat sink compound," a silicon base grease) becomes increasingly important. When you are working with *high power*, it is best to specifically ask for an insulating grease manufactured for this purpose. What you want is a reduction of thermal resistance between the semiconductor and heat sink. Remember, clean your work surface before applying the heat sink compound and be sure your mounting screws are tight.

CHAPTER TEN

Basic Fundamentals for Radio Frequency Solid State Devices and Circuits

The fast-growing field of home satellite TV systems, cable TV systems and direct broadcast satellites, in both the U.S. and Canada, offers opportunities for the progressive service technician who understands radio frequency (rf) semiconductors. Figure 10-1 clearly illustrates that high frequency solid state semiconductor packages are different in appearance from the standard audio frequency transistor packages most of us are familiar with.

This chapter not only includes an introduction to the subject of rf fundamentals needed by anyone dealing with rf equipment and today's semiconductors (such as those shown in Figure 10-1); it also shows you what to do as well as how to do it. The presentation is mostly practical, and therefore essentially nonmathematical.

Frequency Spectrum

The term *frequency* has different meanings for different electronic technicians. For example, a person working with audio systems will usually think of audio frequencies (abbreviated af) as falling roughly between 15 cycles per second (Hz)

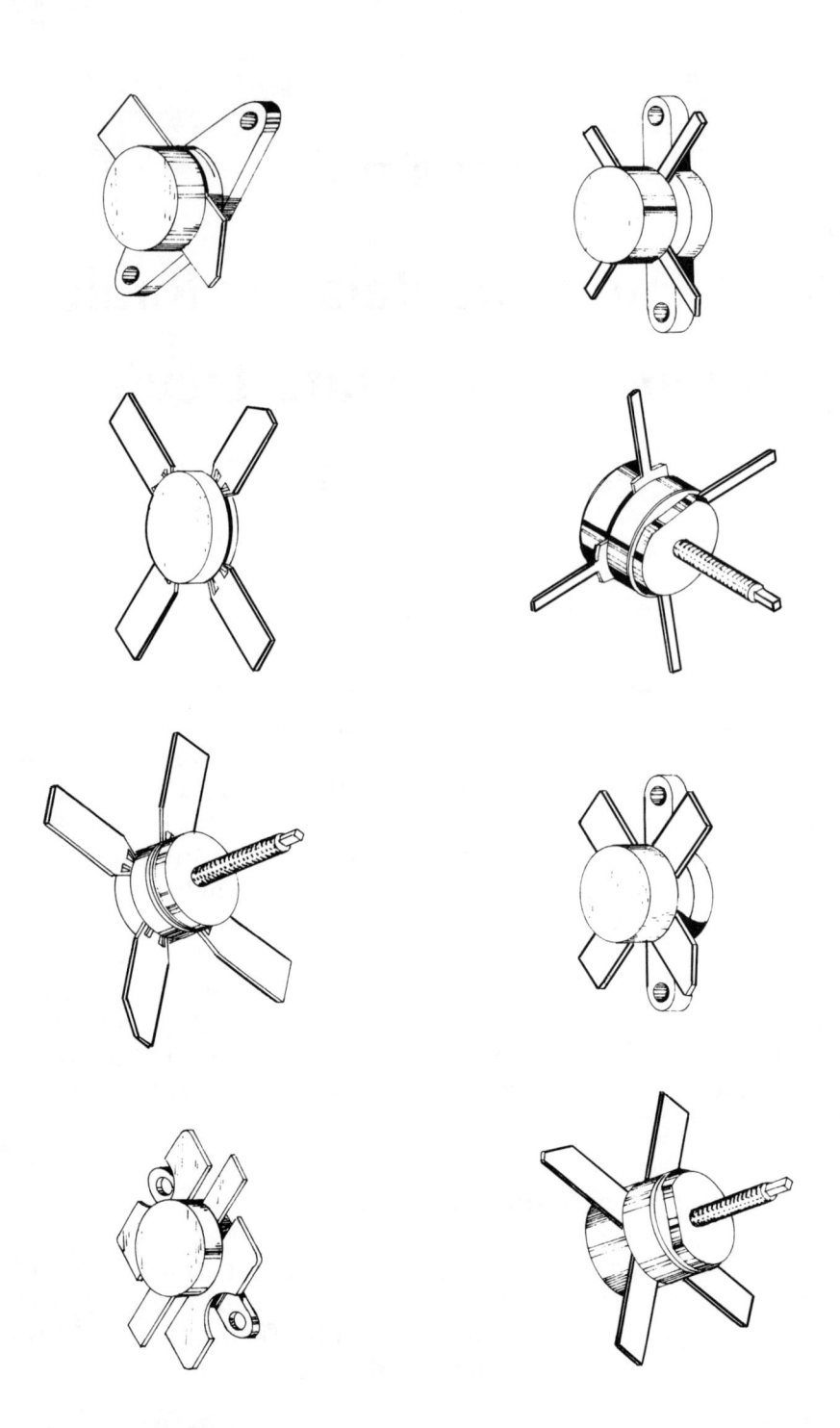

256 RADIO FREQUENCY SOLID STATE DEVICES AND CIRCUITS

Figure 10-1: State-of-the-art packages used for rf power transistors.

FREQUENCY RANGE	DESIGNATION	
30 to 300 hertz	ELF	Extermely Low Frequency
300 to 3000 hetz hertz	VF	Voice Frequency
3 to 30 kilohertz	VLF	Very-Low Frequency
30 to 300 kilohertz	LF	Low-Frequency
300 to 3000 kilohertz	MF	Medium-Frequency
3 to 30 megahertz	HF	High-Frequency
30 to 300 megahertz	VHF	Very-High Frequency
300 to 3000 megahertz	UHF	Ultra-High Frequency
3 to 30 gigahertz	SHF	Super-High Frequency
30 to 300 gigahertz	EHF	Extremely-High Frequency

Table 10-1: Frequency bands.

and 20,000 Hz. But someone working on rf equipment will probably think of a frequency range of something like 10 kHz and 30,000 MHz. Actually, frequencies are divided into many subgroups. Table 10-1 lists the frequency bands. Referring to this table, radio frequency ranges from VLF to EHF (3 kHz to 300 gHz).

Frequency Versus Wavelength

In Chapter Nine, we discussed photo transistors and other opto devices such as photo diodes, etc. It was pointed out that many of these semiconductors operate in the near infrared band. A few years ago, infrared was often described in terms of Hertz. Today, however, you will usually find that infrared is considered in terms of wavelengths. Typically, these range all the way from 300 to 3 microns or micrometers in wavelength. Figure 10-2 shows the wavelength in micrometers (μm) for a pulsed infrared emitting diode, in reference to the constant energy spectral response.

Photosensors such as photo transistors respond to the entire visible radiation range as well as to infrared. The visible spectrum ranges in wavelength from 0.7 to 0.4 μm. Incidentally, it should be pointed out that all silicon photosensors (diodes, transistors, Darlingtons, triacs) show the same basic radiation response to the same wavelength; i.e., they respond to the near infrared peak, as shown in Figure 10-2.

As you can see, when dealing with electromagnetic waves it is often preferable to use wavelength (λ) rather than

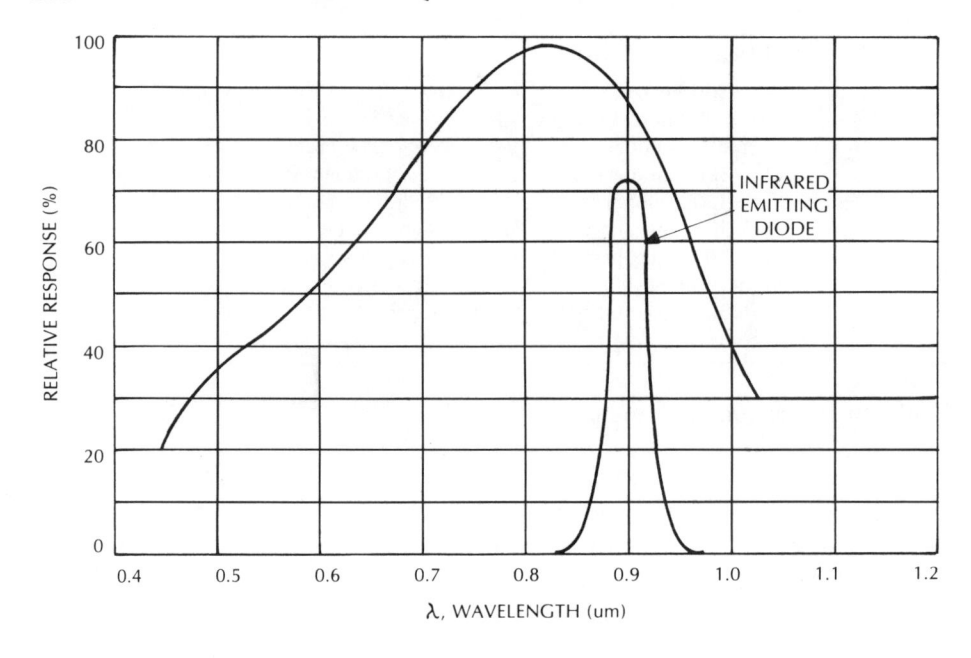

Figure 10-2: Infrared light-emitting diode spectral responses.

frequency. Generally, in situations that use wavelength, it is because wavelength is more easily measured than frequency. In such a case, frequency would be calculated from the measured wavelength using the familiar formula: frequency in Hz = speed of light/wavelength in meters.

RF Semiconductors

Semiconductor devices that are designed to operate in the VHF part of the spectrum and in the so-called *microwave band* are easily identified because they usually require special designs for the terminations (leads) of the transistors, diodes, etc. These designs (see Figure 10-1) are required to minimize the inductance caused by the pin-connecting arrangements.

The devices shown in Figure 10-1 are called *Strip-line Opposed Emitters* (SOE) transistor packages. Figure 10-3 shows how the transistor chip leads are connected to the flat rectangular strips used as connecting leads. This illustration is shown

RADIO FREQUENCY SOLID STATE DEVICES AND CIRCUITS 259

in a submounted package (an SOE package) with the top cap removed so that you can see the inside connections of the transistor.

Impedance (at rf frequencies)

If an impedance mismatch exists between the rf power transistor, the transmission line (for example, the strip-line leads shown in Figure 10-3) and the next stage (the load), some

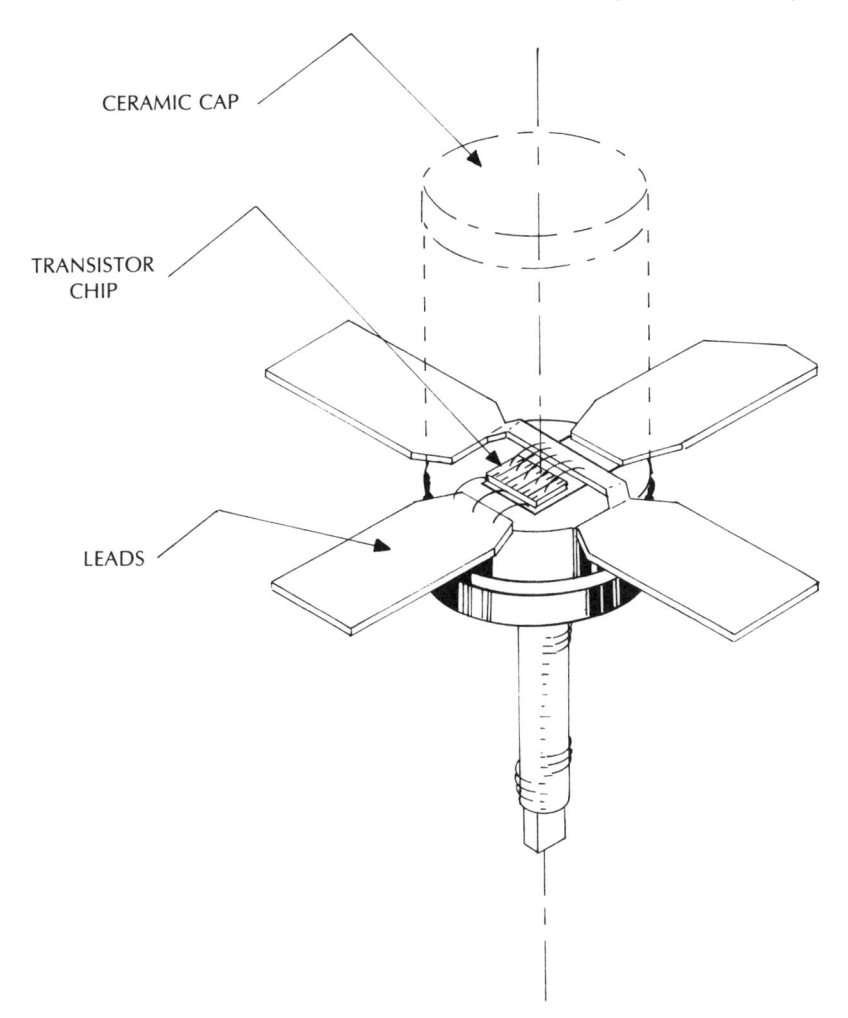

Figure 10-3: Low inductance strip-line leads connected to a transistor chip.

of the transmitted energy will be reflected back into the transmission line. This will set up what is called *standing waves* of voltage and current within the system and, most important, will possibly damage a high power component (and certainly cause a loss of energy).

This all means that, for the best possible energy transfer from stage to stage, *impedance matching* is necessary when you are working with rf systems. For example, if the signal source impedance is 50 ohms (a typical value), the transmission line (coaxial cable, etc.) is 50 ohms, and the input impedance of the load is 50 ohms. The impedance of the system is matched and there will be maximum transfer of energy from the source to the load.

The output impedance of rf power transistors is usually given by all manufacturers, on the transistor data sheets, as a capacitance C_{out}. In general, the capacitive reactance output impedance will more nearly approach a resistive output impedance (which is what you want) in the upper part of the transistor operating bandwidth. However, the load that must be presented to an rf transistor is frequently indicated on the transistor data sheet.

Matching Networks

There are numerous basic impedance-matching networks that can be used when one must connect electronic components having different impedances into the same system. For instance, at sufficiently high frequencies (commercial TV frequencies and higher), matching networks using quarter-wave transformers may be used to accomplish impedance matching. Matching networks normally take the form of filters—both passive and active. As an example of an active filter, an **OP AMP** with capacitive feedback (voltage leads current by a given amount of degrees) can be designed to act like a circuit that will appear as an inductance, or even an entire **LC** network.

In summary, every rf system, no matter what its physical form, will have a definite value of impedance at the point where the transmission line (coax, or whatever) is to be connected to each component (rf transistor amplifier, rf signal generator, or any other rf device). The problem is to transform this impedance to the proper value to match the connecting lines so that all components of the entire system are matched to each other *ohm for ohm*.

As a simple example, your TV antenna has an output impedance of 300 ohms. Therefore, the down lead from the TV antenna to the TV receiver must be 300 ohms. Next, the input impedance to the TV receiver must be 300 ohms. If each of the components is not the same impedance, it must be matched by using a matching network. For instance, if there is a 75-ohm coaxial cable coming down from a 75-ohm antenna, you would need to purchase a 75-to-300-ohm matching transformer from an electronic supply house such as Radio Shack, etc., to match the down lead to the receiver. Remember, impedance matching is important in any rf project you attempt, whether it's connecting test gear, antennas, or rf transistors in a circuit.

Testing Solid State Radio Frequency Systems

You have one slight technical advantage with very low-priced solid state rf systems, be they CB radios, FM two-way radios, aircraft transceivers, marine radio telephones, or ham gear. Their design is simple. For example, budget priced CB radios' front panels contain only volume, squelch, and channel knobs. Generally there is an LED channel readout, and occasionally there's a meter, and sometimes a switch—but that's about it. Low-cost CB radios are simpler, by far, than their more expensive cousins: aircraft transceivers, marine radio telephone, and other sets such as deluxe AM/SSB mobile CB rigs.

When testing complete rf systems such as these, connect for transmitter testing first. Solid state transmitters are easier to service and quicker to diagnose than solid state receivers. Your first step is to measure the transmitter rf power (unmodulated). To do this, if you are working on a CB transceiver or a ham rig, etc., you'll need a test meter. One that you could use is the Power/Modulation/SWR meter sold by Radio Shack. Typically low-cost meters such as these will measure power output in three ranges: 0-5 watts for CB, 0-5 watts and 0-500 for amateur band equipment. Frequently the same set measures percent of modulation from 1 to 120%, and can also be used to check the standing wave ratio (SWR) of the antenna system.

Another piece of test gear that one should have when checking rf systems is a field strength meter. This test instrument will help you operate the rf system with maximum efficiency. Generally, they are quite inexpensive (under $20.00), and they will measure SWR forward and reflected power to 1000 watts. Most of the CB types have a 52-ohm characteristic impedance and operate between 2 and 30 MHz.

It is important that you realize that measuring output power, SWR, or percent of modulation does not require an FCC operator's license. Nor is an FCC license required to work on a transmitter if it is connected to a *nonradiating* dummy antenna. However, if any repairs or changes are made to a transmitter circuit (other than amateur rigs) that affects power out, modulation, or frequency, before it is connected to a radiating antenna it *must* be checked by a licensed technician.

To check communications receivers, you will need the right signal generator and a good diagnostic ability. But IC's, in either a transmitter or a receiver, should not worry you much. For example, a certain input signal (usually given on the equipment schematic or spec sheet) should produce a specified output signal. If it doesn't, check the IC's dc voltages. If they are OK and you still don't get the proper output signal, the IC is probably defective and you should replace it.

Other than the special equipment listed, you will need the standard shop equipment. This includes a voltmeter and probes (both rf and high impedance). Prior to using anything but your voltmeter, check the power supply (perhaps the most important circuit in the transceiver). Before you undertake any other procedure, measure and verify that these voltages (regulated and unregulated) are where they are supposed to be.

Controlled Quality Factor (Q) Transistors

When, as a technician, you encounter statements such as "family of *controlled Q transistors*," it can become confusing because the letter "Q" is used in the electronics field for many things; for example, quantity of electric charge expressed in coulombs, the identification letter for transistors in a schematic diagram, the ratio of reactance of a circuit to the resistance of a circuit, and a measure of sharpness or the frequency selectivity of an electrical system such as a transistor (the one we are discussing when we say *controlled Q transistor*).

Actually, what is desired in controlled Q transistor systems is that the input and output network present a certain impedance (usually 50 ohms), and that the transistor will operate at the desired output power and efficiency over a given bandwidth. The bandwidth of Motorola's controlled Q transistor is about 450 to 512 MHz. A variety of police, taxi, trucking, and public utility maintenance communications systems operate in this part of the UHF spectrum.

Internal matching in the transistor is a transformer network (fabricated inside the transistor package) that is necessary to raise the normally low base impedance of the transistor to the desired 50 ohms, over the entire bandwidth of operation. As an example, in some designs the input/output filters are low-pass networks composed of a shunt capacitor, a series transmission line, and a shunt capacitor. The series transmission line is used in place of an inductor because it may be printed and, of course, reduces cost.

It is very important that you understand that any external circuits must be coupled to a transistor. Two circuits are said to be coupled when voltage or current in one network produces a voltage or current in the other one. For proper coupling of two rf circuits, filter networks frequently are used to match one impedance value to another one. These are often called *matching networks*.

The most important consideration in any coupling network you place between a transistor and a load is the amount of power delivered to the load over the desired bandwidth (Q of the circuit). The circuit you want to couple to the transistor has inductive and capacitive reactance, plus resistance. The reactive effects associated with even small inductance and capacitance place drastic limitations on UHF transistor operation if you do not "cancel them out" by placing the proper coupling network (inserting the right amount of the opposite kind of reactance) between the source (transistor) and a reactive load. It is beyond the scope of this book to cover coupling networks; however, most handbooks of practical electronics reference data and manufacturers' reference data will contain this information.

Another term you will encounter is *insertion loss*. The insertion loss is the ratio of the power delivered to the load with the coupling network in the circuit, to the power delivered to the load without the coupling network.

Introduction to Modern Resonant Circuits

Solid state devices such as tuning diodes (voltage-variable-capacitance diodes) are used as the variable element of a circuit to change (or tune) its *resonant* frequency. To understand what resonance means, let's examine resonant circuits.

In general, all rf circuits (whether within an **IC** or designed using discrete components) are based on the use of resonant circuits consisting of some form of capacitance and inductance connected in series or parallel, as shown in Figure 10-4.

At the resonant frequency (see Figure 10-4), the inductive and capacitive reactance (X_L and X_C) are equal, and the parallel circuit presents a high input impedance, or a low input impedance in the case of the series circuit. However, in either circuit any combination of capacitance and inductance will have some resonant frequency, as the formula shows.

To permit tuning of the resonant circuit over a given frequency range, either the capacitance or the inductance can be variable. Therefore, when we replace the capacitance with a tuning diode, capacitance becomes the variable element in the resonant circuit. In any case, the two basic design considerations for rf resonant circuits are resonant frequency and the **Q** factor that was discussed in the previous section of this chapter.

Usually, resonant circuit **Q** is measured at points on either side of the resonant frequency where the signal amplitude is down either 6dB (0.707 × peak voltage or current), or down 3dB (half-power), in respect to the resonant value of voltage or power, respectively. Figure 10-5 shows current in series-resonant circuits having different Q's (see Figure 10-5 A). Relative impedance of parallel-resonant circuits with different Q's is shown in Figure 10-5 B.

The formula for the **Q** of a circuit that has the various bandwidths shown in Figure 10-5 is

$$Q = F_0/(F_1 - F_2)$$

where F_0 is the resonant frequency and F_1 is the lowest fre-

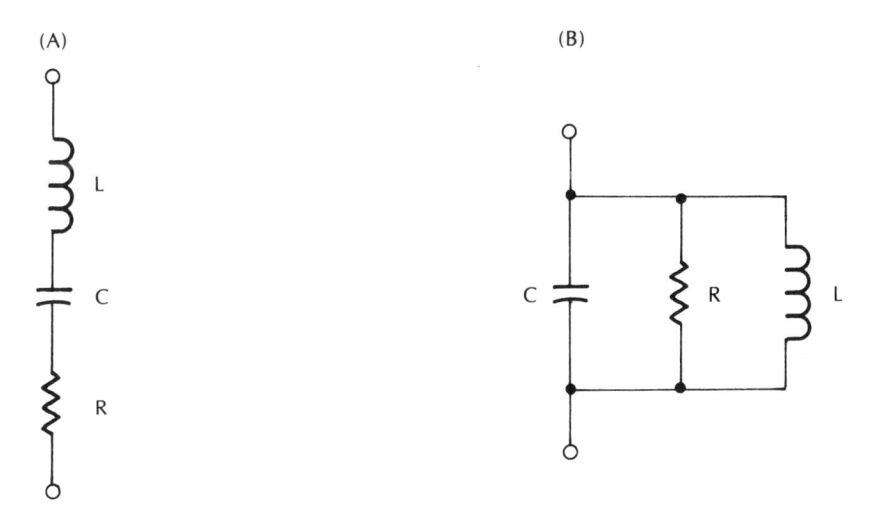

RADIO FREQUENCY SOLID STATE DEVICES AND CIRCUITS

Figure 10-4: A series circuit is shown in (A). The circuit shown in (B) is parallel. Resonant frequency $= \dfrac{1}{2\pi\sqrt{LC}}$, where L is in henrys, C is in farads, and frequency is in Hz.

quency and F_2 is the highest frequency at the 3dB bandwidth points. Or, another formula you can use is

$$\text{bandwidth} = (3\text{dB}) = F_0/Q.$$

If you work out a few example problems, you will find the Q must be increased for an increase in resonant frequency if you desire to retain the same bandwidth. Another point to remember is that you must decrease Q if you wish to increase bandwidth.

In summary, the input and output tuning circuit of an rf solid state amplifier must perform two functions:

1. Tune the amplifier to the desired frequency. Peak up at the resonant freqüency while, at the same time, permitting the half-power bandwidth frequencies to pass at 0.707 × peak amplitude.
2. Match the input and output impedances of the transistor (for example, the control Q transistors we discussed previously, or any other rf transistor, for that matter) to the impedance of the rf signal source and load. If you do not do this, there will be a considerable loss of signal, due to the mismatch and resulting standing wave ratio.

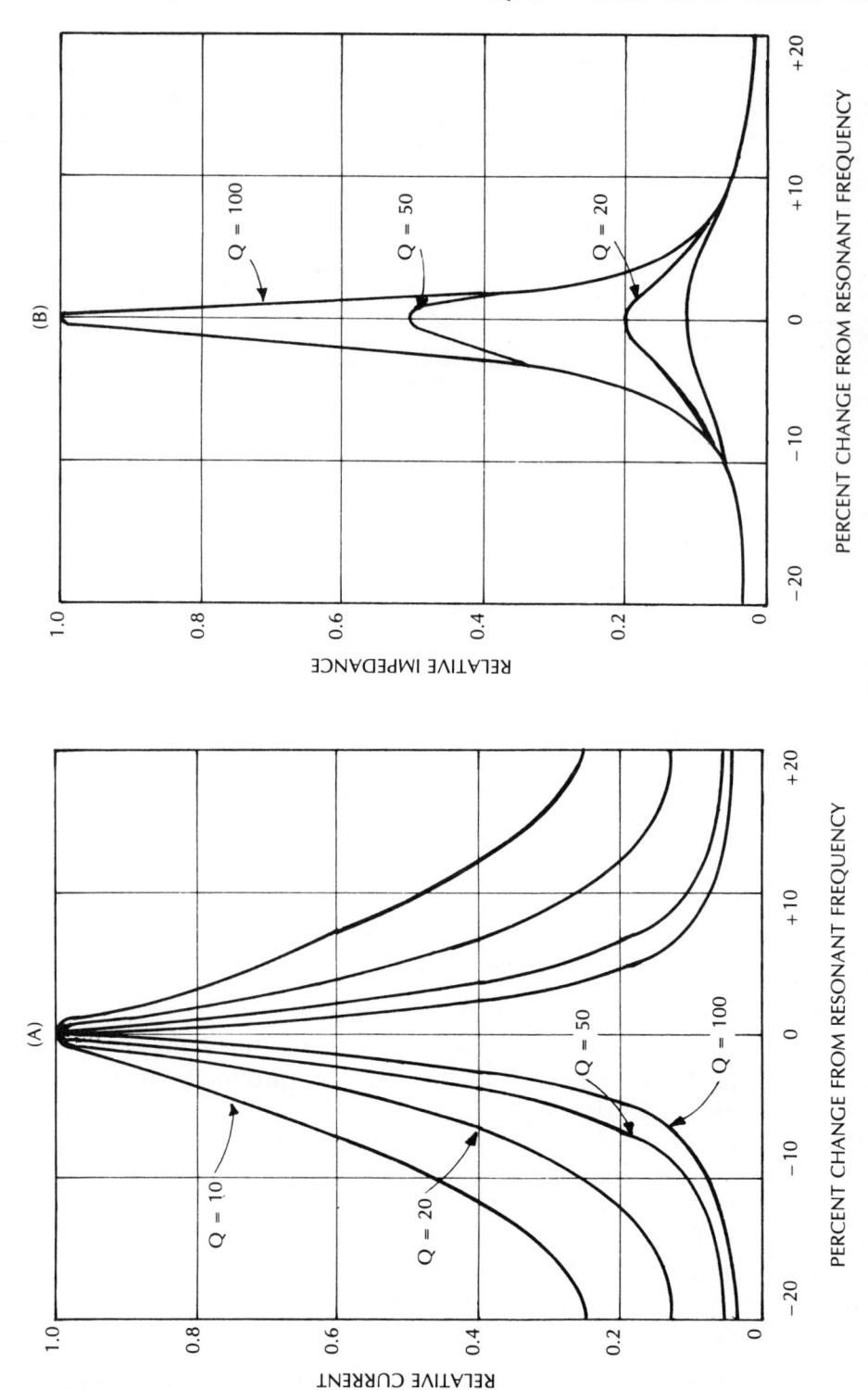

Figure 10-5: Series-resonant circuits having different Q factors but same peak current. Note—the lower value Q, the wider the circuit bandwidth (shown in A). The B section shows parallel-resonant circuit impedance characteristic at different Q's.

Practical Solid State RF Circuits and Devices

Regardless of the application—amateur radio transmitters, AM, FM transmitters, or single sideband CB transceivers—rf semiconductors are common elements in all the transmitters that most of us come into contact with today. These solid state devices are used in oscillators, rf amplifiers, frequency multipliers, and frequency down-converters. For example, one application of a down-converter is a home satellite TV system. This system converts satellite frequencies to the lower commercial broadcast frequencies needed by home TV receivers.

When you select a semiconductor device (such as an rf transistor), the one you choose will depend primarily upon the frequency band in which you want to operate, and the power output needed. For instance, a simple transistor oscillator circuit that meets your frequency stability needs could be used as a transmitter at some low rf frequency, but the power output would be quite low.

As a general rule, the output of a very stable oscillator (usually crystal controlled) is fed into one or more amplifiers to bring the power up to the desired level. The last amplifier is an rf power amplifier, in every case. Figure 10-6 shows a working schematic of a single transistor rf power amplifier.

In this type of rf amplifier, the radio frequency chokes (RFC) should present an inductance reactance impedance of about 1 k ohm to 3 k ohms at the operating frequency. The bypass capacitors' value should be about 0.001 to 0.1 μf. Also, note that one of the capacitors in the output tuning circuit is labeled "Loading," and the other is "Resonant tuning." When tuning a circuit such as this, one capacitor (tuning) is tuned to the resonant frequency and the other (loading), is adjusted for proper impedance match to the load placed on the rf output line. To test an rf amplifier such as the one shown in Figure 10-6, you can use the test setup shown in Figure 10-7 for adjusting the amplifier.

The rf amplifier should be connected as shown in Figure 10-7. In general, you first adjust the input until you read mini-

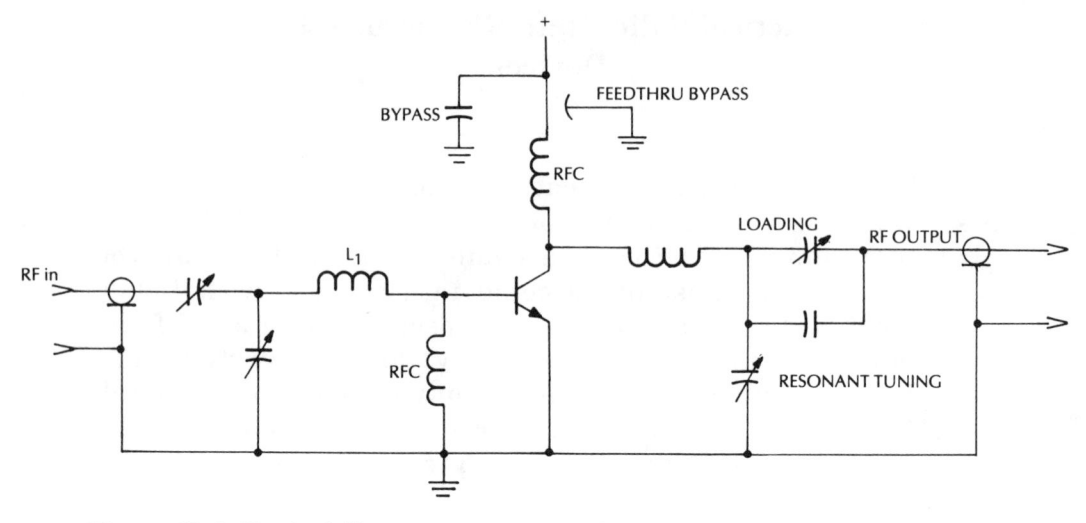

Figure 10-6: Typical discrete component rf power amplifier circuit. The selection of the transistor and components would depend on operating frequency and power input/output requirements of the transmitter design.

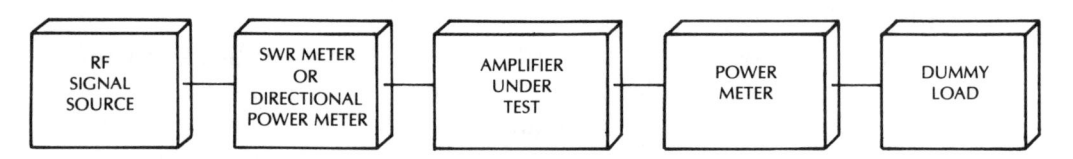

Figure 10-7: Test setup for adjusting an rf amplifier.

mum reflected power (SWR as low as possible), and then adjust the amplifier circuits until you read the highest output power possible on the power meter. This should be done with only a small amount of power being supplied by the rf signal source, and it will probably require tuning the amplifier output circuit capacitors several times. Also, in many cases a small cooling fan is needed to cool the power transistor heat sink during the test—especially if you are going to test for maximum power out. There are many rf power transistors available for application in just about any area one could wish for. Table 10-2 lists a few with different frequency and power rating.

RADIO FREQUENCY SOLID STATE DEVICES AND CIRCUITS

Device Type	P_{in} Input Power Watts	P_{out} Output Power Watts	G G_{PE} Power Gain dB Min	V_{CC} Supply Voltage Volts	Package
MRF466	0.1	3.0 PEP	15	12.5	TO-220
MRF406	1,25	20 PEP	20	12.5	211-07
MRF421	10	100 PEP	10	12.5	211-08
		1.5–30 MHz HF/SSB TRANSISTORS			
MRF476	0.10	0.5	10	12.5	TO-39
MRF 450A	4.0	50	11	12.5	145A-09
MRF458	5	80	12	12.5	211-11
		14–30 MHz, CB/AMATEUR TRANSISTORS			

Table 10-2: Low-voltage rf power transistors.

Radio Frequency Measurements

In the beginning of this chapter, it was shown that when dealing with electromagnetic waves it is often preferable to use wavelength rather than frequency; for instance, when dealing with optoelectronic devices. However, when working with rf circuits that are to be used with any of the rf power transistors listed in Table 10-2, it is usually more convenient to measure frequency rather than wavelength.

By and large, measurements in the rf spectrum require special instruments, and the frequency counter is probably the best low-cost, all-around instrument offered today. These low-priced counters often provide frequency measurements in both UHF and VHF bands; in addition they may be hand-held types for in-the-field work, or they may be more sophisticated bench types for in-shop use. In most cases, these counters can be directly connected to the circuit under test. Or, for frequency counts without a direct connection, you can use an antenna to take a count of the frequency being radiated from the device under test.

Special RF Measurement Precautions

If you have not tried it, you will soon find out that radio frequency tests are quite a bit more demanding than audio frequency tests. Sales catalogs such as Heathkit tell you all about how accurate their frequency counters are, but they don't tell you about problems you can encounter when using the instrument at the workbench. For example, how easy it is to get into trouble due to stray coupling between test leads or between circuit components. Or how easy it is to lock on to a harmonic rather than the fundamental frequency. You will find that even the smallest value of stray reactance (for instance, a resistor wire lead can easily become a small inductor at UHF frequencies), is one of the factors that will make high frequency measurements more exacting. The following precautions will go a long way toward making your day at the workbench more pleasant when rf testing:

1. Keep all wire leads as short, straight, and as large a diameter wire as possible.
2. Use a common ground point for all test instruments, circuits, etc. That is, avoid using a separate ground point for each rf system, whether it is a frequency counter, transmitter, receiver, or any other piece of the test setup.
3. For proper operation, use bypass capacitor and rf chokes (RFC's) in your circuit. The bypass capacitors and RFC's are usually called for in any rf amplifier circuit. See Figure 10-6.
4. Place your hand close to an unshielded oscillator circuit and you will see it change frequency of operation (assuming that you have a frequency-measuring instrument connected to an operating rf circuit). This is called *body capacitance,* and it can detune or reduce signal levels in an unshielded rf circuit. The problem may be caused by improper grounding and/or inadequate shielding.
5. If you are experiencing *drift* (a slow change in frequency), it is probably caused by trying to test equipment before it has had time to warm up properly. The equipment under test, and all test instruments, must be at the same temperature (usually room temperature) and *stabilized* at that temperature.
6. Another common problem is *electrical noise pickup.* This perplexing problem can be held to a minimum if

you always use *short leads, proper power line filters,* and *rf shielding.* In some cases where radio interference is a problem, your best bet is to work in a shielded booth, called a *screen room.* This is a small room completely inclosed (sides, top, and bottom) with a small mesh metal screen. However, today's counters are generally housed in an interference-free cabinet so that such extremes are not necessary.

7. Don't overdrive the circuit under test. This problem can be prevented by using loose coupling between the circuit under test and test instrument. For best results, always use the loosest practical coupling. For example, if you are using a frequency counter with antenna attached, vary the distance between the counter antenna and the circuit you are testing until you have the least signal pick-up you can have and still be able to make a frequency count. This will help assure that you are measuring the fundamental frequency, and improve the accuracy of your measurement.

8. You should calibrate rf test instruments at regular intervals. Many instruments have self-contained frequency standards that can be used for calibration.

A Collection of Formulas Frequently Needed in RF Work

Conductance, Susceptance, and Admittance Formulas

conductance $G = 1/(R^2 + X^2)$ answer in mhos

susceptance $\beta = 1/X_L$ (when $R = 0$) answer in mhos

admittance $\gamma = 1/\sqrt{(R^2 + X^2)}$ answer in mhos

Capacitance and Inductance: Reactance Is Known

$C = 1/2\pi F X_C$ and $L = X_L/2\pi F$

Power Factor Formulas (pf)

pf = $\cos \theta$, where θ is the angle of lead or lag

pf = watts/(current)(voltage)

Inductance and Capacitance Needed for Resonant Frequency When Frequency and One Value Are Known

$$L = 1/(4\pi^2)(Fr^2) C$$
$$C = 1/(4\pi^2)(Fr^2) L$$

Impedance Formula

$$Z\underline{/\emptyset} = \sqrt{R^2 + X^2} \quad \tan^{-1} X/R \text{ (for series circuit)}$$
$$Z\underline{/\emptyset} = RX/\sqrt{R^2 + X^2} \quad \tan^{-1} R/X \text{ (for R and X in parallel)}$$

Decibel Formulas

$$dB = 10 \log P_2/P_1 = 20 \log E_2/E_1$$
$$= 20 \log I_2/I_1 \text{ (when input/output impedance are equal)}$$

$$dB = 10 \log P_2/P_1 = 20 \log E_2 \sqrt{Z_1}/E_1 \sqrt{Z_2}$$
$$= 20 \log I_2 \sqrt{Z_2}/I_1 \sqrt{Z_1} \text{ (when input/output impedances are not equal)}$$

Frequency Modulation Index

modulation index = frequency deviation / modulation frequency.

CHAPTER ELEVEN

How to Select and Use High Frequency Diodes, Transistors, Integrated Circuits, and Modules

In the past, electronic experimenters needed to remember only a few practical semiconductor characteristics in order to breadboard countless projects for various applications. However, digital circuits and things such as personal computers have changed all of this. No longer can a few semiconductor devices be memorized and used in almost every situation.

Not only the semiconductor devices themselves, but also the circuits have changed over wide latitudes, for different applications. Thus, the need for this chapter is very evident to anyone who works with today's high-speed circuits, where cutting nanoseconds is the name of the game.

Introduction to Electronic Tuning and Control Applications

In Chapter Ten, you were introduced to resonant circuits and learned that the resonant frequency of a circuit can be

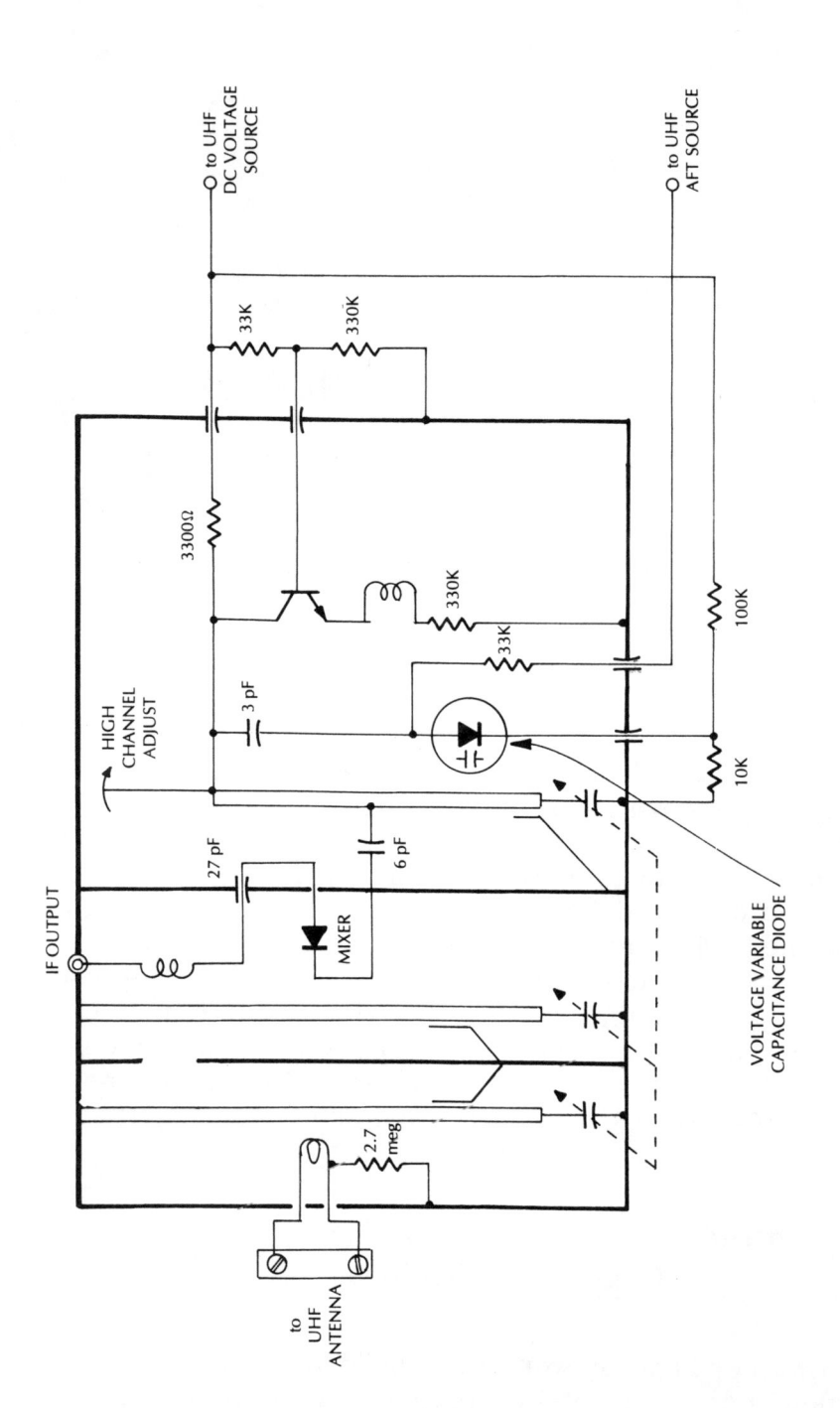

Figure 11-1: Schematic of a television receiver UHF tuner that uses a voltage variable capacitance diode and UHF mixer diode.

DIODES, TRANSISTORS, INTEGRATED CIRCUITS, MODULES 275

shifted by changing either the inductance or the capacitance in the circuit. In the same chapter, you also learned that a voltage variable capacitive diode can be utilized in an oscillator circuit to shift the resonant frequency of the oscillator. To see how this can be done, let's examine a UHF oscillator circuit in a piece of electronic gear we all know—a TV set. See Figure 11-1 for the schematic of a UHF television receiver tuner that is tuned by use of a voltage variable diode.

Tuning Diode

The voltage variable capacitance diode that may be called a *varactor diode* (especially if it is used in AM tuning), *silicon capacitor, voltage controlled capacitor*, etc., is a two-terminal solid state device whose capacitance varies with the applied voltage. Notice that this capacitance diode is in the UHF oscillator circuit and is utilized by the AFT (automatic fine tuning) circuit to adjust the UHF oscillator for optimum tuning. The AFT voltage applied to the capacitance diode determines the capacitance, and therefore the operating frequency of the UHF oscillator.

Mixer Diode

The *mixer diode* must also be a high frequency device because it must mix the incoming frequency (off the UHF television antenna) with the UHF oscillator frequency, to produce the intermediate frequency (IF) that is amplified in the TV receiver IF stages. As you can see from the schematic, this diode is also a two-terminal solid state device.

There are several different circuit arrangements that utilize voltage variable capacitance diodes. However, tuning in these tuner circuits is accomplished by utilizing the characteristics of voltage variable capacitor diodes. This causes them to behave as capacitors when they are reverse biased. The amount of reverse bias voltage applied to the diode will determine its capacitance.

It is now common to find these diodes used for tuning the antenna input, rf amplifier, mixer, and oscillator circuits. Most modern TV's and other receivers select a channel by use of push-buttons which apply the proper voltage for a particular channel to the voltage variable capacitor diodes. The improvement of manufacturing of semiconductor devices is the primary reason behind the greatly improved performance of modern tuners, whether in TV, CB, or test instruments.

Hot-Carrier Diodes

Some of the names you will encounter when working with high frequency diodes are tuning, hot-carrier, and PIN diodes. We have discussed the tuning diode, so let's now take up the hot-carrier diode. The silicon hot-carrier diodes (Schottky barrier diode), MBD101 and MBD102, manufactured by Motorola, are designed primarily for UHF mixer applications but are also suitable for use in ultra-fast switching circuits.

These diodes (there are different packages; see Figure 11-2) have extremely fast turn-on and turn-off times, excellent diode forward and reverse characteristics, lower noise characteristics, and wider dynamic range. Typically, they have very low capacitance—less than 1.0 pF @ zero volts—and a low noise figure of −7.0 dB max. @ 1.0 GHz.

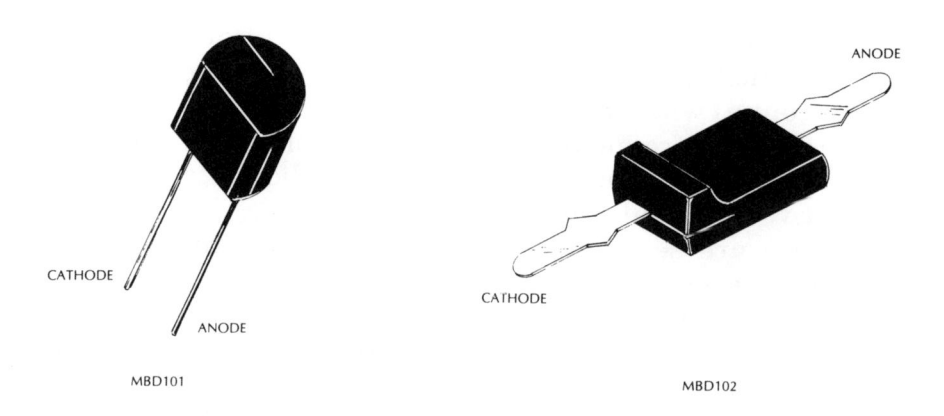

Figure 11-2: Silicon hot-carrier diode packages. (A) inexpensive plastic package. (B) low-impedance mini-L package for high volume requirements.

PIN Switching Diodes

A silicon switching diode may be purchased in the same type package as illustrated in Figure 11-2. The anode and cathode pins will also be as shown in that figure. However, a less expensive package looks like a small carbon resistor (body and two connecting leads); for example, the MPN3500 series, which is designed for VHF band switching applications and general switching circuits.

Notice that these devices will function properly at VHF band frequencies. The storage time for the PIN diode is long enough that it completely fails to rectify at UHF (microwave) frequencies and above. When used in the microwave part of the rf spectrum, a PIN diode behaves like a variable resistor with its value controlled by a dc bias current. Therefore, you will find that it will work well as a variable microwave attenuator. In fact, there are several commercial microwave attenuators that are designed using PIN diodes. These devices are usually called *PIN-diode attenuators* and are a two-part network consisting of two or more PIN diodes controlled by a driven circuit.

Selecting the Right Transistor for Your Application

As has been explained, most of the *needed* information for a particular transistor can be obtained from the data sheet that comes with the transistor when you purchase it. However, this is not true in every case. If you are working with extremely high frequency circuits, such as personal computer circuits, where high frequency switching characteristics are of particular importance, you may have to test a transistor using the actual working circuit you expect it to perform in.

Nevertheless, you will always have to refer to the data sheet to begin, because you will find that each manufacturer uses different ways to present the information you need. However, they all present about the same data: maximum ratings and electrical characteristics. As you can imagine, it would be all but impossible to present all data sheet formats in this chapter (or in a book, for that matter); therefore we will use Motorola's presentation of transistor specifications.

Table 11-1 is the data sheet for an MPS-U02 that is

NPN SILICON ANNULAR AMPLIFIER TRANSISTOR

Designed for general purpose, high-voltage amplifier and driver applications.

High Power Dissipation — P_D = 100W @ T_C = 25°C

Complement to PNP MPS-U52

MAXIMUM RATINGS

Rating	Symbol	Value	Unit
Collector-Emitter Voltage	V_{CEO}	40	Vdc
Collector-Base Voltage	V_{CB}	60	Vdc
Emitter-Base Voltage	V_{EB}	5.0	Vdc
Collector-Current Continuous	I_C	800	mAdc
Total Power Dissipation @ T_A = 25°C	P_D	1.0	Watt
Derate Above 25°C		8.0	mW/°C
Operating and Storage Junction	T_j, T_{stg}	−55 to +150°	°C

THERMAL CHARACTERISTICS

Characteristic	Symbol	Value	Unit
Thermal Resistance, Junction to Case	$R_{\theta JC}$	12.5	°C/W
Thermal Resistance, Junction to Ambient	$R_{\theta JA}$	125	°C/W

ELECTRICAL CHARACTERISTICS (T = 25°C unless otherwise noted)

Characteristic	Symbol	Min	Max	Unit
OFF CHARACTERISTICS				
Collector-Emitter Breakdown V (I_C = 1.0mAdc, I_B = 0)	BV_{CEO}	40	—	Vdc
Collector-Base Breakdown V (I_C = 100μAdc, I_E = 0)	BV_{CBO}	60	—	Vdc
Collector Cutoff Current (V_{CB} = 40Vdc, I_E = 0)	I_{CBO}	—	100	nAdc

Table 11-1: Typical transistor data sheet (MPS-U02).

a fairly typical amplifier transistor. For example, it has numerous substitute types: GE-63, TR-72, 2SC1384, PTC144, SK3199, ECG188, WEP3020, and RT-156, which are all general-purpose replacements.

This transistor is listed as an NPN *silicon annular amplifier transistor*. In case the word "annular" is new to you, it means a *mesa* transistor in which the semiconductor

DIODES, TRANSISTORS, INTEGRATED CIRCUITS, MODULES 279

ON CHARACTERISTICS				
DC Current DC Current Gain (I_C = 10mAdc, V_{CE} = 10Vdc) (I_C = 150mAdc, V_{CE} = 10Vdc) (I_C = 500mAdc, V_{CE} = 10 Vdc)	h_{FE}	50 50 30	— 300 —	— — —
Collector-Emitter Saturation (I_C = 150mAdc, I_B = 15mAdc)	$V_{CE}(sat)$	—	0.4	Vdc
Base-Emitter Saturation V (I_C = 150mAdc, I_B = 15mAdc)	$V_{BE}(sat)$	—	1.3	Vdc
DYNAMIC CHARCATERISTICS				
Current Gain-Bandwidth Product (I_C = 20mAdc, V_{CE} = 20Vdc, f = 100MHz)	f_T	100	—	MHz
Output Capacitance (V_{CB} = 10 Vdc, I_E = 0, f = 100kHz)	C_{ob}	— —	— 20	pf

Table 11-1 (continued)

materials are arranged in concentric circles about the emitter. A mesa transistor is one in which the base and emitter semiconductor materials appear as plateaus above the collector region. A non-mesa semiconductor is called a *planar* device (for example, a planar diode).

Maximum Ratings

The first specifications listed in Table 11-1 are *maximum ratings*. The maximum collector emitter voltage is listed as V_{CEO}. Actually, this is a test voltage rather than an experimenter/designer operating voltage. V_{CEO} usually tells you the collector emitter breakdown voltage with the base circuit open. Of course, you would not normally operate a transistor this way in a circuit. Nevertheless, you can use this specification. The 40 Vdc figure can be considered the *absolute maximum* voltage that you should use as a collector voltage during circuit experiments.

A few points to keep in mind are

1. In general, operate a transistor with a collector voltage less than the supply voltage (there are exceptions to this rule; for example, in certain rf circuits). But

there are very few exceptions in low-frequency analog applications.

2. *Never breadboard a circuit design and then apply power if the source you are using is a higher voltage than the maximum voltage rating of the transistor, even with a resistor in the collector lead!*

The next two ratings are collector base voltage (V_{CB}), 60 Vdc max, and emitter base voltage (V_{EB}), 5.0 Vdc max. Again, these are test voltages rather than operating voltages. For example, in actual practice, you will usually have current flowing through the emitter base junction at all times. Therefore, the *real* voltage drop across this junction will be about 0.5 to 0.7 volts for a silicon transistor (about 0.2 to 0.4 volts for a germanium transistor).

In most practical breadboarding situations, you will have to select a bias (V_{EB} voltage) on the basis of what your input signal to the transistor is to be, rather than the spec sheet's given value. In any case, you should always consider any input signal that you wish to apply to the emitter base junction, in addition to what you set as a no-signal operating bias.

Collector Current

In Table 11-1, the collector current, continuous (I_C), is listed as 800 mA. It is important that you realize *collector current will increase with temperature*. This rating (and all the rest) is specified at 25°C (77°F). Notice that the next line, "Total Power Dissipation," is given as (@) $T_A = 25°C$.

RULE: *Do not operate any transistor at or near the maximum rating given in the maximum rating section of the spec sheet.*

Total Power Dissipation

In practical breadboarding (designing and prototyping a circuit), it is the *total power dissipation* (P_D) ratings that are of major concern. Note that there are two given: T_A (operating) and T_C (collector). T_A, P_d is to be derated above 25° at 8.0 mW per degree Celsius, or 8.0 mW/°C. The other Total Power Dissipation @ $T_C = 25°C$, should be derated above 25°C at 80 mW/°C. For example, assume that the collector operates at 40 V and 800 mAdc. This results in a power dissipation of 32 watts—far above the 10 watts specified for T_C at 25°C.

Operating and Storage Junction Temperature Range

How much power a transistor can dissipate is closely associated with the transistor's temperature range. As shown in Table 11-1, P_D must be derated above 25°C for each degree above the value given. The temperature range (T_j, T_{stg}) given is −55 to +150°C. In most cases, the high end (+150°C) is the most important. This is because, as a rule of thumb, collector current increases with temperature. That will, in turn, cause an additional rise in temperature, followed by an increase in collector current. Result: *thermal run-away!* End result: the transistor will burn out and you will have to purchase another.

An external heat sink is required for maximum power output for all but the smallest semiconductor devices (power dissipation rating of less than 1 watt). If the area of the heat sink is insufficient to dissipate the heat generated within the semiconductor device (for example, an **IC**), in many instances an automatic thermal shutdown circuit will turn off the **IC** and prevent it from being damaged.

Thermal Characteristics

Notice that the thermal characteristics given in Table 11-1 are both listed as *Thermal Resistance*. Transistors designed for power applications (as is the MPS-U02 we are discussing), have these ratings specified to indicate power dissipation capability. Thermal resistance can be defined as the increase in temperature of the transistor *junction* with respect to either case ($R_{\theta jc}$) or ambient temperature ($R_{\theta ja}$), or with respect to some other reference, divided by the power dissipation, or °C/W, as shown.

OFF Characteristics

The OFF characteristics given in Table 11-1 are strictly test values. For example, collector emitter breakdown voltage

(BV_{CEO}) is the breakdown voltage (BV), with the collector (C) and emitter (E) connected to a voltage source and the base left open (O), or BV_{CEO}.

In this example, it is listed as 40 Vdc min. with the transistor non-biased. From this, we get *OFF characteristics*. Notice that each of the following includes a zero as the last symbol that indicates the transistor lead left open. For instance, BV_{CBO} indicates the emitter is open, and I_{CBO} also shows the same condition. However, as was explained before, test ratings such as these can be used for design or test purposes. The given values can be considered as the *absolute maximum* values you should apply to the device.

ON Characteristics

The symbol, in this case, is h_{fe} (also known as *current transfer ratio*). You will see, by referring to Table 11-1, that the dc current gain for this transistor is given for three different conditions (the first being I_C = 10 mAdc, V_{CE} = 10 Vdc). The current transfer ratio of the transistor, under these conditions, is listed as an h_{fe} of a minimum of 50. Also, note that when I_C is equal to 500 mAdc @ V_{CE} = 10 Vdc the h_{fe} drops off to a minimum of 30. In other words, the transistor will not amplify as well at the larger collector current values. To repeat: *do not operate any transistor at, or near, its maximum current rating!* Although direct current characteristics are important for selecting proper bias circuits, they are primarily test values.

Dynamic Characteristics

Table 11-1 shows two transistor characteristics under the heading Dynamic Characteristics: Current Gain—bandwidth product, and Output Capacitance, both of which are important if you intend to use this transistor in high frequency circuits. Under the test condition given (I_C = 70 mAdc, V_{CE} = 20 Vdc, f = 100 MHz), the minimum current gain bandwidth of this transistor is 100 MHz. As a general rule, gain bandwidth product is the frequency (100 MHz, in this case) at which gain drops to a very low level. In some cases, this drop will be down to unity (0 dB).

The other output capacitance (under the given test conditions) is a maximum of 20 pF. The lower this value (C_{ob}), the better. In general, capacitance will shunt the signal, resulting in a loss of amplification and a change of the expected circuit impedance.

CB/Amateur High Frequency Transistors

High frequency characteristics are especially important to the person working with citizen band (CB) and amateur radio transistors. To properly match impedances, two components must be considered. These are output resistance versus frequency and output capacitance versus frequency. Usually this information is provided on data sheets, in graphical form, over the frequency operating range of the transistor. Figure 11-3 shows a graph of these two transistor characteristics for the MRF475. This transistor is designed primarily for use in single sideband linear output application in CB and other rf equipment operating to 30 MHz.

The reason this information is presented by means of graphs or curves, rather than in tabular form, is that you need to know these resistance and reactance values (either capacitance or inductance) over a wide range of frequencies (1.5 to 30 MHz, in this example), not at some specific frequency (unless you happen to be testing a transistor for a specific frequency only).

There are two terms you will encounter, when reading rf transistor spec sheets, that may not be familiar: *carrier power (P_C)* and *peak envelope power (PEP)*. Carrier power is speaking of the power available at the output terminals of the rf transistor (in its test circuit) when the transistor output terminals are connected to the normal load (usually 50 ohms). In this example, carrier power is typically 4.0 watts (continuous), with V_{CC} = 13.6 Vdc, and connected to a 50 ohm load. Note that this is usually expressed as under *continuous* power output.

Peak envelope power is the *average power* supplied to some form of load (antenna, dummy load, etc.) by the transistor (with its test circuit), during one rf cycle at the highest crest of the modulation envelope, taken with the transistor and test circuit connected to its normal load. Again, this is 50 ohms in our example. However, the value for PEP is frequently derived

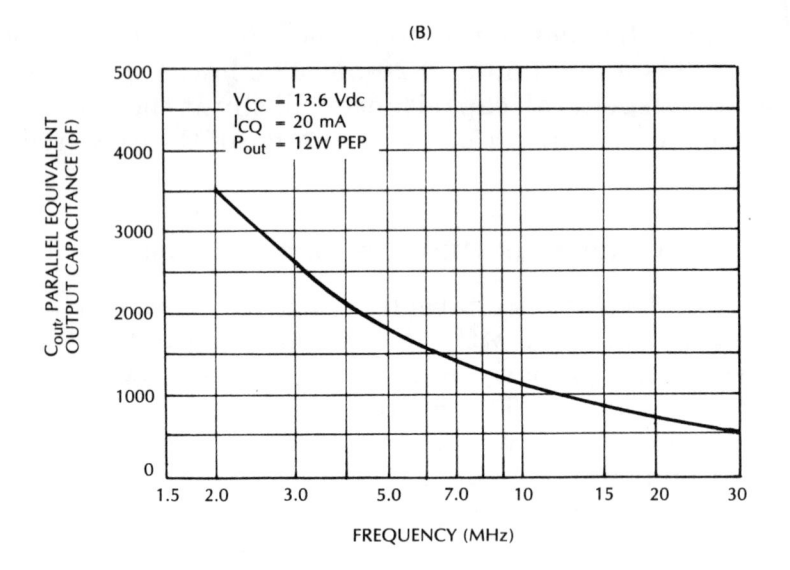

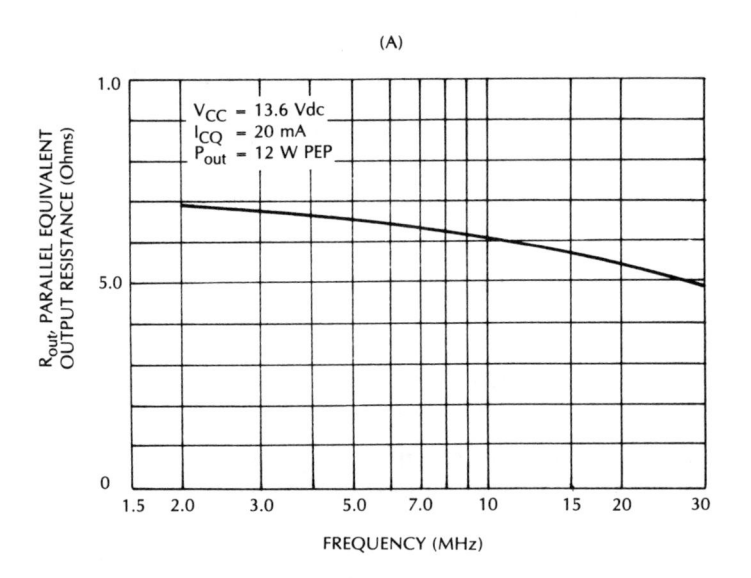

Figure 11-3: (A) Output resistance versus frequency. (B) Output capacitance versus frequency, for a MRF475. This device is a NPN silicon rf power transistor rated at 4.0 watts continuous wave (CW) @ 30 MHz.

under test conditions by doubling the original test setup power input, to simulate driver modulation.

When you divide PEP by P_C, subtract 1 and then multiply your answer by 100. This results in the percentage of *up-modulation*. In formula form, this is

$$\text{percentage up-modulation } [(PEP/P_C) - 1] \times 100.$$

This percentage of modulation is given in spec sheets at a certain P_C value (4.0 watts in the transistor spec sheets we are discussing), and, in this spec sheet, equals 100%. Percentage up-modulation, and other tests, can be measured and/or calculated by using the common emitter test circuit shown in Figure 11-4. See Table 11-2 for parts list.

If you have both an rf ammeter and an rf voltmeter, the power output is the product of the two readings regardless of the value of the load impedance, *provided the load you are using is a pure resistance*. This applies to a so-called *key-down* (operating) CW test circuit.

Measuring the output of a single-sideband (SSB) test circuit is a bit more complicated. Your best bet, if you don't have a regular rf test set available, is to use an oscilloscope.

For example, let's say that you are modulating a SSB test setup with a single tone, and carrier insertion is used. In this case, adjust both until you get all test patterns of equal heights, as shown on your scope. Turn the audio gain up on the SSB driver unit (it would be connected to the rf input shown in Figure 11-4), until the highest waveform is seen on your scope—*without distortion*.

Next, use an rf voltmeter and measure the peak voltage of the modulation pattern (the value from zero carrier to peak carrier). Now, convert this peak value to rms value by multiplying by 0.707, squaring, then divide by the value of your load (R_L). This will give you the peak envelope power of a SSB excited test circuit such as the one shown in Figure 11-4. The formula for reading PEP on a SSB setup is

$$PEP = (E_{peak} \times 0.707)^2 / R_L.$$

where R_L is the load resistance you use to terminate the test circuit.

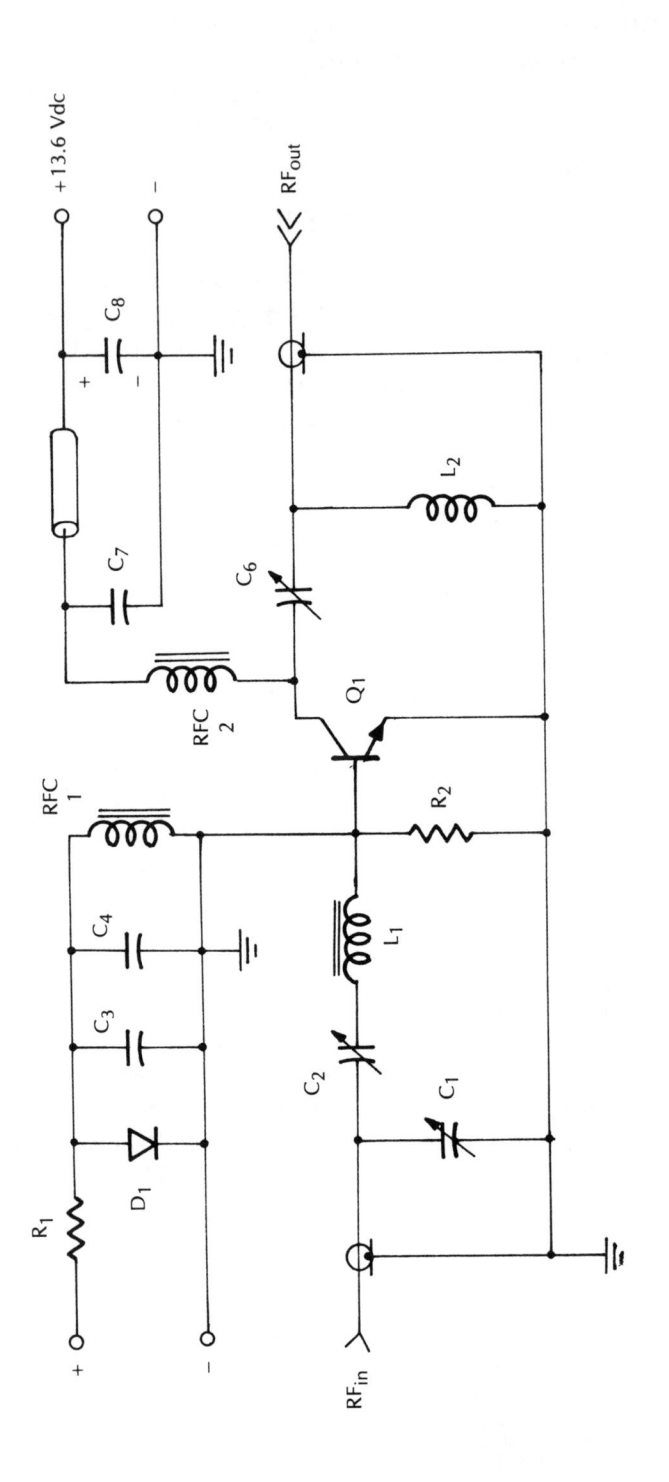

Figure 11-4: Common emitter test circuit for measuring carrier output power, transistor efficiency, power gain, etc., of the rf power transistor MRF475. See Table 11-2 for parts list.

DIODES, TRANSISTORS, INTEGRATED CIRCUITS, MODULES

C1, 2, 6	ARCO 466 Trimmer Capacitors.
C3	1000 μF, 3.0 V_{dc} Electrolytic.
C4, 7	0.1 μF, Disc Ceramics.
C8	100 μF, 15 V_{dc} Electrolytic.
R1	10 ohm, 5.0 W Resistor.
R2	10 ohm, 1.0 W Resistor.
L1	2.2 μH Molded Choke.
L2	4 turns #18 AWG wire, 1/2" I.D., 5/16" long.
RFC1	10 μH Molded Choke.
RFC2	15 turns #20 AWG Wire on 5.6 k ohm 1.0 W Carbon Resistor
RFC3	5 Ferroxcube, #56–590–65/3B, Beads on #18 AWG Wire
D1	1N4997.
Q1	MRF475.

Table 11-2: Parts list for test circuit shown in Figure 11-4.

Driving rf Output Transistors

When a single transistor circuit is not capable of providing the required output power to another stage, an additional transistor amplifier should be added. Another reason to use more than one amplifier stage is heat distribution. The use of two transistor amplifiers rather than a single unit splits the delivered power between the two stages.

As an example of how this can be done, let's use two transistors designed for 12.5 volt amplifier applications, in the 27 MHz CB radio band. The driver transistor could be a MRF8003. This transistor is rated at 0.5 watts, with a minimum gain of 10 dB, and an efficiency of 50%. A 27 MHz amplifier circuit schematic is shown in Figure 11-5. See Table 11-3 for parts list.

Notice that the transistor used in the schematic in Figure 11-5 is rated for only 0.5 watt. This transistor amplifier

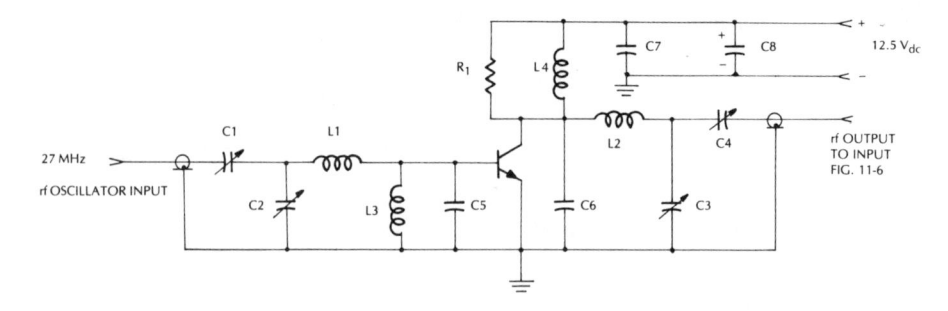

Figure 11-5: Schematic diagram for a 27 MHz transistor (MRF8003) amplifier driver stage. See Table 11-3 for parts list.

Parts List for Figure 11-5
C_1, C_2, C_3, C_4 9.0 - 180 pF ARCO 463 or equiv.
C_5 – 25 pF UNDERWOOD.
C_6 – 100 pF UNDERWOOD.
C_7 – 1000 pF UNDERWOOD.
C_8 – 10 μF ELECTROLYTIC.
$L_1 L_2$ – 0.47 μH Molded Coil.
L_3 – VK 200-20/48 RFC.
L_4 – 16 Turns No. 26 wire closewound on R_1.
R_1 – 390 ohms, 2 W.

Table 11-3: Parts list for circuit illustrated in Figure 11-5.

is suitable for use as a driver for the rf power transistor MRF8004, rated at 3.5 watts, 27 MHz. Figure 11-6 shows a schematic diagram for this amplifier. (See Table 11-4 for parts list.) It can be used as the final amplifier to a CB rig. Incidentally, no provisions for modulation are shown. Therefore it would be impossible to use these stages in a transmitter without including some form of modulating circuit. Of course, you would also have to have a 27 MHz oscillator in conjunction with the driver and power amplifier circuits.

Some form of output indicator should be used for loading (delivering the desired amount of power to each stage, and to the pure resistive load—dummy load—of the final stage). The

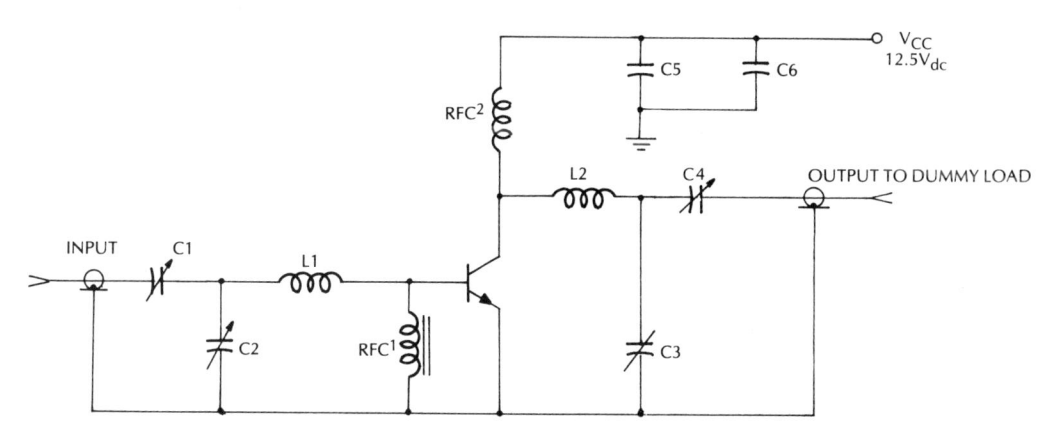

DIODES, TRANSISTORS, INTEGRATED CIRCUITS, MODULES

Figure 11-6: Schematic diagram for a 27 MHz transistor (MRF8004) amplifier, that can be used following the driver stage shown in Figure 11-5.

output indicator may be an rf ammeter, rf voltmeter, or the forward position of an SWR meter. Where no modulation is applied, adjust the loading circuit (the variable capacitors) until you see no further increase at the output metering instrument. *Do not key (apply operating power) for more than 30 seconds at a time while making the loading adjustments.* Basically, the foregoing is also true for a SSB rf circuit—*unmodulated conditions.*

FM Mobile Communications Transistors

There are two general frequency bands utilized in FM mobile radios, for which Motorola manufactures transistors. There are 27 to 50 MHz (low-band FM transistors) where longer distance mobile communication is desirable, and 66 to 88 MHz, midband transistors.

An example of a low-band FM transistor that can be used in rf amplifier circuits operating up to 50 MHz is the MRF402. Although this is a low output power transistor (1 watt @ 50 MHz), it does not require special mounting. It has a TO-39 case rather than the typical rf type packages shown in Figure 10-1.

Figure 11-7 shows a test circuit and parts list that can be used to check the MRF402 transistor. The loading procedures

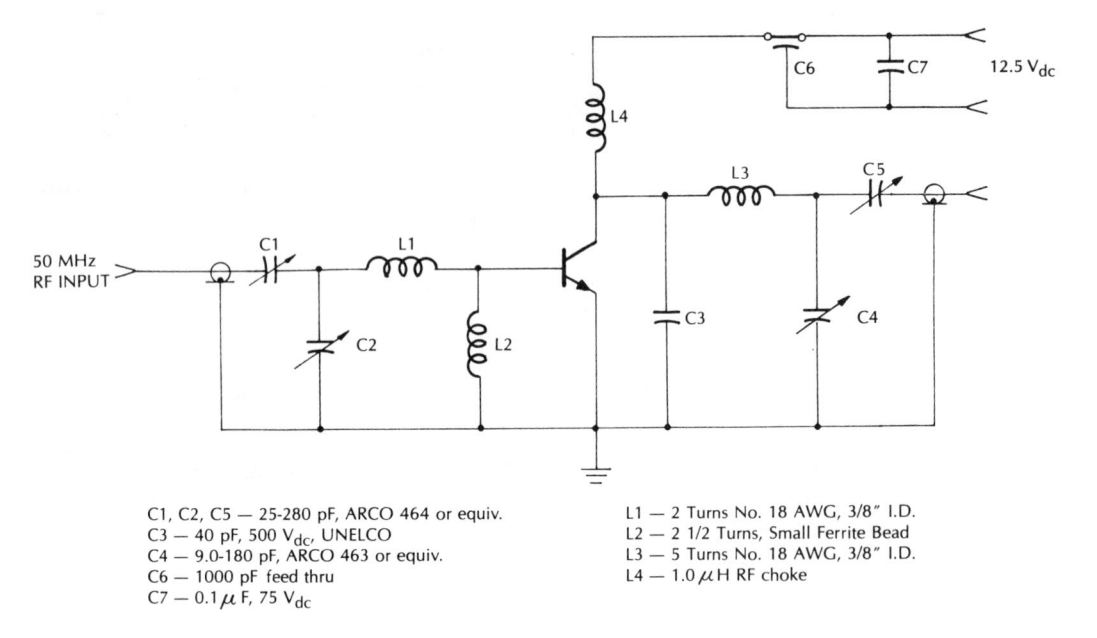

C1, C2, C5 — 25-280 pF, ARCO 464 or equiv.
C3 — 40 pF, 500 V$_{dc}$, UNELCO
C4 — 9.0-180 pF, ARCO 463 or equiv.
C6 — 1000 pF feed thru
C7 — 0.1 μF, 75 V$_{dc}$

L1 — 2 Turns No. 18 AWG, 3/8" I.D.
L2 — 2 1/2 Turns, Small Ferrite Bead
L3 — 5 Turns No. 18 AWG, 3/8" I.D.
L4 — 1.0 μH RF choke

Figure 11-7: Schematic diagram and parts list for a MRF402 test setup.

Parts List for Figure 11-6
C$_1$, C$_2$ – 9.0 pF ARCO 463 or equiv.
C$_3$, C$_4$ – 5.0 - 80 pF RCO 462 or equiv.
C$_5$ – 0.02 μF Ceramic disc.
C$_6$ – 0.1 μF Ceramic disc.
RFC$_1$ – 4 Turns #30 Enameled Wire wound on Ferroxcube Bead Type 56-590-65/3B.
RFC$_2$ – 26 Turns #22 Enameled Wire (2 layers - 13 turns eachZ) 1/4" inner diameter.
L$_1$ – 0.22 c Molded Choke.
L$_2$ – 0.68 c Molded Choke.

Table 11-4: Parts list for power amplifier circuit.

are the same as explained in the last section. However, it bears repeating. *Do not apply operating power for more than 30 seconds at a time while adjusting the capacitors for maximum permissible power-out to the resistive load.*

Marine Radio Transistors / Modules

The frequency range of these transistors and modules is from 156 to 162 MHz. A transistor that is provided in a T-39 grounded emitter package, and will operate at 4 watts up to 175 MHz, is the MRF237. The grounded emitter TO-39 package is designed so that the transistor has high gain (12 dB minimum @ 175 MHz) and very good total power dissipation (8.0 watts @ T_c of 25°C).

A VHF power amplifier module that operates in the same frequency range as the MRF237 transistor is the MHW613. Actually, this module operates at 150 to 174 MHz, with an output power of 30 watts. Figure 11-8 shows the package, with pins identified, for this rf power amplifier module. The MHW613 is designed for 12.5 volts VHF power amplifier applications in industrial and commercial equipment.

If you are working with equipment that utilizes a VHF power amplifier module such as the MHW613 or an equivalent, there are a few points to keep in mind:

1. Input impedance at pin 4, and output impedance at pin 1, is 50 ohms. In other words, that is the impedance you must match when connecting external circuits to the module.
2. Pin 3 is a gain control pin that may be using either manual or automatic output level control.
3. The maximum rf input power to pin 4 is 500 mW. Do not exceed this rating if you are testing one of these modules. Normally, you should use about 300 mW with

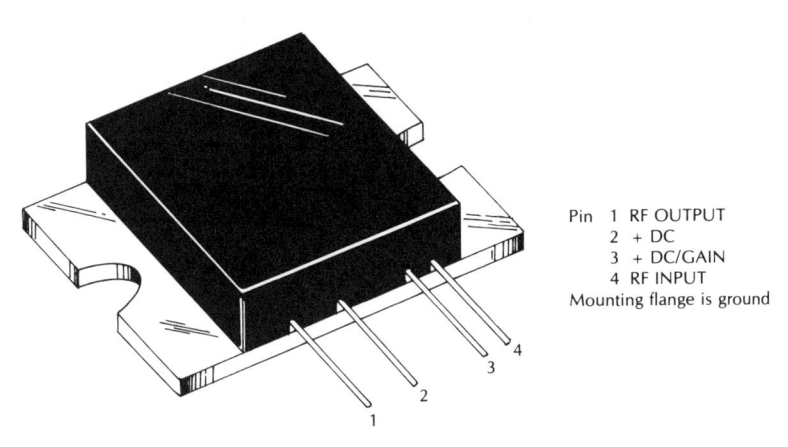

Pin 1 RF OUTPUT
2 + DC
3 + DC/GAIN
4 RF INPUT
Mounting flange is ground

Figure 11-8: RF power amplifier module (30 W @ 150 – 174 MHz), MHW613.

pin 2 set for 30 watts output power.
4. Do not exceed +16 Vdc on pins 2 and 3 during any test. You should use near 12.5 Vdc on these pins.
5. Due to the high gain of these units, decoupling networks are a must on both pins 2 and 3. These are basically low-pass filters and are generally included on the spec sheets.

Typically, marine radios are required to reduce their power-out to below 1 watt during in-harbor operations. You will find that most equipment designed around the MHW613 rf module achieves this by removing the dc voltage from P_2 (leaving P_2 open). Pin 3 will usually have 12.5 Vdc, and power input is normally achieved at 300 mV. Under these conditions, the module power-out will be between 0.3 watt to something under 1 watt.

CHAPTER TWELVE

30 Electronic Projects Using State-of-the-Art Solid State Devices

New developments in semiconductor electronic components by manufacturers such as Motorola have opened a broad range of projects to the electronics technician/experimenter. However, an important step when planning a project or experiment is the availability of circuit components. Admittedly, resistors, capacitors, and quite a few solid state devices are easily obtainable from electronics retail stores such as Radio Shack.

Nevertheless, in many instances semiconductor components are expensive and difficult to obtain. Therefore, you will find that the circuits shown in this chapter were selected because the solid state devices are relatively low cost and readily available through any authorized Motorola Semiconductor Distributor. These distributors are located in almost every major city in the United States and Canada.

It is important that the reader realize that the following circuits were not designed or authorized by the Motorola Corporation. Each circuit and IC was selected by the author to help the reader gain experience in the field of electronics. Because of this, each reader may have to adjust the circuit component values to his needs (for example, if you use a 6-, 12-, or 18-volt power transformer, the circuit component values could be different for each of the three voltages). In general, you should not experience any difficulty with the following projects and experiments provided reasonable care is taken. Or, to put it another way, when in doubt, monitor the circuit currents and voltages. Just remember, excessive current is usually caused by not enough resistance or too much voltage.

Project 12.1: Low-Cost Rectifier Circuit for Solid State Projects

Parts List for Project 12.1:

D_1, D_2, D_3—Rectifier, silicon diode, rated at 6 ampere average rectified current @ 50 volts. MR 750 or equivalent.

T_1—Transformer, 120 Vac primary. For 6V-output, use a power transformer rated at 6V or higher. The current reading should be at least 3 A if the circuit is to be used with the 3 A regulator shown in Project 12.2.

Comments:

You can choose the transformer for either a 6-, 12-, or 18-volt output. However, whichever you select, it must be remembered that the actual output voltage will be 1.414 times the secondary of the transformer (no load condition). This is

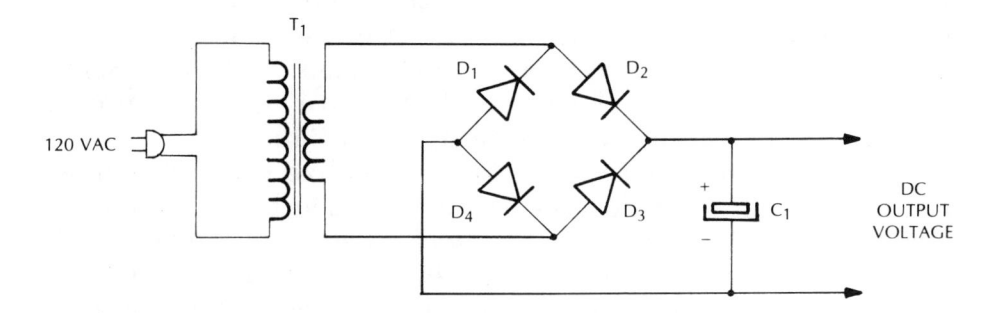

Figure 12-1: Schematic for a basic solid state rectifier.

STATE-OF-THE-ART SOLID STATE DEVICES

because the output filter capacitor (C_1) will charge up to the peak voltage of the transformer secondary. Any load that is placed on the output terminals will cause the output voltage to decrease.

Project 12.2: Practical 5-Volt, 3-Ampere Regulator

Parts List for Project 12.2:

C_1—Capacitor, 0.22 µF.
C_2—Capacitor, 10 µF.
Q_1—Transistor, MJE370 or similar.
R_1—Resistor, 2 ohm, 8 watt.
R_2—Resistor, 1 ohm, 5 watt.
V_{reg}—Voltage regulator, LM109.

Comments:

Providing 5 volts at 3 ampere, this on-board type regulator will work well with the rectifier circuit shown in Project 12.1. This is a positive voltage regulator and it has several all-important circuits: current limiting, internal short-circuit protection, and internal thermal overload protection. It is especially useful for powering single board computer projects that can easily draw 1, 2, or 3 amperes.

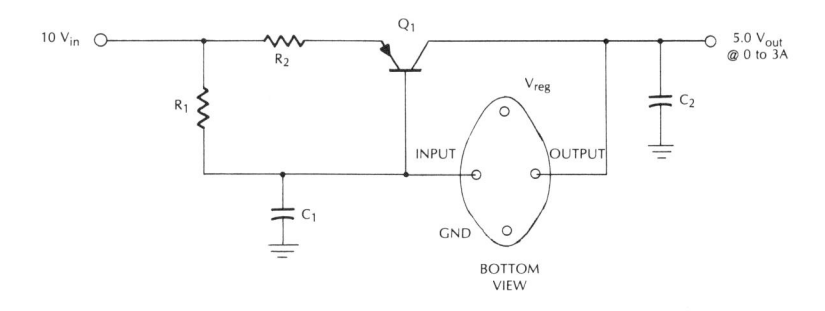

Figure 12-2: Wiring diagram for a 5-V regulator (LM109).

Also, because of the few parts needed, the entire circuit is easily mounted right on the same breadboard that you used to mount the preceding rectifier circuit.

Project 12.3: Overvoltage Protection Circuit

Parts List for Project 12.3:

C_1—Capacitor, 0.1 µF.
D_1—Diode, zener. 1N5232.
F_1—Fuse, 3A (if used with load rated at 3A).
R_1—Resistor, 10 ohm.
R_2—Resistor, 220 ohm.
SCR—Silicon controlled rectifier, 2N2573.

Comments:

Several "crowbar" SCR circuits that provide overvoltage protection are available in self-contained modules. However, using the circuit diagram and parts list given here, you can build your own. Whichever way you go, it can be well worth your time to build or buy one of these circuits. It just may save you

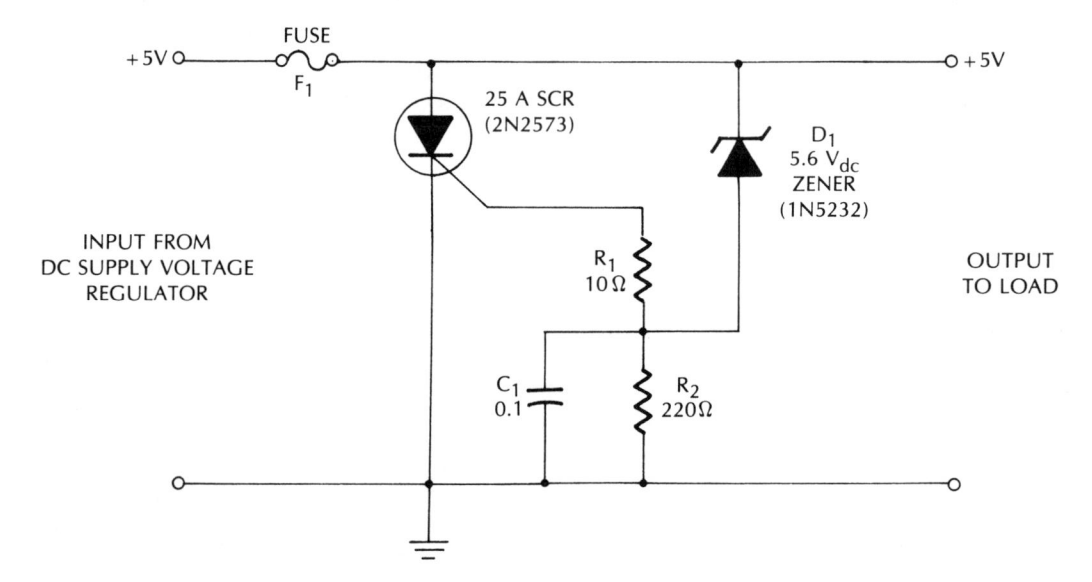

Figure 12-3: Overvoltage protection circuit (crowbar).

STATE-OF-THE-ART SOLID STATE DEVICES   297

hundreds of dollars in case there is a sudden rise in supply voltage. See Chapter Seven for more details on this circuit.

Project 12.4: Using a Power Transistor as a Rectifier

Parts List for Project 12.4:

C_1—Capacitor, electrolytic, 2200 μF, working Vdc (WVdc) depends on the peak output voltage expected. For solid state work, 35 V is usually satisfactory.

Q_1—Power transistor. Any general purpose PNP power transistor that has a continuous maximum current rating of at least 25% more than the expected load current. For example, a 2N4929 rated at 0.5 A, 2N3244 rated at 1 A, 2N4234 rated at 3 A, etc. Each of the listed power transistors uses a TO-39 package.

T_1—Transformer, 120 Vac primary. For a 17Vdc (peak) output, you must use a transformer with a 12-volt secondary. For example, a Radio Shack catalog No. 273-1513 rated at 120 Vac primary, 12 Vac secondary @ 5 A load current.

Comments:

What you are doing in this experiment is forming a rectifying diode by use of the collector base junction of a power transistor. The emitter is not used. A power transistor can be used in this manner in any rectifier configuration: half-wave, full-wave, or bridge rectifier circuit. Just remember,

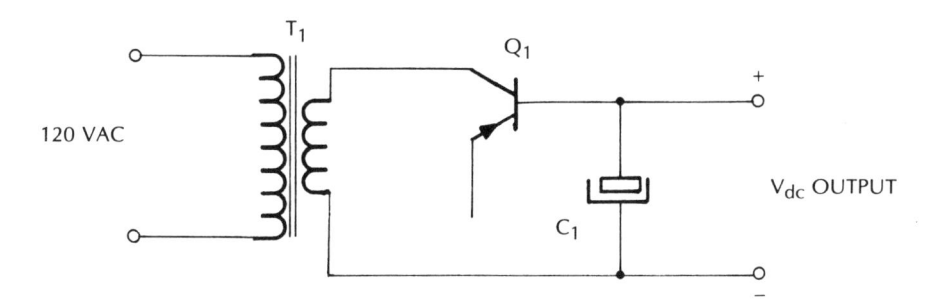

Figure 12-4: Schematic for using a power transistor as a rectifier.

always use the collector as the anode of the rectifier, and the base as the cathode. *If you use a NPN transistor, it will reverse the output polarities.*

Project 12.5: Lab-Type Power Supply for Shop Use

Parts List for Project 12.5:

C_1—Capacitor, electrolytic, 2200 µF, working volts 35 Vdc, axial. Radio Shack catalog No. 272-1020.

C_2—Capacitor, electrolytic, 100 µF, working volts 35 Vdc, axial. Radio Shack catalog No. 272-1016.

D_1—Bridge rectifier, V_{RRM} volts 50. Motorola 3N246 or equivalent.

D_2—Zener diode, 9.1 volts. Motorola 1N4696 or equivalent.

Q_1—Transistor, NPN, Motorola 2N110 for a case power dissipation of 5 W @ 25°C with a maximum current of 1 A. *Our suggested current load for this power supply is 250 mA.*

R_1—Resistor, 560 ohm, ½ W.

T_1—Transformer. For about 9 Vdc output voltage, use a 12 V output transformer. Radio Shack catalog number 273-1385 transformer may be used *if the load current does not exceed 300 mA.*

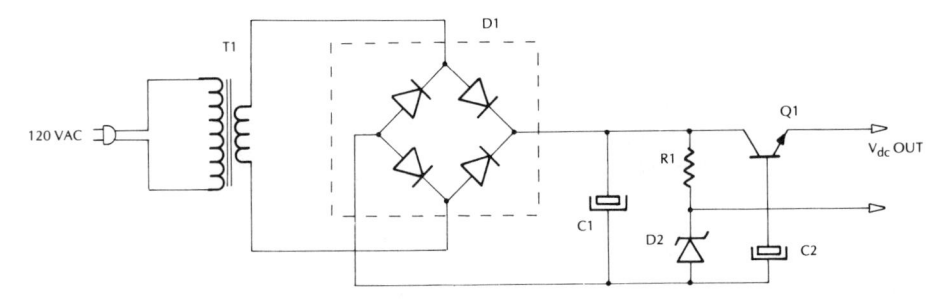

Figure 12-5: Wiring diagram for a full-wave bridge rectifier and active filter.

Comments:

The zener diode (D_2) and transformer you select sets the output voltage of this power supply. For example, Radio Shack's transformer, catalog NO. 273-1384, a 6.3 V transformer, and Motorola's zener diode 1N918A (5.1 V, 1.5 W) should produce about 5 volts on the output. However, you might have to adjust the value of resistor R_1 when using these components. See the next project for a voltage regulator.

Project 12.6: Simple Fixed Output Voltage Regulator (MC78 Series)

Parts List for Project 12.6:

C_1—Capacitor, 0.33 µF.

C_2—Capacitor, 0.1 µF.

Voltage regulator, MC78L05C.

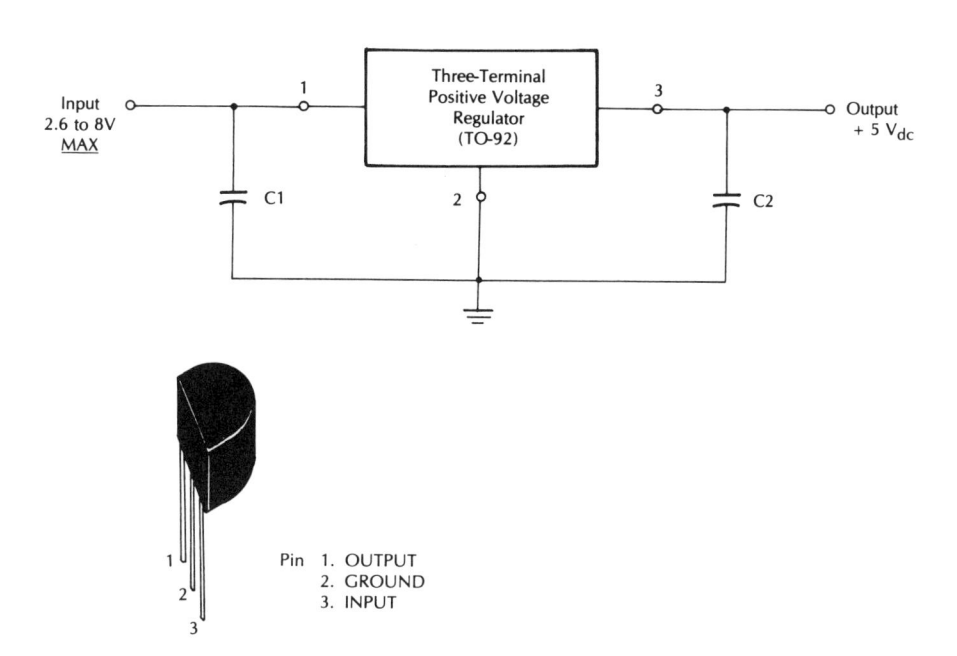

Figure 12-6: Wiring diagram for a fixed voltage regulator (MC78 series).

Comments:

The voltage regulator IC used in this project has internal current limiting and thermal shutdown protection. As an experiment, you should be able to disconnect both capacitors and see very little difference in the output voltage.

However, C_1 is needed if the wire leads are long between the power supply and regulator. The purpose of C_2 is to improve transient response; especially important in digital work where transient voltages can be quite a problem.

Project 12.7: How to Double the Output Voltage

Parts List for Project 12.7:

C_1, C_2—Capacitors, electrolytics, 2200 μF. Working voltage equals at least twice the *peak* ac voltage on the transformer secondary.

D_1, D_2—Silicon diode, 0.5 A or larger. Select diodes with a peak-inverse voltage of at least twice the output voltage of the doubler.

T_1—Transformer. 120 Vac primary. Secondary voltage equals desired output voltage times 0.5.

Comment:

The dc output voltage of this circuit will be approximately the *sum* of the two voltage readings you measure across capacitors C_1 and C_2.

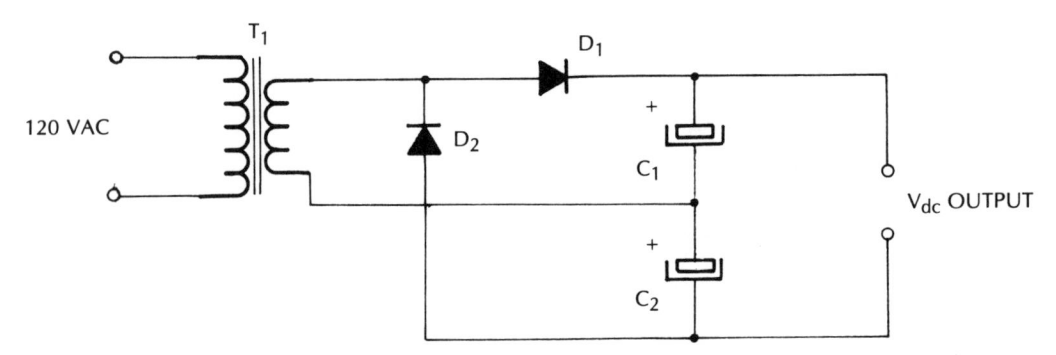

Figure 12-7: Schematic diagram for a voltage doubler.

Project 12.8: Easy-to-Make High Impedance Microphone

Parts List for Project 12.8:

B1—Battery, 9V, transistor type, Radio Shack No. 23-464 or a 9V battery eliminator, Radio Shack Cat. No. 270-1552.
C1—Capacitor, electrolytic, 10 F, 35 working volts (WVdc), Radio Shack Cat. No. 272-1013.
C2—Capacitor, 0.47 F, 35 WVdc, Radio Shack Cat. No. 272-1417 (or equivalent).
Q1—Transistor, NPN, Radio Shack Cat. No. 276-2014.
R1—Resistor, 270 Kohm, ½ watt.
R2—Resistor, 27 Kohm, ½ watt.
S1—Switch, SPST, Radio Shack Cat. No. SPST 275-602.
SP—Speaker, 8 ohms, Radio Shack Cat. No. 40-245 (2 inch).

Comments:

This circuit (as shown) will serve in place of almost any high impedance microphone. Actually, the circuit is both an audio amplifier and an impedance matching device. What it is doing is providing an impedance match between the 8-ohm input and the desired high impedance load (loads of 7 k ohms or higher).

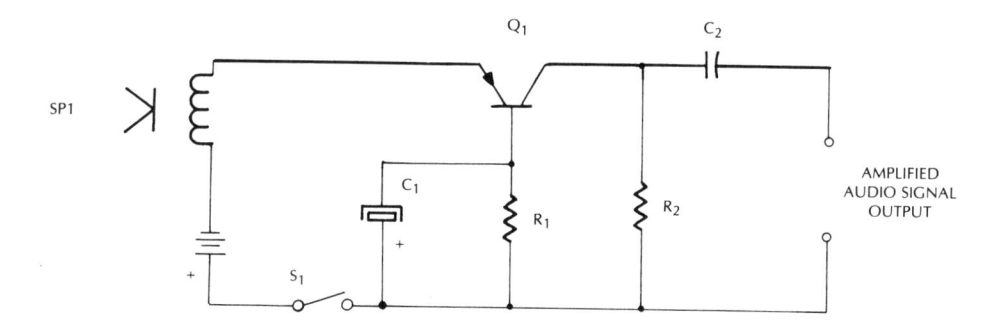

Figure 12-8: Wiring diagram for an easy-to-make high-impedance microphone.

Project 12.9: Using a MC1741 OP AMP to Interface a High Impedance to a Low Impedance

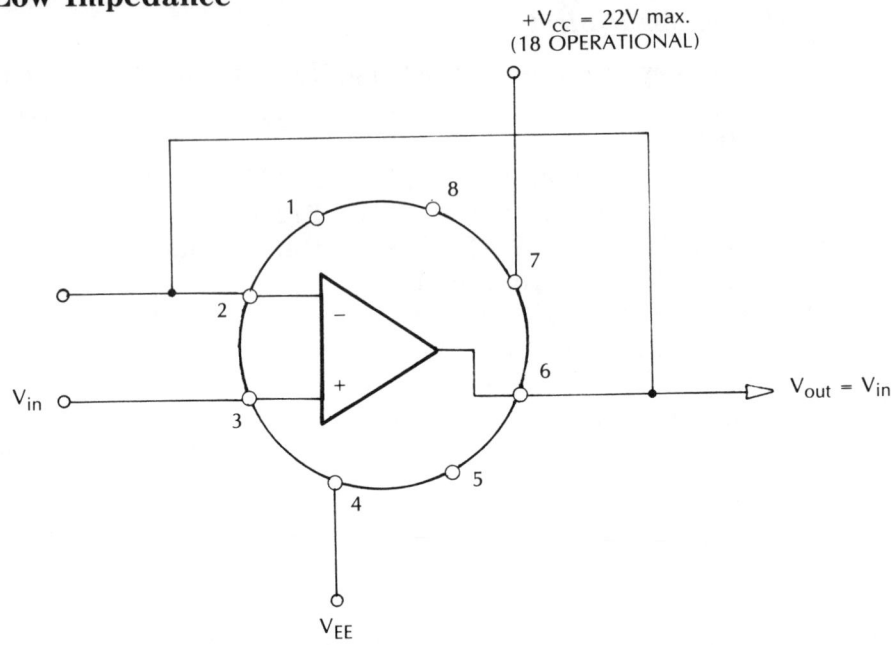

Figure 12-9: Wiring diagram for using a MC1741 OP AMP follower as an impedance-matching device.

Comments:

The MC1741 general purpose operational amplifier requires no frequency compensation and is short-circuit protected. *Important:* An MC1741C **IC**, which is exactly the same type of **OP AMP**, has a *maximum* power supply voltage (V_{CC}) of only 18 volts on V_{CC} and V_{EE}. Typical *operating* voltage for the MC1741 shown in Figure 12-9 is 18 volts. Incidentally, even though this **IC** has provisions for offset voltage compensation (pins 1 and 5), leave these pins unused in this project.

Project 12.10: Building a Voltage Level Detector Using a MC1741C OP AMP

Parts List for Project 12.10:

I_C—General purpose **OP AMP** MC1741C.

STATE-OF-THE-ART SOLID STATE DEVICES

LED—Light-emitting diode. See Comments

R₁—Potentiometer, 50 k ohms (sets the voltage detection threshold level). Radio Shack catalog No. 271-219.

R₂—Resistor, 1 k ohm. See Comments.

Comments:

It is necessary to limit current to LED's. Resistor R_2 in Figure 12-10 is the current-limiting resistor in this circuit. Usually, continuous forward current (I_F) in LED's is from 5 to 40 mA. The forward voltage drop (V_F) ranges from 1.65 to 2.2 volts. The value of resistor R_2 for the various LED's can be calculated by using the formula

$$R_2 = (V_{CC} - V_F)/I_F.$$

The operation of the circuit is this: any time the input voltage exceeds the threshold voltage you set, using the adjustable resistor (R_1), the LED will glow. However, the threshold voltage setting cannot be set to detect a voltage input greater than 9 volts.

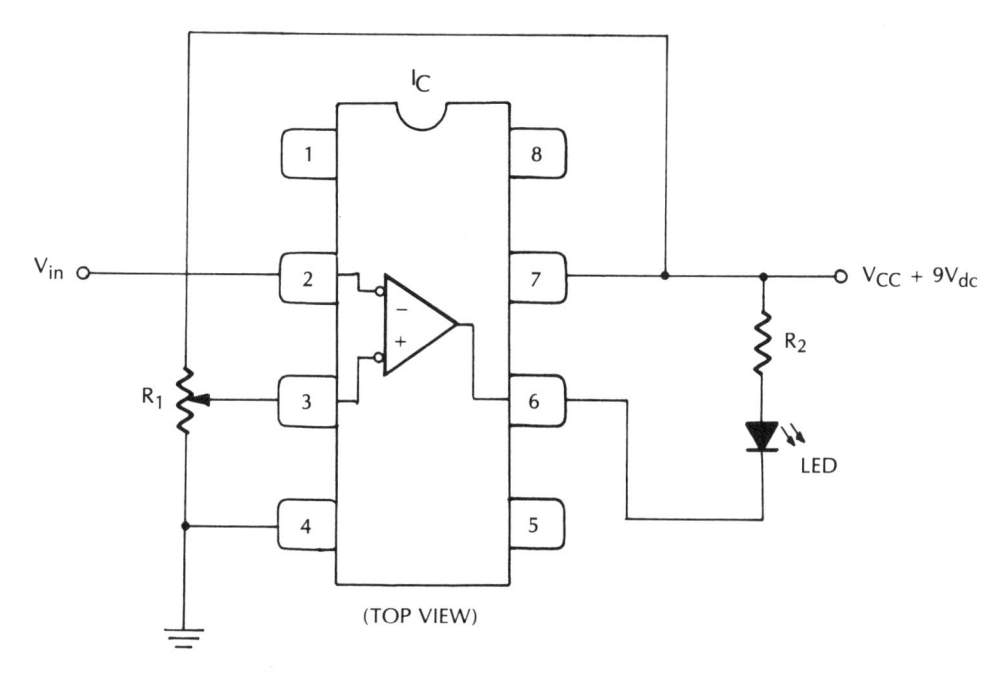

Figure 12-10: Wiring diagram for a voltage level detector.

Project 12.11: How to Connect and Test a High Performance Dual Operational Amplifier

Parts List for Project 12.11:

C_1, C_2—Capacitor, 0.1 μF.

R_1—Resistor, 1 k ohm.

R_2—Resistor, 100 k ohm.

I_{C1}—Dual operational amplifier MC1747CP2.

Comments:

This **IC** is actually two MC1741 **OP AMP**'s in a single 14-pin dual-in-line package. This type of **IC** can be used just

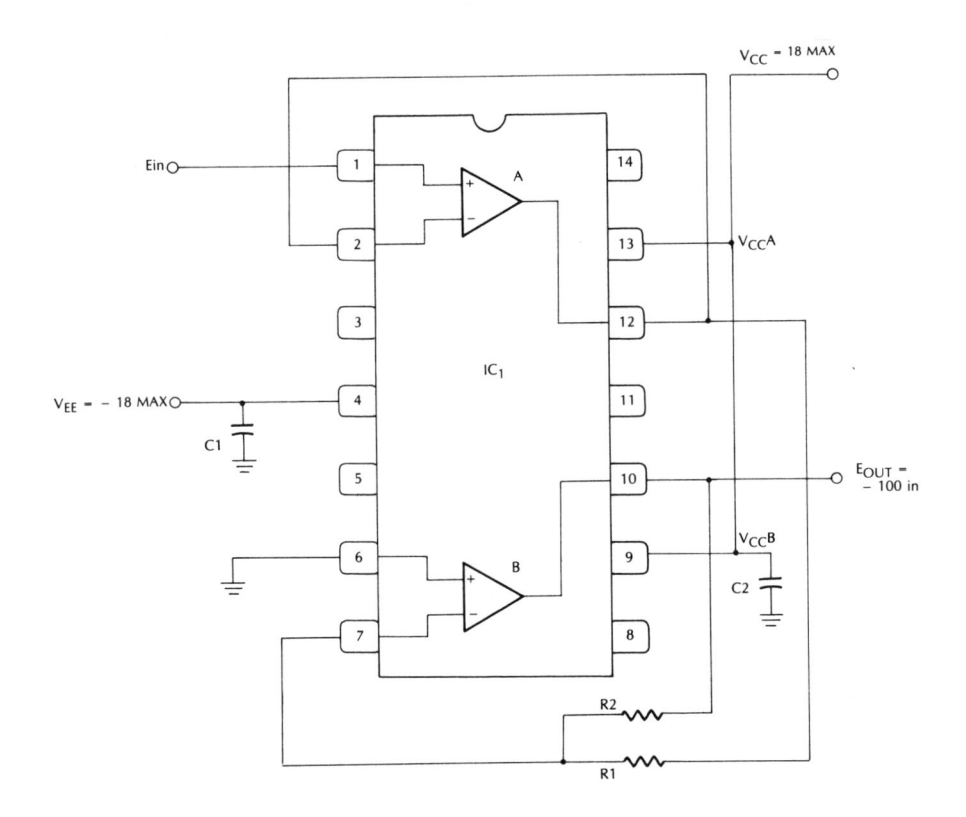

Figure 12-11: Wiring diagram for the high performance ($E_{out} = 100 \times E_{in}$) dual operational amplifier MC1747.

STATE-OF-THE-ART SOLID STATE DEVICES 305

about anywhere two or more single **OP AMP's** of this type are required. However, there are quite a few quad **OP AMPs** available that could possibly (where several **OP AMP's** are required), save you space, time and money. See Figure 1-20, Chapter One.

Project 12.12: How to Construct a Dual Flasher

Parts List for Project 12.12:

C_1, C_2—Capacitor, 10 μF.

I_{C1}—Dual operation amplifier MC3401.

R_1—Potentiometer, 500 k. Radio Shack catalog No. 271-221.

R_2—Resistor, 10 k.

LED—Light-emitting diode. Radio Shack catalog No. 276-041, or any similar device. Also, see Project 12.10 Comments.

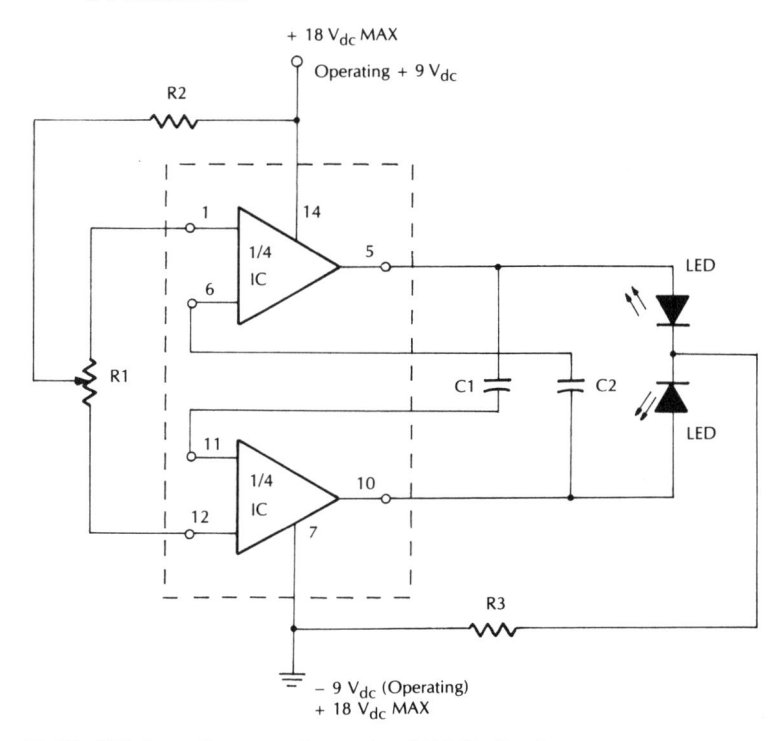

Figure 12-12: Wiring diagram for a dual LED flasher.

Comments:

Although there are four individual **OP AMP**'s in this **IC**, they all use a common single positive power supply to provide the supply voltage to pin 14. The common ground is pin 7. The output pins are 4, 5, 9, and 10. We have used output pins 5 and 10, which are the outputs for **OP AMP**'s 2 and 4 in the **IC**. You can choose any combination of two **OP AMP**'s you wish. The **OP AMP**'s inputs and outputs are

INPUTS	OUTPUTS
pins	pins
3(−), 2(+)	4
6(−), 1(+)	5
8(−), 13(+)	9
11(−), 12(+)	10

Important: If you accidentally reverse the polarities on pins 14 (V_{CC}) and 7 (ground), it may damage this **IC**. Also, *do not* short any of the output pins to either V_{CC} or ground. This **IC** has no short protection!

Project 12.13: Audio Amplifier

Parts List for Project 12.13:

B—Battery, 9 V, or dc power supply set at 9 Vdc.

C_1—Capacitor, 0.1 µF.

C_2—Capacitor, 15 pF.

C_3—Capacitor, 1 µF.

C_4—Capacitor, 200 pF.

C_5—Capacitor, 0.05 µF.

IC—½ W, audio amplifier MC1306 or equivalent.

R_1—Potentiometer, 5 k (volume control). Radio Shack catalog No. 271-217.

R_2—Resistor, 1 Meg.

R_3—Resistor, 1 k.

R_4—Resistor, 10 k.

Speaker, 8 ohms.

STATE-OF-THE-ART SOLID STATE DEVICES 307

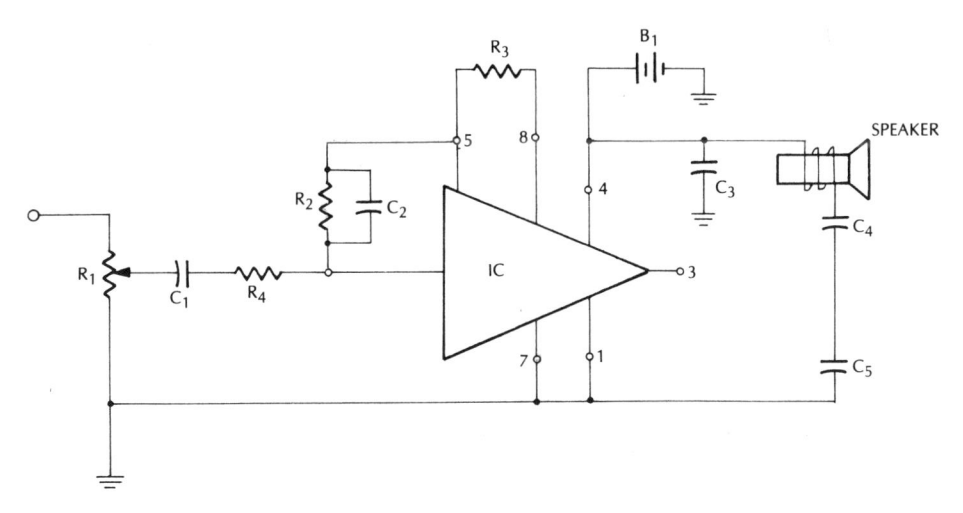

Figure 12-13: Power audio amplifier that will direct-drive an 8-ohm speaker.

Comments:

The **IC** used in this amplifier contains a complementary power amplifier and preamplifier designed to deliver ½ watt into an 8-ohm speaker, with a 3.0 mV (rms) typical output. Both gain and bandwidth are externally adjustable. See Figure 6-1 in Chapter Six for another wiring diagram using the MC1306P.

Project 12.14: Lamp Dimmer Using a Power MOSFET

Parts List for Project 12.14:

FET—Power MOS. See Comments.

R_1—Potentiometer, 1 Meg ohm. Radio Shack catalog No. 271-229.

R_2, R_3—Resistors, 1 Meg ohm. Radio Shack catalog No. 271-059.

Comments:

Motorola distributors will probably have the power **MOSFET** listed as a TMOS® device. Radio Shack stores will usually have their **MOSFET**'s listed as a VMOS solid state de-

308   STATE-OF-THE-ART SOLID STATE DEVICES

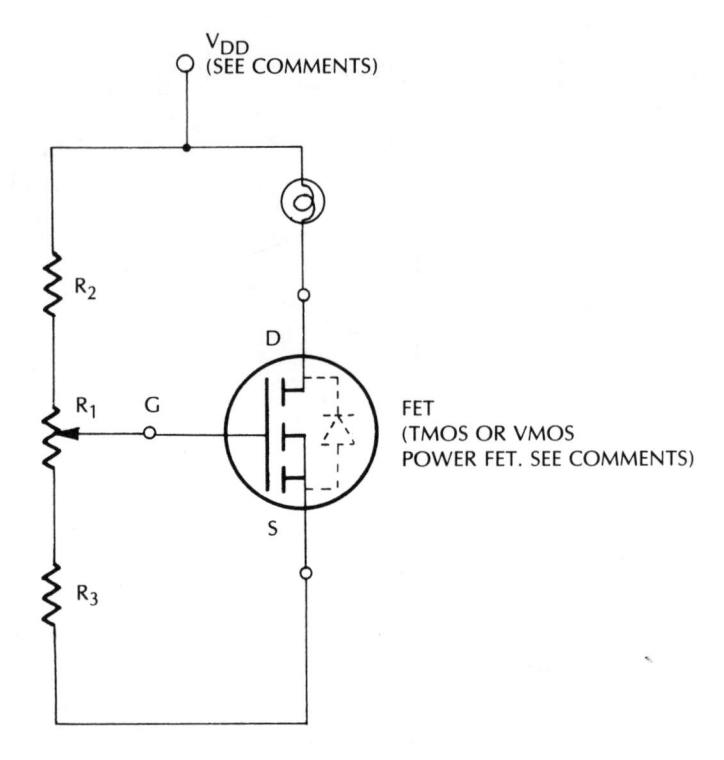

Figure 12-14: Wiring diagram for a lamp dimmer, using a TMOS transistor.

vice. A typical part number for Motorola will start with MTM and then have an identifying number; for example, MTM474. Radio Shack will probably have one of these listed as a VN something. For example, VN67, VN10, etc.

The source voltage (V_{DD}) will also be different for the various transistors. For instance, you can use a V_{DD} voltage between 6 and 12 Vdc for the VN67. On the other hand, Motorola's TMOS power **FET**'s will permit several hundred volts to be applied between the drain and source (V_{DSS}).

The diode shown by the dashed lines is included in the Motorola design. However, other manufacturers may not include such a source-to-drain diode. With an N-channel enhancement mode **TMOS FET**, the diode conducts when the source (S) is positive with respect to the drain (D).

STATE-OF-THE-ART SOLID STATE DEVICES

Project 12.15: Variable Timing Circuit (Clock)

Parts List for Project 12.15 *(continued on next page):*

C_1, C_5—Capacitor, 0.01 μF.
C_2—Capacitor, 10 μF.
C_3—Capacitor 1 μF.
C_4—Capacitor 0.1 μF.

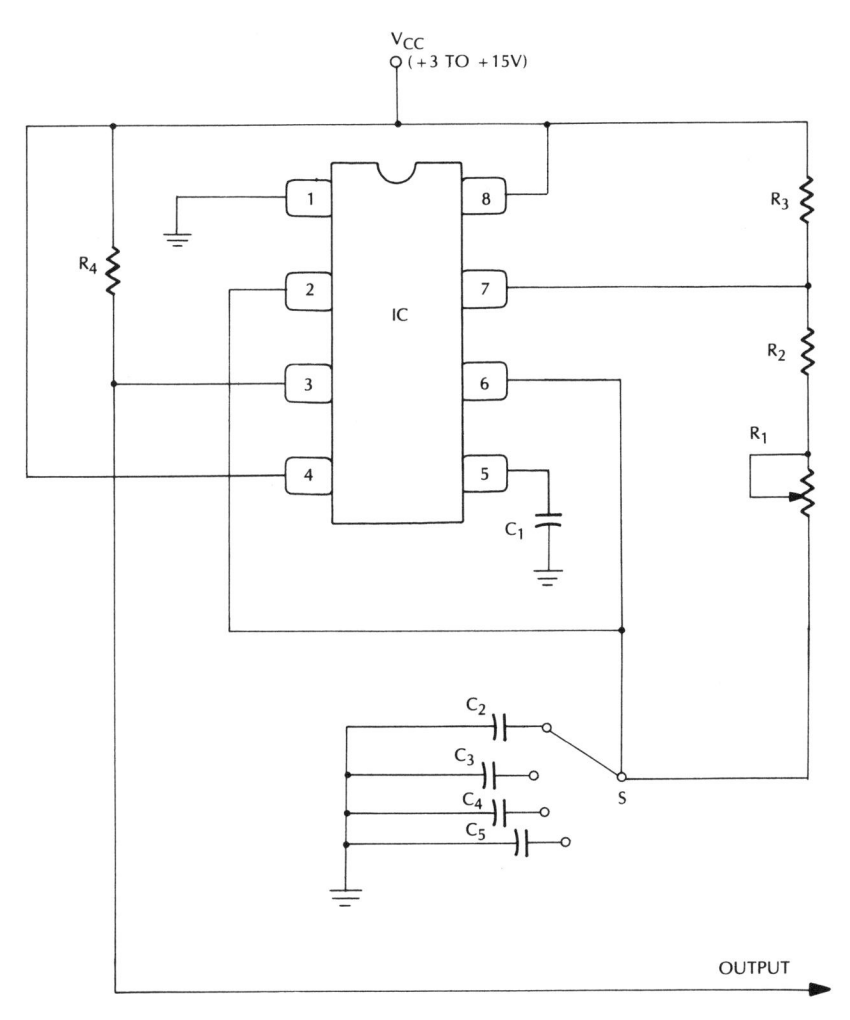

Figure 12-15: Wiring diagram for a variable timing circuit that can be used to clock digital circuits.

IC—Timer, 555, 7555 or equivalent.

R_1—Potentiometer, 500 k, Radio Shack catalog No. 271-221 (Thumbwheel).

R_2, R_3—Resistors, 470 ohms.

R_4—Resistor, 1 k

S—Switch, rotary. Single pole, 4-position (or more) Radio Shack catalog No. 275-1385.

Comments:

There are two adjustments that can be used to adjust the timing of this clock circuit: a frequency range switch (S) and a fine adjust (R_1). The range of output frequencies for each capacitor setting is

C_2 = approx 5 to 60 Hertz
C_3 = approx 10 to 600 Hertz
C_4 = approx 100 to 6 or 7 kHz
C_5 = approx 1000 to 70,000 Hertz.

A regular 14-pin wire wrapping DIP socket (Radio Shack No. 276-1993) can be used to mount the **IC** (using only 4 pins on each side), or a solder mounted 8-pin socket such as Radio Shack's catalog No. 276-1995 could be used, if it is more convenient.

Pin 3 of this **IC** (555/7555) is the output pin. The fine adjust control (R_1) adjusts the clock signal to the desired frequency.

Project 12.16: Contact Debouncer

Parts List for Project 12.16:

IC—Hex buffer (noninverting), MC14050B or equivalent.

R_1—Resistor, 100 k.

S_1—SPDT pushbutton. Radio Shack catalog No. 275-1547 (normally open).

Comments:

The pushbutton switch (when momentarily depressed) will force the circuit into one state or the other. The feedback

STATE-OF-THE-ART SOLID STATE DEVICES

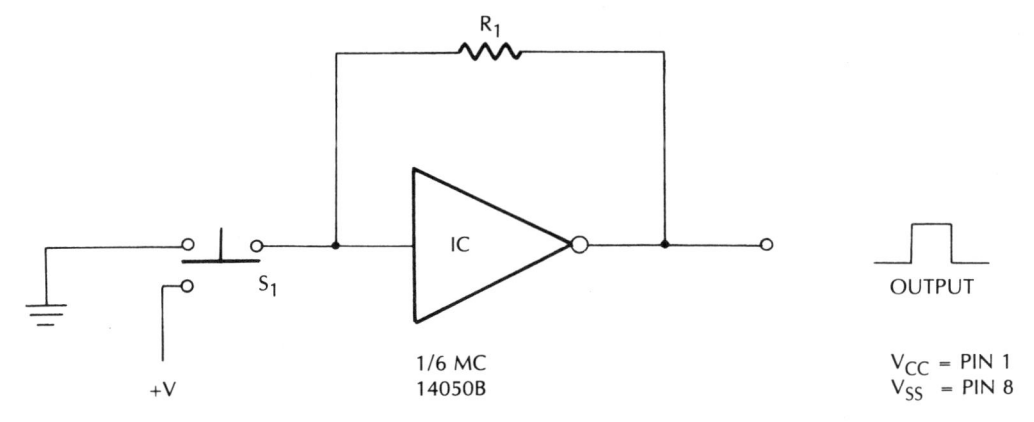

Figure 12-16: Simple contact debouncer circuit.

resistor (R_1) will hold the output in either a high or a low state. You can extend the circuit to six debounced switches by utilizing all six buffers contained in the IC, and adding the other needed feedback resistors and pushbutton switches.

Project 12.17: Square-Wave Generator

Parts List for Project 12.17 (continued on next page):

C—Capacitor, 0.01 µF.

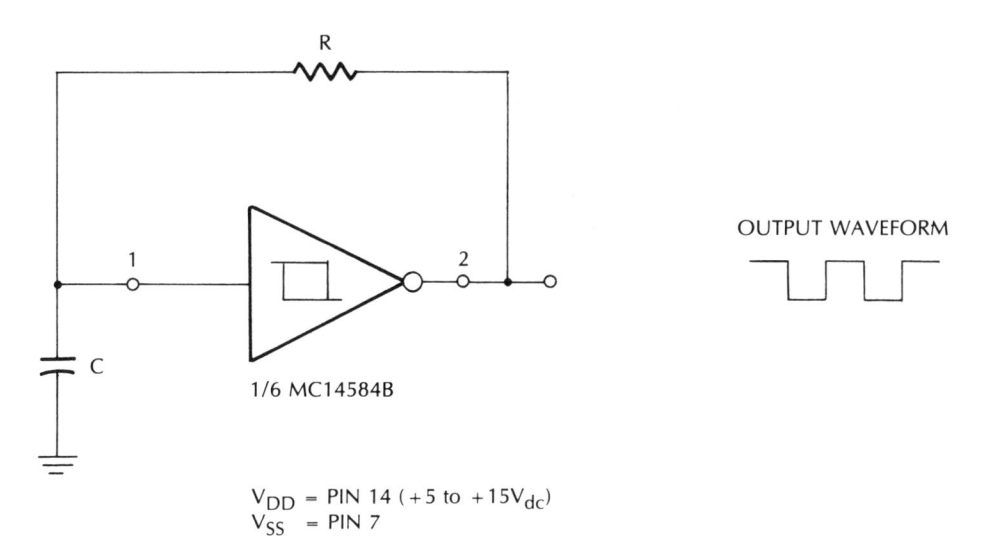

V_{DD} = PIN 14 (+5 to +15V_{dc})
V_{SS} = PIN 7

Figure 12-17: Wiring diagram for square-wave generator.

R—Resistor, 330 k.

I_C—Hex Schmitt trigger MC14584 or equivalent.

Comments:

With the values shown, you should have an output frequency of about 1 kHz. However, the frequency can easily be altered by selecting different values for C_1 and R_1.

Project 12.18: Touch-Controlled Flip-Flop

***Parts List for Project 12.18** (continued on next page):*

I_C—MC14011B.

R_1, R_4—Resistors, 100 k.

R_2, R_3—Resistors, 4.7 Megohm.

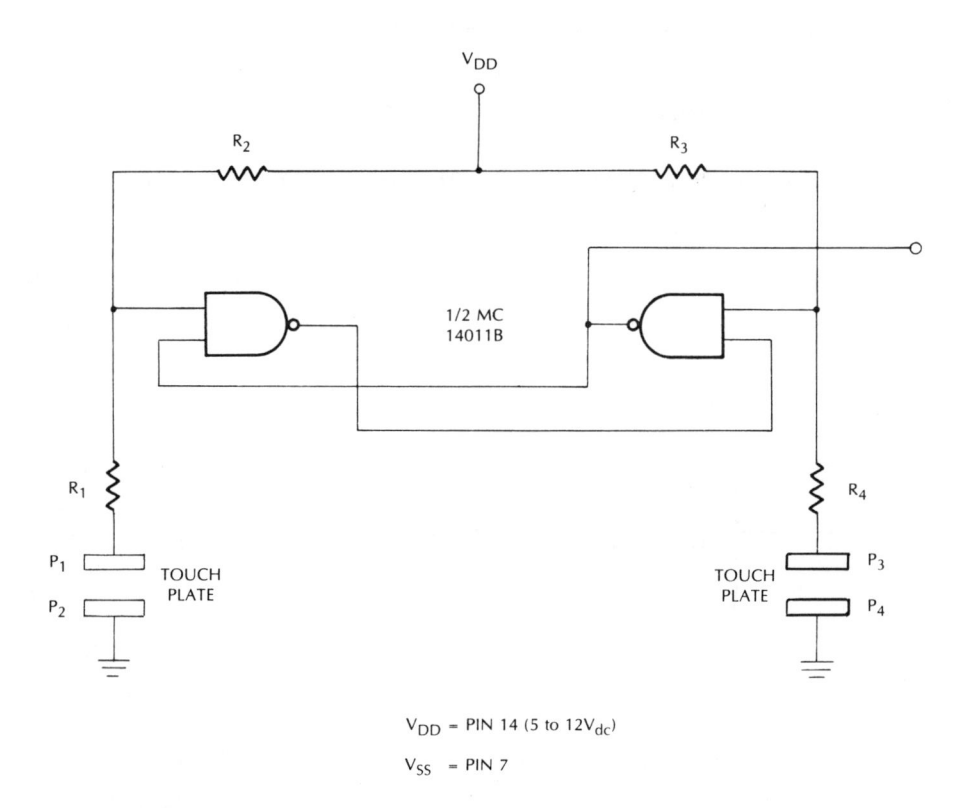

V_{DD} = PIN 14 (5 to 12V_{dc})

V_{SS} = PIN 7

Figure 12-18: Wiring diagram for touch-controlled flip-flop.

STATE-OF-THE-ART SOLID STATE DEVICES 313

Touch plate—Any conductive material cut into four small pieces (P_1, P_2, P_3, P_4). Each touch plate consists of two pieces with a small gap between each piece.

Comments:

To operate the circuit, first touch plates P_3 and P_4 at the same time (with one finger). This should produce a high on the output. Next, touch plates P_1 and P_2 (again, both at the same time), and the output should drop to a low; that is, clear the circuit.

Project 12.19: CMOS Quad 2-Input NAND Gate (MC14011B)

Testing Unit as a Control Gate:

Step 1. Connect all pins *except* pins 1, 2 and 3 (the first gate in Figure 12-19) to either pin 7 or 14.

Step 2. Connect V_{DD} and V_{SS} (+3 to 15 Vdc) to pins 14 (+V_{DD}) and pin 7 (−V_{SS}). A 9-volt battery serves as a power source, if desired.

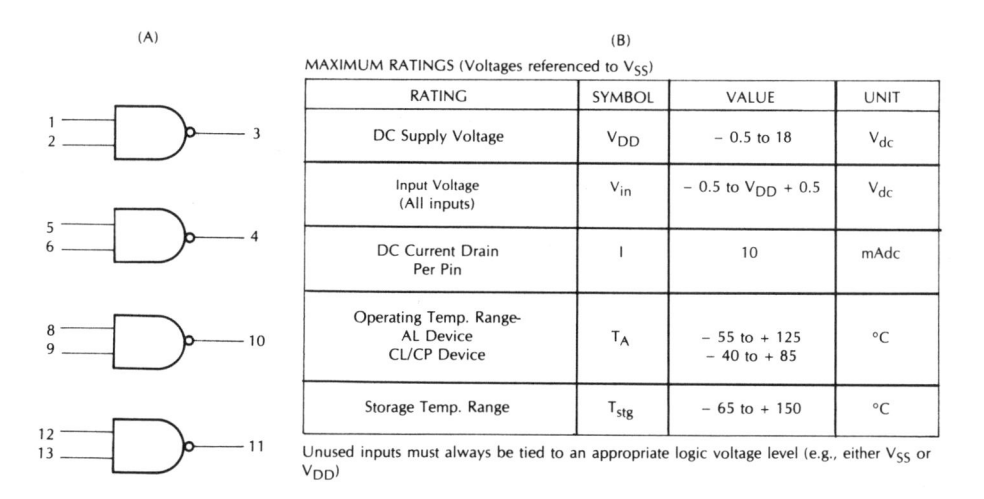

(A)

(B)
MAXIMUM RATINGS (Voltages referenced to V_{SS})

RATING	SYMBOL	VALUE	UNIT
DC Supply Voltage	V_{DD}	− 0.5 to 18	Vdc
Input Voltage (All inputs)	V_{in}	− 0.5 to V_{DD} + 0.5	Vdc
DC Current Drain Per Pin	I	10	mAdc
Operating Temp. Range- AL Device CL/CP Device	T_A	− 55 to + 125 − 40 to + 85	°C
Storage Temp. Range	T_{stg}	− 65 to + 150	°C

Unused inputs must always be tied to an appropriate logic voltage level (e.g., either V_{SS} or V_{DD})

Figure 12-19: (A) Logic diagram for the quad 2-input NAND gate MC14011B. (B) Maximum ratings for this IC.

INPUTS		OUTPUTS
Pins		Pin
1	2	3
L	L	H
L	H	H
H	L	H
H	H	L

Table 12-1: Truth table for testing a MC14011B single control gate.

Step 3. Use truth table 12-1, using the listed inputs and resulting outputs, to check the gate (L = low, H = high).

Comments:

There are four individual **NAND** gates in this **IC**. Use the same procedure to check the other three. *Important:* Always connect all *unused* pins to either pin 7 or pin 14 during each gate test.

Project 12.20: Constructing an Inverter Using a MC14011B

Testing the MC14011B Wired as an Inverter:

Step 1. Connect all pins *except* pins 1, 2, and 3 to either pin 7 or pin 14.

Step 2. Connect V_{DD} and V_{SS} to power source.

Step 3. Use truth table 12-2 to check the inverter gate (L = low, H = high).

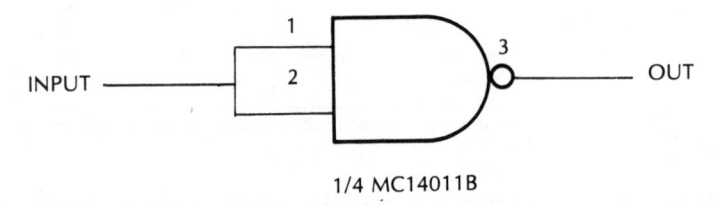

Figure 12-20: Wiring diagram for an inverter.

STATE-OF-THE-ART SOLID STATE DEVICES 315

INPUTS		OUTPUTS
Pins		Pin
1	2	3
L	L	H
H	H	L

Table 12-2: Truth table for testing the inverter shown in Figure 12-20.

Comments:

By using the wiring diagram shown in Figure 12-20 and connecting pins 1 and 2, 5 and 6, 8 and 9, 12 and 13, it is possible to construct four separate inverters with this single **IC**.

Project 12.21: Building an AND Gate Using a MC14011B

Testing the IC Wired as an AND Gate:

Step 1. Connect all pins *except* 1, 2, 3, 4, 5, and 6 to either pin 7 or 14.

Step 2. Connect V_{DD} and V_{SS} to the power source. See maximum ratings for this **IC** (given in Figure 12-19).

Step 3. Use truth table 12-3 to check the **AND** gate (L = low, H = high).

Comments:

A formula for an **AND** gate states that "A AND B = C" (pins 1 AND 2 = A AND B in Table 12-3). The symbol for multi-

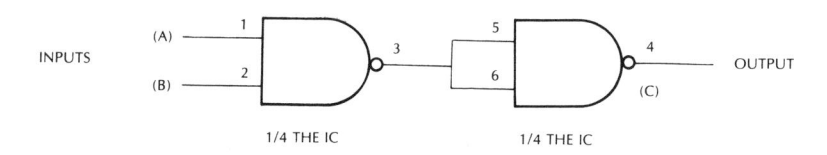

Figure 12-21: Wiring diagram for constructing an AND gate using a MC14011B.

INPUTS		OUTPUTS
Pins		Pin
1 (A)	2 (B)	3 (C)
L	L	L
L	H	L
H	L	L
H	H	H

Table 12-3: Truth table for testing the AND gate shown in Figure 12-21.

plication (·) stands for AND in Boolean algebra. The table shows the resulting output condition for all possible inputs and the formula is A · B = C, or (using pin numbers) 1 · 2 = 4.

The formula for the previous project (12.20) is A + B = C. To put it another way, it takes two lows to produce a high. On the other hand, it takes two highs to produce a low. By referring to Table 12-2, you can see that this is true.

Project 12.22: Using the MC14011B to Build an OR Gate

Testing the OR Gate:

Step 1. Connect pins 11, 12, and 13 to pin 14 (V_{DD}).

Step 2. Connect V_{DD} and V_{SS} to the power source. See maximum ratings for this **IC** (given in Figure 12-19).

Step 3. Use truth table 12-4 to check the **OR** gate (L = low, H = high).

Comments:

You should find that any high input (input A or B) to an **OR** gate will produce a high on the output. The symbol for addition (+) stands for OR in Boolean Algebra.

STATE-OF-THE-ART SOLID STATE DEVICES 317

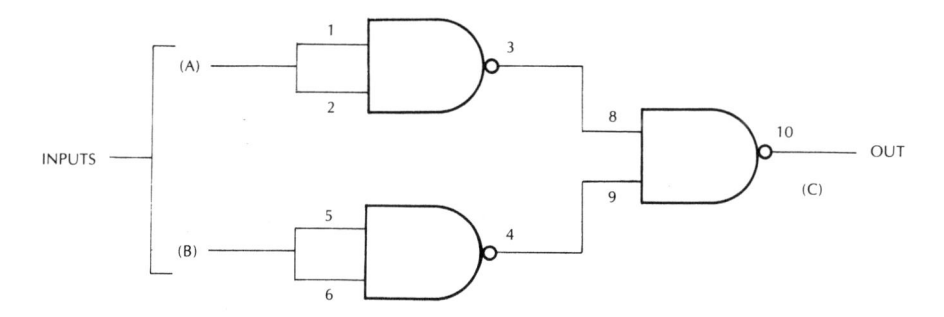

Figure 12-22: Wiring diagram for constructing an OR gate using a MC14011B.

INPUTS		OUTPUT
Pins		Pin
1, 2 (A)	5, 6 (B)	10 (C)
L	L	L
L	H	H
H	L	H
H	H	H

Table 12-4: Truth table for an OR gate. The formula for an OR gate states that "A OR B = C" (A + B = C).

Project 12.23: Combination Gates

Testing the Combination Gate:

Step 1. Connect all unused pins to either pin 14 (V_{DD}) or pin 7 (V_{SS}).

Step 2. Connect V_{DD} and V_{SS} to the power source. See maximum ratings for this IC (given in Figure 12-19).

Step 3. Use truth table 12-5 to check the combination gate (L = low, H = high, X = don't care).

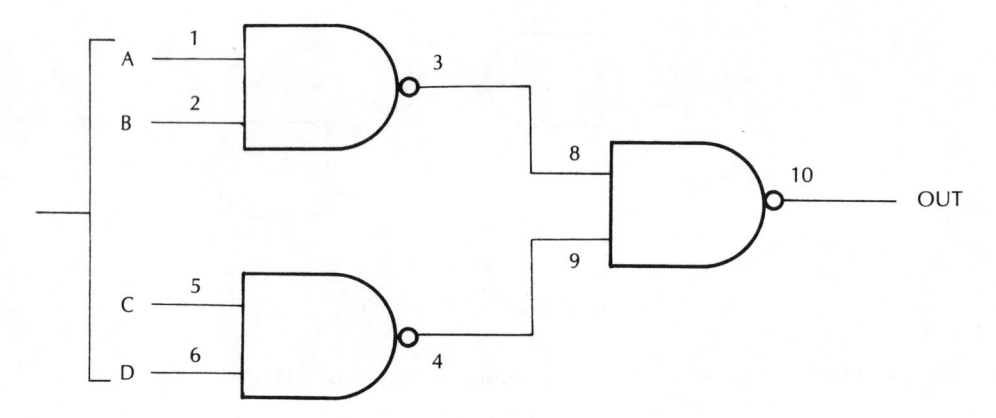

Figure 12-23: Wiring diagram for building a combination gate (AND-OR) using a MC14011B.

	INPUTS	OUTPUTS
	Pins	Pin
1,2	5, 6	10
X, X	H, H	H
H, H	X, X	H
H, H,	H, H	H

Table 12-5: Truth table for combination gate (AND-OR). The formula for this combination gate is output at pin 3 = A · B, output at pin 4 = CD, output at pin 10 = AB + CD.

Comments:

All types of logic gates, **NOR** (MC14001B), **OR** (MC14071B), and **AND** (MC14081B), can be combined together to form switching arrangements to perform certain operations or functions. We have just described the **AND**-to-**OR** gate network. However, you can build other combinations using different multiple gate IC's. It is well worth your time to run each of the gate projects because these gates are the building blocks of all digital logic circuits.

STATE-OF-THE-ART SOLID STATE DEVICES

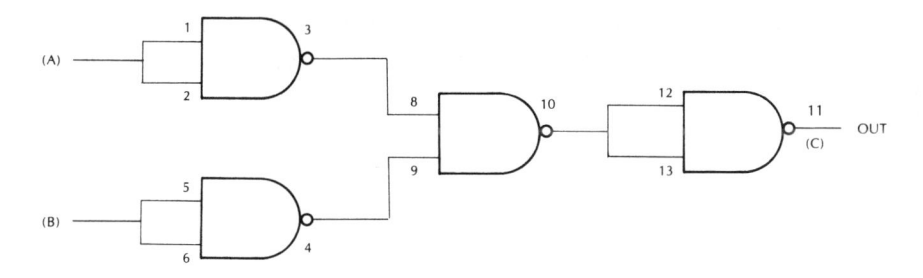

Figure 12-24: Wiring diagram for constructing a NOR gate using a MC14011B.

INPUTS		OUTPUTS
Pins		Pin
1, 2 (A)	5, 6 (B)	10 (C)
L	L	H
L	H	L
H	L	L
H	L	L

Table 12-6: Truth table for a NOR gate.

Project 12.24: Using the MC14011B to Build a NOR Gate

Testing the NOR Gate:

Step 1. Connect power to V_{DD} and V_{SS}. Do not exceed the maximum ratings given in Figure 12-9.

Step 2. Use truth table 12-6 to check the inputs and outputs (L = low, H = high).

Comments:

The **NOR** gate is actually an **OR** gate (see Project 12.22, Figure 12-22) followed by an inverter, and may be read as

"NOT-OR," hence the term "**NOR.**" By referring to Table 12-6, you will note that when both inputs are low, the output of the **NOR** gate will be high. The formula reads "A Not and B Not = C" (A + B = C). Incidentally, when the inputs of the NOR gate are connected together, it becomes an inverter.

Project 12.25: Building an Exclusive-OR Gate Using a MC14011B

Testing the Exclusive-OR Gate:

Step 1. Connect power to pins 14 (V_{DD}) and 7 (V_{SS}).

Step 2. Use truth table 12-7 to check the inputs and output (L = low, H = high).

Comments:

Some digital circuits require an **OR** gate to produce a high on the output when only one input is a high. This type of **OR** gate is called an *Exclusive OR gate*. The formulas given below (Table 12-7), read "A and B Not or A Not and B = C." The truth table shows that only one input with a high will produce an out of a high.

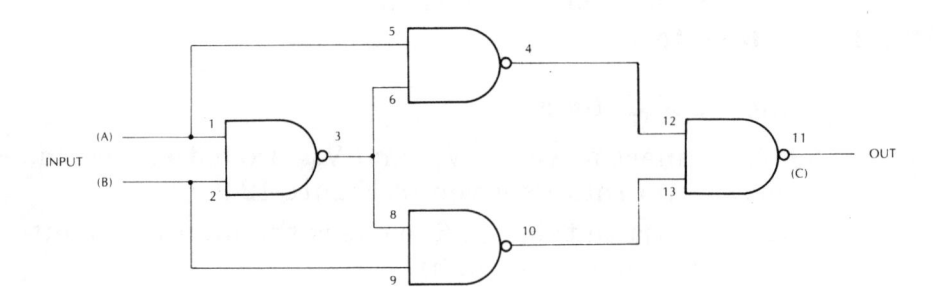

Figure 12-25: Wiring Diagram for the Exclusive-OR Gate.

STATE-OF-THE-ART SOLID STATE DEVICES

INPUTS		OUTPUTS
Pins		Pin
1, 5 (A)	2, 9 (B)	11 (C)
L	L	L
L	H	H
H	L	H
H	H	L

Table 12-7: Truth table for the Exclusive-OR gate. Formula: AB + AB = C.

Project 12.26: Building a 4-Input NAND Gate Using Two MC14011B's

Testing the 4-Input NAND Gate:

Step 1. Connect all unused pins to either pin 14 (V_{DD}) or pin 7 (V_{SS}).

Step 2. Connect the power source to V_{DD} and V_{SS}.

Step 3. Use truth table 12-8 to check the inputs and output (L = low, H = high, X = don't care).

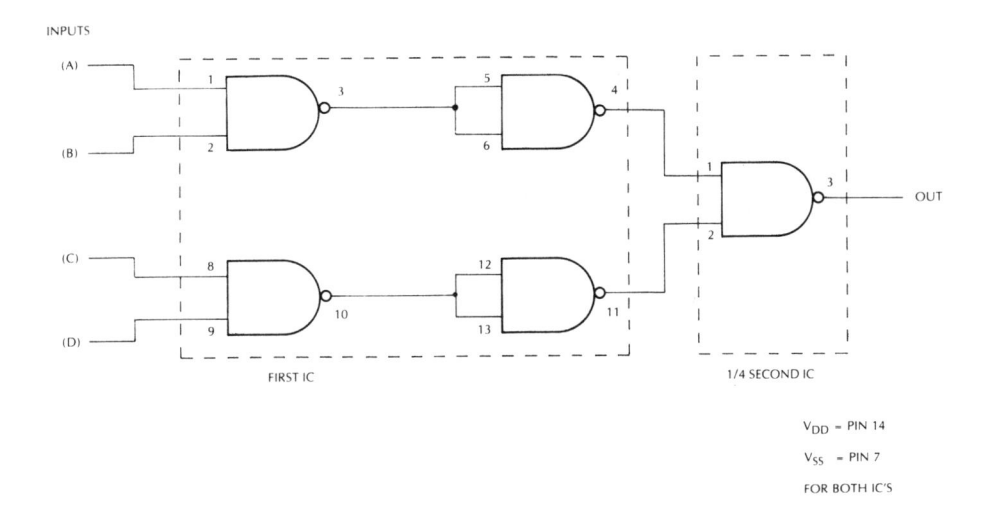

Figure 12-26: Using two MC14011B's to construct a 4-input NAND gate.

INPUTS				OUTPUT
Pins				Pin
1	2	8	9	3
L	X	X	X	H
X	L	X	X	H
X	X	L	X	H
X	X	X	L	H
H	H	H	H	L

Table 12-8: Truth table for a 4-input NAND gate.

Comments:

This project uses a quad 2-input **NAND** gate to construct a single 4-input **NAND** gate. However, there are 4-input **NAND** gates in a single package; for example, the dual 4-input **NAND** gate, MC140128. Of course, this **IC** uses exactly the same truth table for both of its gates.

Other quad 2-input gates that you can experiment with are the **NOR** gate MC14001B, the **OR** gate MC14071B, and the **AND** gate MC14081B. In fact, there is an entire line of this type of gate called "**CMOS** B-series gates," with 2-inputs, 3-inputs, 4-inputs, and 8-inputs.

Project 12.27: Divide-by-4 Counter Using Dual J-K Flip-Flops (MC14027B)

Testing the Counter:

Step 1. Unused inputs (pins 2 and 14) must always be tied to an appropriate voltage level (V_{SS} or V_{DD}).

Step 2. Connect the power source (+3 to 15 Vdc) to pin 16 (V_{DD}) and pin 8 (V_{SS}).

Step 3. Place a clock signal at the input shown (pins 3 and 13 tied together), and the divide-by-4 count should start.

STATE-OF-THE-ART SOLID STATE DEVICES 323

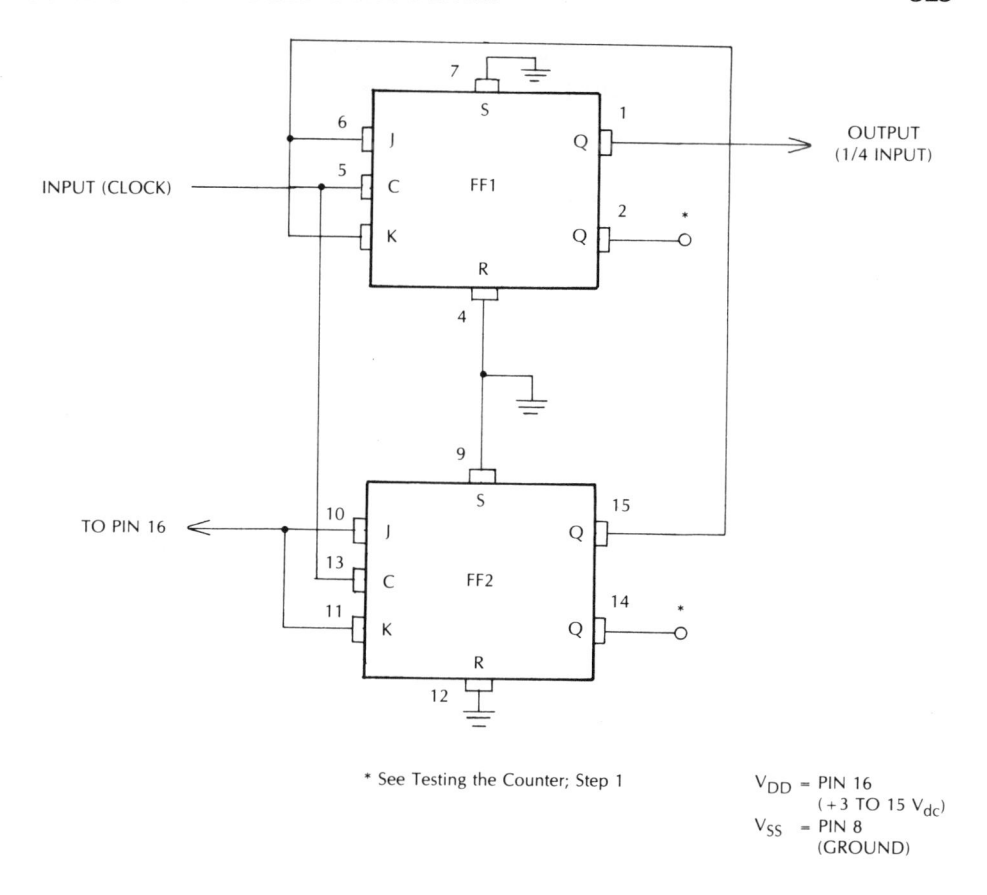

Figure 12-27: Wiring diagram for a divide-by-4 counter using a dual J-K flip-flop (MC14027B). Both flip-flops contained within the IC are used in this project (FF$_1$ and FF$_2$).

Comments:

To set (S input, pins 7 and 9) or reset (R input, pins 4 and 12), you can place a logic high on either of these inputs. They are shown grounded (a logic low) in Figure 12-27.

For proper operation, it is recommended that the input signal voltage and output signal voltage be kept so that V$_{SS}$ is always less (V$_{in}$ or V$_{out}$) than V$_{DD}$. Also, you will find a truth table for this **IC**, shown in Chapter Two (see Table 2-2).

Project 12.28: Building a 4-Bit Serial Shift Register Using Two Dual J-K Flip-Flops (MC14027B's)

Testing the Serial Shift Register:

Step 1. Tie all unused input pins to either V_{DD} or V_{SS}.

Step 2. Connect the power source (+3 to 15 Vdc) to pins 16 (V_{DD}) and 8 (V_{SS}).

Step 3. After you have the two IC's properly wired and power applied, activate the clock and place a signal on the data input line. See Figure 2-18 and Table 2-3 in Chapter Two, for shifting and timing sequence for a shift register.

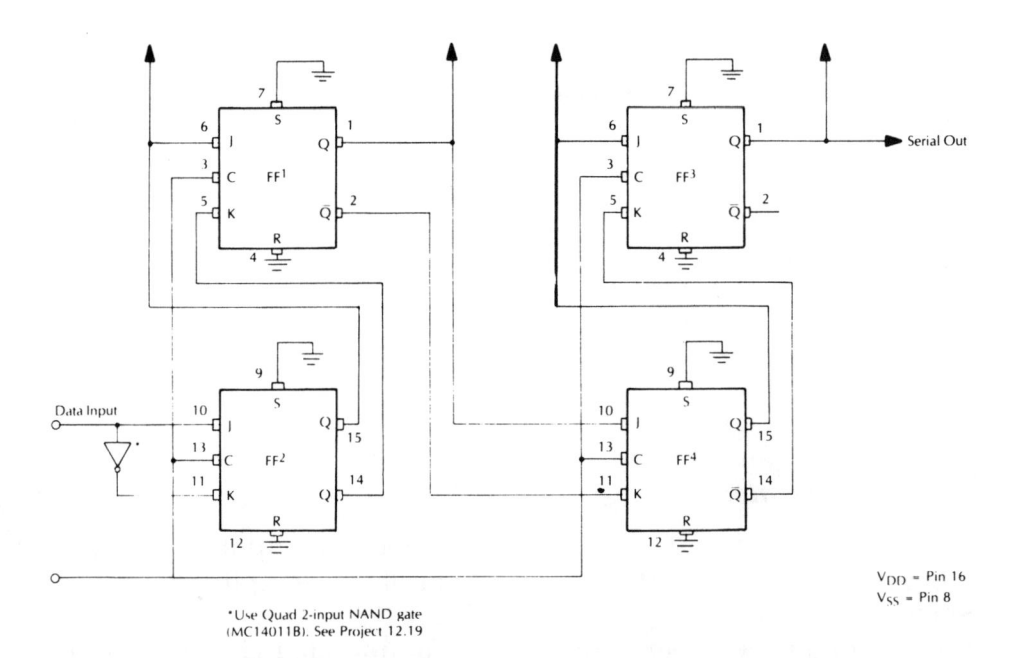

*Use Quad 2-input NAND gate (MC14011B). See Project 12.19

V_{DD} = Pin 16
V_{SS} = Pin 8

Figure 12-28: Wiring diagram for a serial shift register. FF1 and FF2 are a single MC14027B dual J-K flip-flops, as are FF3 and FF4.

STATE-OF-THE-ART SOLID STATE DEVICES

Comments:

As has been pointed out, the number of flip-flops determines the amount of data a register can store. In this case, the word length is 4 bits. Therefore, it is a 4-bit shift register. It follows that two more of these dual J-K flip-flops could be wired in the same manner, and used in conjunction with this project, to make an 8-bit register. Four more could be used to construct a 16-bit, and so forth.

Project 12.29: Phototransistor Light-Operated Relay

Parts List for Project 12.29 (continued on next page):

C_1—Capacitor, 0.1 µF.
Q_1—Phototransistor, MRD300 or similar device.
Q_2—Bipolar transistor, MPS3394 or equivalent.

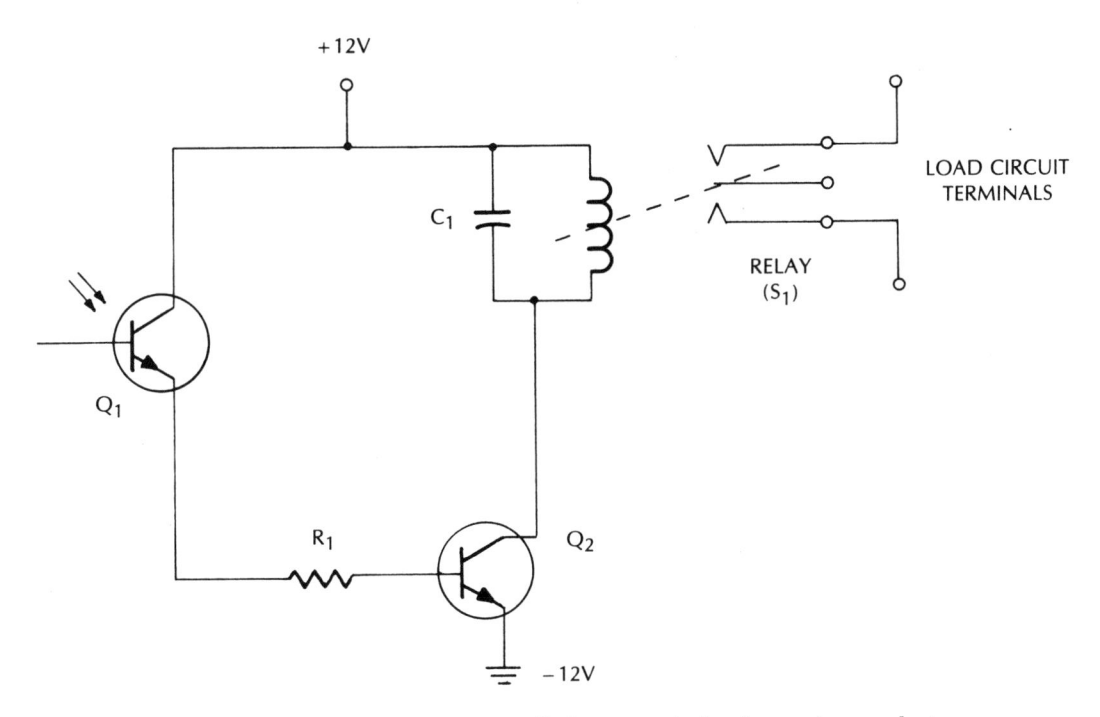

Figure 12-29: Wiring diagram for a light-operated relay using a phototransistor (MRD300).

R_1—Resistor, 1.5 k.

S_1—Relay, miniature SPDT (high sensitivity) designed for transistor circuits with a low current drain. Contacts rated @ 1A @ 125 Vac, more or less, depending on the circuit the relay is to control.

Testing the Phototransistor Light-Operated Relay:

Step 1. After wiring the circuit as shown in Figure 12-29, apply the source voltage (about +10 or 12 Vdc).

Step 2. Energize the phototransistor. Illumination for the phototransistor should be fairly strong. The brighter the light source, the greater the current flow within the circuit.

Comments:

The base current of Transistor Q_2 depends on the illumination of phototransistor Q_1. In turn, the value of the collector current (and, of course, the relay current) flowing through transistor Q_2 is controlled by the Q_2 base current.

This current (Q_2 collector current) must be sufficiently large to activate your selected relay. For example, if the relay requires 10 mA to close the points, Q_2's collector current should be at least 10 mA at saturation. *Note:* The components chosen for this project must be rated to handle the relay current. The component used in Figure 12-29 should work with a relay rated at about 5 or 10 mA. But the biasing (current-limiting) resistor R_1 may have to be increased or decreased, if you choose different solid state components and/or relay.

Project 12.30: Diode Transistor Coupler (Optoisolator)

Parts List for Project 12.30:

I_{C1}—Opto coupler/isolator, 4N26.

R_1—Resistor, 47 ohms.

R_2—Resistor, 2.2 k. *Note:* Collector current = 5 mA.

STATE-OF-THE-ART SOLID STATE DEVICES 327

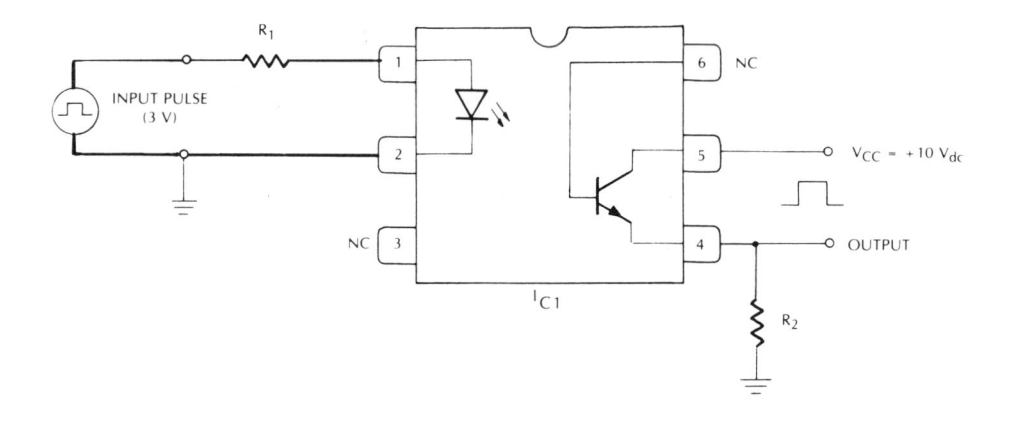

Figure 12-30: Wiring diagram for testing an opto coupler/isolator (4N26).

Testing the Optoisolator:

Step 1. After wiring the device as shown in Figure 12-30, apply the source voltage to pin 5 and ground.

Step 2. Apply an input pulse and you should see an in-phase output pulse (using the wiring diagram shown in Figure 12-30). See Comments.

Comments:

You can place an LED in the circuit between pin 2 and ground to monitor the input pulse. The output will pulse either in phase with this LED, or out of phase, depending on where you place the resistor R_2.

If you place R_2 in the transistor emitter circuit (as shown), the output will be in phase with your input signal. Or place R_2 in the transistor collector circuit and you should find that the output is 180° out of phase, in respect to the input.

Larger input signals may be used. However, resistor R_1

will have to be increased in value (if this is done) to insure that the diode is not subjected to currents over its rating. Refer to the coupler spec sheet for the current rating of the diode in the coupler you purchased. Also, additional amplification of the input signal can be achieved by adding a transistor amplifier and connecting its input (through a capacitor) to the output of the coupler (at pin 4).

INDEX

A

Accumulator, 96
ACIA:
  asynchronous communications interface adapter, 118
  control lines, 120
  registers, 119
A/D converter, 206
Adapter:
  asynchronous interface, 118
  controller IC's, 119
Address:
  accumulator, 98
  decoder, 209
  direct, 98
  extended, 99
  immediate, 98
  indexed, 98
  inherent, 100
  relative, 99
Addressing modes, 95, 126
Amplifier:
  audio power, 38
    schematic, 307
  Darlington, 162
    photo, 249
  differential, 34
  driver (rf), 288
  linear, 33, 35
  operational (OP AMP), 181
  phonograph, 156
  preamplifier, 174
  test, 268
  troubleshooting, 178
    audio, 157
  VHF power module, 291
Analog-to-digital conversion, 206
AND gate, 30, 37
Architecture, 96
ASCII, 82, 84
Assembler program, 94, 131
Asynchronous communications, 118

Audio:
  amplifier (schematic), 307
  circuit power requirements, 166
  frequency, 255
  oscillations, 179
  output-transformerless (OTL), 158
  phonograph amplifier, 156
  troubleshooting, 157
  wideband measurements, 169

B

Bandwidth (Q), 263
Battery charger (solar), 202
BCD-to-7 segment decoder, 76
Beta (transistor), 222
Bias:
  transistor, 24
    troubleshooting, 179
Bilateral trigger (DIAC), 243
Binary words, length of, 129
Biose phototransistor, 248
Breadboarding, 143
Bus:
  extender, 224
  handshake, 225
  interface, 224

C

Cable:
  audio, 144
  coax, 138, 144
Capacitance:
  color code, 18
  formula, reactance known, 271
  reactance, 16
  resonant frequency, 272
  voltage variable, 274
Capacitor:
  bypass, 267
  characteristics, 18

Capacitor: *(cont'd.)*
  filter (dc power supply), 183
  minimum value (audio output), 164
  reactance, 19
  silicon, 275
  testing, 179
  tolerance calculation, 138
  working with, 134
Carrier power, 283
Characteristics, transistor, 281
Circuit:
  analog, 34
  bandwidth, 265
  digital, 33
  linear, 33
  loading, 14
    test, 142
  resonant, 264
  parallel, 10, 265
  Q, 264
  series, 9, 264
  tuning, 265
Clock:
  circuit, 309
  timer, 121
CMOS:
  gates, 41, 44, 46
  IC, 43
  interface (LED), 214
  precautions, 51
Coax cable, capacitance measurement, 139
Code:
  ASCII, 81
  binary, 80
  capacitance, 18
  hexadecimal, 81
  mnemonic, 94, 128
  octal, 80
Communications, asynchronous, 118
Compensation, OP AMP, 212
Complementary:
  power transistors, 159
  voltage regulator, 189
Computer:
  addressing, 95
  basic system, 87
  control, 87
  language, 131
  micro, 85
  numbering systems, 79
  programming, 90
  signal descriptions, 88
Conductance, susceptance, admittance formulas, 271
Conversion:
  analog-to-digital, 206
  digital-to-analog, 206
Converter A/D, 206

Counter:
  divide-by-4, 322
  IC, 61, 67
  opto, 247
  program, 97
CPU, 28
Crossover point, 17
Crowbar:
  circuit, 192
  voltage protector, 153
Current:
  converting values, 3
  definition, 4
  measurement, 12

D

Darlington:
  amplifier, 162, 174
  dual, 175
  octal array, 136
  photo amplifier, 249
  quad driver, 238
  transistor, 235
Data bit stream:
  parallel, 118
  serial, 118
dBm, 172
Debouncer circuit, 310
Debugging, 146
Decibel:
  formulas, 272
  measurement, 169
Decoder:
  address, 209
  BCD-to-decimal/binary-to-octal, 73
  display, 75
Design:
  fundamentals (audio), 163
  power requirements, 166
DIAC, 243
Differential amplifier input, 34
Digital:
  CMOS, 38
  ECL, 38
  logic levels, 36
  to analog conversion, 206
  TTL, 38
  voltmeter, 206
Dimmer lamp, 307
Diode:
  bidirectional thyristor, 245
  hot carrier, 276
  light-emitting (spectral response), 259
  mixer, 275
  PIN, 23, 276
    switching, 277
  switching, 23

INDEX

Diode: *(cont'd.)*
  tuning, 275
  tunnel, 23
  varacter, 23
  zener, 23
    dc power supply, 184
Display:
  decimal, 214
  interface, 217
  multiplexing, 217
  output pins (IC), 217
  7-segment, 217
Distortion, audio, 167
DTL, 231
Drivers:
  computer, 231
  line, 231

E

ECL gate, 42
Emitter follower, 236
Encoder, priority, 72, 74
EPROM, 71, 113
Exclusive-OR gate, 320

F

Fan-out, 214
Feedback, audio, 163
Filter:
  active (OP AMP), 260
  power supply (dc), 183
  zener diode, 184
Flag, 123
Flasher, LED, 305
Flip-flops:
  D type, 57
  J-K, 57
  touch control, 312
Flow chart, 93
Formula:
  Impedance, 212
  phase angle, 212
  rf, 271
FPROM, 113
Frequency:
  bands, 257
  modulation index formula, 272

G

Gate:
  AND, 30, 37
  CMOS, 41
    driving LED's, 214
  combination (test), 317
  ECL, 42

Gate: *(cont'd.)*
  exclusive-OR, 320
  NAND, 30, 39
    4-input, 313
    testing, 313
  NOR, 30, 40, 319
  OR, 30, 37
    test, 316
  TTL (driving LED's), 214
Ground:
  artificial, 182
  testing rf, 270
  true, 182

H

Hexadecimal:
  #, 128
  $, 128
Hot carrier (diode), 276

I

IC:
  digital, 29
  LED driver, 214
  linear, 28
Impedance:
  control Q transistor, 263
  dynamic input (IC), 211
  formula for, 212, 272
  input measurement of, 172
  matching, 260, 302
  output, 172
  rf power transistor, 259
Inductance:
  formula, reactance known, 271
  resonant frequency, 272
Inductors, working with, 134
Input, differential amplifier, 34
Inputs/outputs:
  8080 IC, 108
  1802 IC, 107
Insertion loss, rf, 263
Instruction:
  jump, 129
  microprocessor IC, 131
  set, 129
Integrated circuits:
  CPU, 28
  ULSI, 28
  VLSI, 28
Interface:
  bus, 223
  CMOS-to-CMOS, 215
  IC, 205
  LED-to-CMOS, 220
  memory (MPU), 227

332 INDEX

Interface: *(cont'd.)*
  MOS and DTL, TTL, 214
  output (IC), 214
  peripherals, 231
  7-segment display, 217
  TTL-to-LED, 215
Interrupt, 106, 130
Inverter, 314
I/O:
  based system, 208
  port address decoder, 209
    interface, 211
  signals, 113
Isolator:
  optical, 245
  opto, 247
  transistor assembly, 253

J

JFET, 48

L

Lamp dimmer, 307
Language:
  BASIC, 131
  Computer, 131
Latch, 116
LED:
  current limiting, 220
  driving (TTL, CMOS), 214
  flasher, 305
  infrared, 245
  interface, 214
  resistor, current limiting, 303
Light operated relay, 325
Line voltage regulation, 190
Linear:
  amplifier, 33
  circuits (breadboarding), 143
Loading circuit, 14
Logic:
  elements, 52
  levels, 36
  monitor, 147
  pulser, 145

M

Mask-programmable, 114
Matching networks, rf, 263
MCU:
  comparison, 110
  6805 IC, 109
MECL, 114
Measurement:
  ac, 13

Measurement: *(cont'd.)*
  capacitance, 139
    coax cable, 138
  carrier output power, 285
  current, 12
  dc, 10
  decibel, 169
  digital multimeter, 11
  impedance (dynamic output/input), 172
    IC, 211
  inductance, 139
  SWR, 262
  THD, 168
  transistor efficiency, 285
    gain, 285
  triangular, 13
  watts, 13
Memory:
  adding, 112
  addresses, 96
  controller, 226
  IC, 68
  instructions, 95
  interface (MPU), 227
  mapping (A/D converter), 209
Meter scales:
  linear, 140
  10-ohm center scale, 142
  square law, 140
Microcomputer IC packages, 29
Michrophone (schematic), 301
Microwaves, 258
Mixer, diode, 275
Mnemonic code, 94, 128
Mode, addressing, 95
Modem, 118
Modulation:
  measurement of, 261
    SSB, 285
  percentage, 285
Module (VHF), 291
MOSFET, 41, 44, 47, 49, 239
MPU:
  breadboarding, 209
  instructions, 114
  mounting hardware, 160, 177
Multiplexing, 217
Multitester:
  linear scale, 140
  square law scale, 140
Multivibrator:
  astable, 53
  monostable, 54

N

NAND gate, 30, 39
NMOS, 47
NOR gate, 30, 40, 319

INDEX   333

Numbering systems:
  binary, 80, 82
  computer, 79, 82
  hexadecimal, 81, 82, 84
  octal, 80, 82

O

Ohms-per-volt, 14, 141
OP AMP:
  compensation, 212
  impedance matching, 302
  testing, 304
  voltage level detector, 302
Operational amplifier, 181
Opto coupler/isolator, 247
Optoisolators, 245, 326
OR gate, 30, 37
Oscillations (audio), 179
Oscilloscope, wideband measurement with (dB), 169
OTL:
  complementary system, 165
  Darlington, 163
  stacked circuit, 161
Output-transformerless (OTL), 158

P

Packages (rf power transistors), 266
Parallel circuit, 10
Peak envelope power, 283, 284
Peripherals:
  driver testing, 223
  driving, 221
  interface, 231
Phase angle, 211
Photo amplifier, Darlington, 249
Photosensors, 248
Phototransistor, 247, 325
Photovoltaic, 201
PIN diode, 23, 276
Ports, input/output, 16
Power:
  audio carrier, 284
    dissipation, 166
    in, 166
    measurement, 171
    peak envelope, 284
    requirements, 166
  digital supply, 150
  factor formula, 271
  supply, dc, 298
    automatic current control, 191
    battery, 149
    bipolar, 143, 181
    bridge, 151
    complementary voltage regulator, 189

Power: *(cont'd.)*
  filter, 183
  full-wave, 151
  line voltage regulation, 190
  performance curve (hi-fi), 150
  protection techniques, 191
  regulated, 149
  schematic, 294
  series pass transistor, 188, 192
  single-ended, 182
  switching device, 193
  switchmode, 193
  troubleshooting, 197, 201
Probe:
  logic monitor, 147
  pulser, 145
Program:
  assembler, 130, 131
  machine language, 130
  source language, 130
Programming, basics, 129
PROM, 71
protection:
  crowbar circuit, 192
  overvoltage, 296

R

Radiation frequency response, 248
RAM, 68, 85, 87, 111, 112
Reactance:
  capacitive, 16, 19
  inductive, 20
  stray, 270
Receivers (line), 231
Rectifier:
  bridge, 23, 199
  fast recovery, 23
  full-wave, 199
  general purpose, 21
  half-wave, 199
  schematic, 294
  Schottky, 23
  voltage-current relationship, 22
Register:
  bidirectional, 64, 66
  condition code, 97
  control, 120
  dynamic, 62
  index, 96
  input capture, 125
  output compare, 123
  receiver data, 120
  shift, 61, 324
  stack pointer, 130
  static, 62
  status, 119
  transmit data, 120

334

INDEX

Regulator:
  complementary, 189
  5-volt, 295
  fixed output voltage, 299
  line voltage, 190
  photovoltaic, 201
  switchmode, 192
  voltage, 186
  zener diode, 184
Relay (light operated), 325
Resistance:
  color code, 5
  definition, 4
  parallel, 7
  thermal, 281
  tolerance, 6
Resistor:
  bias (audio circuit), 179
  current limiting (LED), 220, 303
  measurements, 136
  working with, 134
Resonant frequency:
  circuit, 264
  formula, capacitance, 272
    inductance, 272
rf:
  choke (RFC), 267
  power measurement, 261
ROM, 68, 85, 87, 113

S

Schematic:
  power supply, lab type, 298
  rectifier, 294
  regulator, voltage, 295
Schottky barrier diode, 276
Screen room 271
Series circuit, 9
Shift register, serial, 324
Signal:
  digital, 36, 37
  flag, 123
  tracing point-to-point, 148
Sine wave measurement, 13
Single-sideband (SSB), 285
Sink current (TTL), 222
Software, 91, 127
Solar battery charger, 202
Soldering iron heat control, 137
Speaker:
  impedance/output power, 164
  output power vs supply voltage, 157
  paralleling, 157
  series, 156
Square-wave:
  generator (circuit), 311
  measurement, 13
Stack pointer, 97, 130

Standing wave, 260, 262
Strip line, 258
Subroutines, 129, 130
Switches, silicon bidirectional (SBS), 244
Switching diode, 23, 277
Switchmode:
  power supply, 193
    basic configurations, 194,
    functional design, 196
    step-down, 196
    step-up, 197
  regulator, 192

T

Temperature, 134, 281
Test:
  capacitors, 179
  coax cable, 139
  digital IC's, 145, 147
  divide-by-4 counter, 322
  equipment, 145
  gates, AND, 315
    combination, 317
    exclusive-OR, 320
    NAND, 313
    NOR, 319
    OR, 316
  inverter, 314
  linear IC's, 148
  OP AMP, 304
  optoisolator, 327
  peripheral driver, 223
  rf, 27, 261, 288
  shift register (serial), 324
THD measurement, 167
Thyristors, 241
Timer, clock, 121
Timing:
  circuit, variable, 309
  signal, clock, 122
    MPU, 121
    RAM, 121
    ROM, 121
Total harmonic distortion, (THD), 149, 167
Transformer, quarter-wave, 260
Transistor:
  base current calculation, 26
  beta, 222
  bias, 24
  bipolar, 24
  CB/amateur, 283
  characteristics, 281, 282
  circuit design, audio, 163
  collector current, 27
  complementary, 158, 176
  control Q, 262
  Darlington, 235

INDEX

Transistor: *(cont'd.)*
  driver (rf), 289
  driving, 287
  experiment (photosensor), 247
  FM mobile, 289
  frequency band, 267
  HEX FET, 240
  isolating, 252
  low voltage, 269
  marine radio, 190
  matching networks, 263
  mesa, 278
  MOS and bipolar, 240
  mounting hardware, 160
    guide, 250
    TO-126 plastic package, 251
  packages, 25
    strip line, 258
  photo, 247, 325
  programmable unijunction (PUT), 243
  rating (max.), 279
  rectifier, 297
  rf power, 188, 192
  silicon annular, 278
  specifications, 277
  temperature, 281
  troubleshooting in-circuit, 280
  TTL interface, 221
  VCC calculation, 28
Triac driver (optically coupled), 245
Triangular waveform measurement, 13
Troubleshooting:
  audio circuits (breadboard), 178
  digital, 147
  rectifier, bridge, 199, 201
    full-wave, 199, 201
    half-wave, 199, 201
  rf, 270
  solid state power supply, 197
    bridge rectifier, 199
    half-wave, 199
TTL sink current, 222
Tuning:
  diode, 275

Tuning: *(cont'd.)*
  resonant circuits, 273
  rf, 267
Tunnel diode, 23

U

Unijunction transistors (UJT), 243
UJT's, programmable, 243

V

Varactor diode, 23, 275
VLSI, 28
VMOSFET, 47
Voltage:
  ac, 2
  converting values, 3
  dc, 2
  doubler (schematic), 300
  Kirchoff's law, 9, 10
  level detector, 302
  regulator, complementary, 189
    IC, 186
    10-amp, 151
Voltmeter (DVM), 206

W

Waveform:
  sine-wave, 2, 13
  square-wave, 13
  triangular, 13
Wavelength:
  infrared, 258
  rf, 257
Wideband measurements (dB), 169

Z

Zener diode, filter (dc power supply), 184